from the leader
in
digital audio...

THE SONY BOOK OF DIGITAL AUDIO TECHNOLOGY

BY H. NAKAJIMA,
T. DOI, J. FUKUDA &
A. IGA OF SONY CORPORATION

A SPECIAL PRINTING FOR THE SONY DIGITAL AUDIO CLUB
Marc Finer and Takeshi Yazawa—editors

FIRST EDITION

SECOND PRINTING

Printed in the United States of America

Library of Congress Cataloging in Publication Data

Dijitaru odio gijutsu nyumon. English.
Digital audio technology.

Translation of: Dijitaru odio gijutsu nyumon.
Includes index.
1. Sound—Recording and reproducing—Digital techniques. I. Nakajima, Heitaro, 1921-
II. Title.
TK7881.4.D5413 1983 621.389'3 82-6020
ISBN 0-8306-2451-1 AACR2
ISBN 0-8306-1451-6 (pbk.)

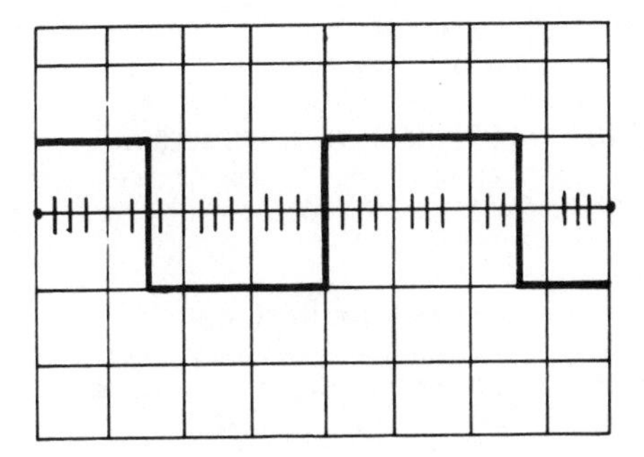

Contents

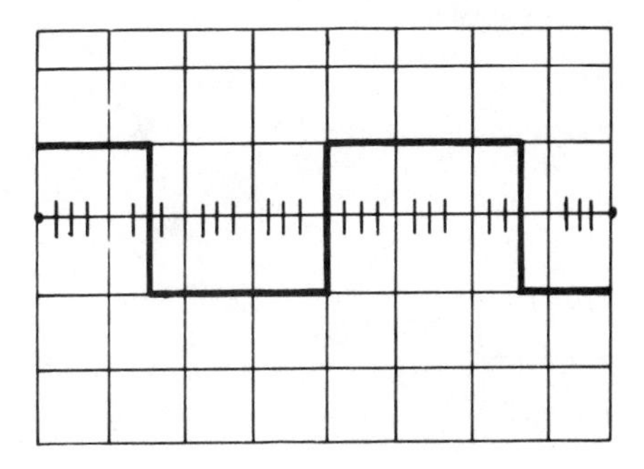

Introduction

The development of digital audio technology promises to create a new standard of excellence in the recording and reproduction of sound. Already major strides have been made toward implementing digital audio in a variety of different ways. PCM systems (Pulse-Code Modulation) are being used in a variety of applications and the future of digital audio looks very bright indeed.

The purpose of this book is to introduce the reader to the techniques being used to develop the science of digital audio technology. This technology is, today, at the forefront of state-of-the-art electronics research and development. Much that has been accomplished will stand the test of years, and much will undoubtedly change as digital audio continues to be developed. But, this much we know for certain—digital audio holds the promise of the ultimate in the true fidelity of reproduction of sound.

Chapter 1

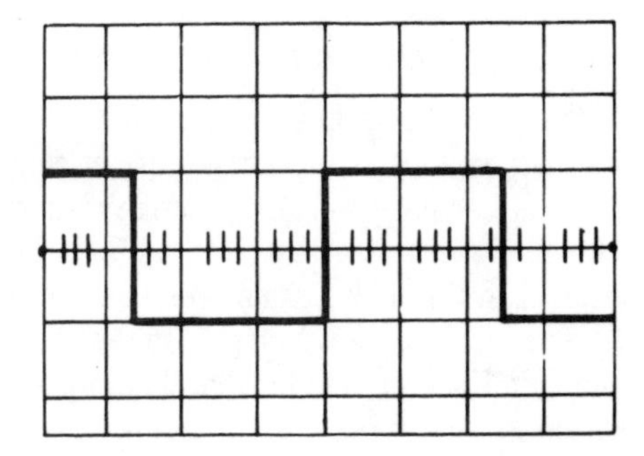

The Road to Super Hi-Fi Recording

The true beginning of progress towards the ultimate aim of recording and reproduction of an audio signal faithfully was the shift from mono to stereo. The spark which started the drive to stereo was the industrial open reel tape recorder running at 38 cm/s with a tape width of 6 mm.

THE BEGINNING OF STEREO RECORDING

In 1948, a monophonic tape recorder was developed by Sony specifically for broadcast use as a master recorder, but in 1956 production of the stereo 45-45 record began. Then the race to produce a stereo record/reproduction tape recorder led to a simplification of techniques which in turn allowed the development of equipment for domestic use. In 1963, FM stereo broadcasting became a reality, and in the same year the first compact stereo cassette player was produced: stereo had arrived. In the two decades since then, attention has been centered on trying to force ever greater sound quality improvements from software, from the three basic sources available: tapes, discs and FM broadcasts. The whole prosperity of the audio business is based upon these alternatives. And behind the undoubtedly great improvements in sound quality lies the open reel tape recorder, which in its essentials is still fundamentally the same as when first invented. Until now, the open reel tape recorder was used as a master for original software and for mother tapes for second generation recordings because it provides good sound quality, and as well as being easy to use for record and reproduction, editing can be carried out with a minimum of bother.

PROBLEMS ENCOUNTERED DURING REPRODUCTION

After 1965 the three basic music sources—records, tape, and FM broadcasts—had become established. Just at this time when FM broadcasting had become standard practice rather than an experimental technique, we began to try to picture the future developments possible in the audio chain. We tried to deduce what roles these future developments would play in the standard recording chain, as shown in Fig. 1-1, and in what ways methods of sound collection, recording, and reproduction might evolve. In other words, there was from the beginning an impetus to clearly define which parts of the audio chain ought to be redesigned, and even which parts should be developed afresh. A whole year was spent in developing and introducing standardized measurement techniques for the then existing audio chain and defining parameters for the appreciation of musical quality, so that from the point of view of both physics and subjective the measurement technique could be both accurate and standardized [1] [2]. With reference to actual measurement of performance, we investigated the concept of dynamic range, which indicates the actual power and spread of the signal, and distortion (including crosstalk) in the widest sense, which shows the extent to which undesirable elements are being added to the original signal.

If we examine the dynamic range of various parts of the audio chain, by looking at the measurement results for each piece of equipment which can be measured electronically, we will find the results shown in Fig. 1-2. The level of distortion which determines the lower limit of dynamic range can be estimated from the affection of system noise and lower frequency component of the noise, by examining the presence or absence of the weighting by the A-curve. As for the upper limit of dynamic range, two levels of data, distortion 1% and 3%, show the affection of system nonlinearity.

The strength and extent of the pick-up sound from a microphone is determined by the combination of the microphone sensitivity and the quality of the microphone pre-amplifier[3], but it is possible to maintain a dynamic range in excess of 90 dB by setting the levels carefully. However, the major problems in microphone sound pick-up are the various types of distortion inherent in the recording venue, which cause a narrowing of the dynamic

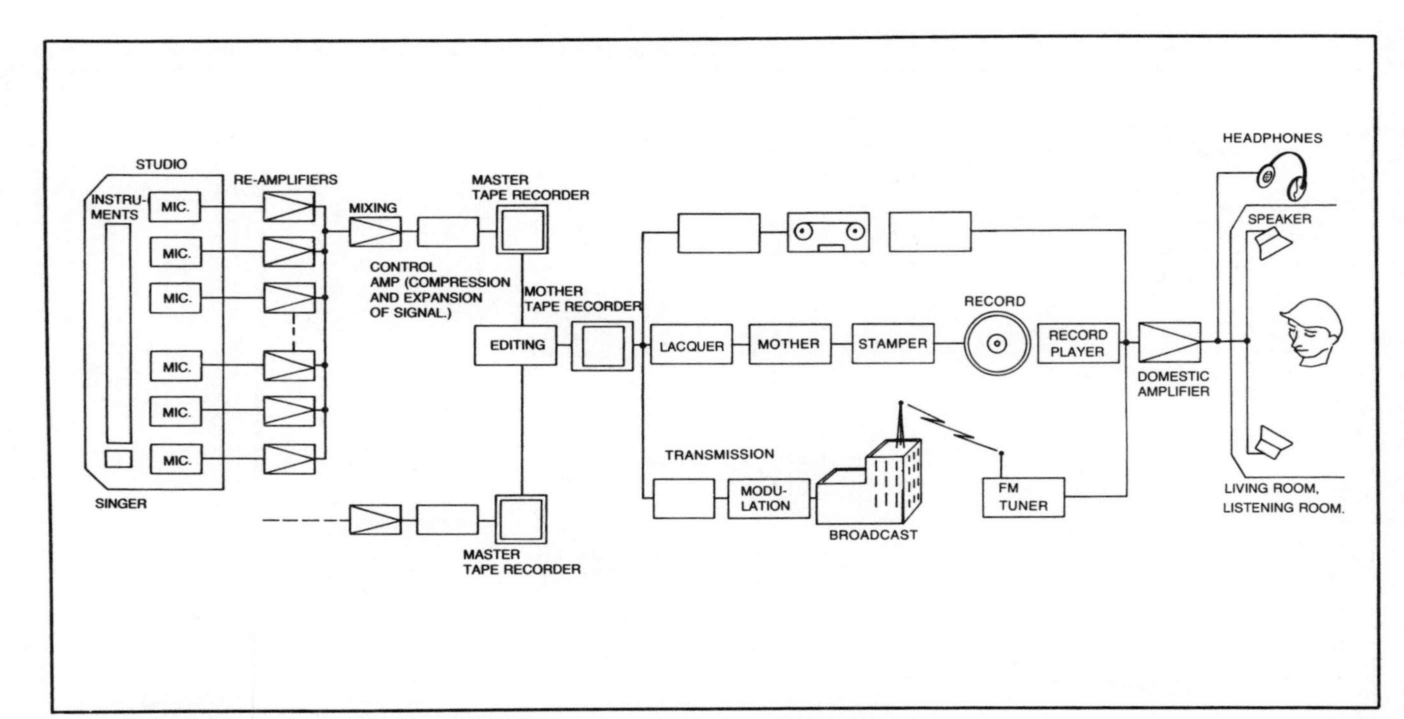

Fig. 1-1. The audio chain from record to reproduction.

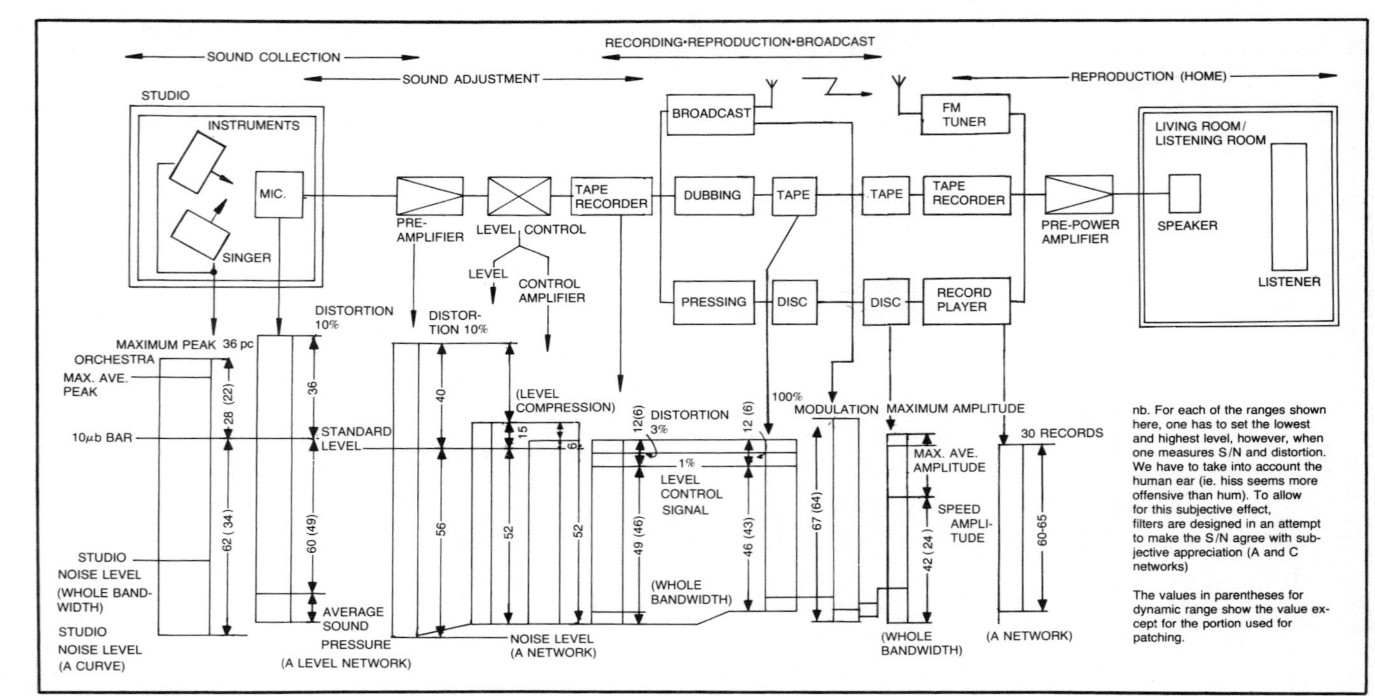

Fig. 1-2. A diagrammatic representation of the dynamic range of the audio chain.

range. For instance, there is the noise level in the studio, particularly when low frequency noise is caused by artists or technical staff moving about, or the noise due to air currents and breath; and all types of electrically induced distortion.

With regard to the pre-mixing and level amplifiers, in the ideal situation where the diagram was a representation of reality, the engineer would experience no great problems. However, depending on the equipment used for level adjustment, the dynamic range is bound to be limited. Because the dynamic range for both low and high levels is limited by using equalization, it is necessary to take this into account, and to stipulate the parameters generally found in actual use. It is, however, the secondary sources, such as tape, disc, and FM broadcasts which, more than anything else, limit the dynamic range available from the recording chain. As a result, control amplification and level compression become necessary, and this effects the sound quality of the audio chain. Furthermore, if one considers the fact that the master tape and mother tape are interposed between any measurement results and the production of the three secondary sound sources, one will appreciate that the narrow dynamic range available from conventional tape recorders becomes a sort of "bottleneck" which affects the whole process. Then again, if one considers the audio reproduction chain, it is clear that the output power of the speakers and the ambient noise level in the listening room also limit the reproduction quality, although at this moment the dynamic range of a speaker is wider than that of a tape recorder.

When considering sound quality, and distortion (in the broadest sense), it is inevitable that one must rely on listening tests, but a number of problems are associated with this. In relation to microphones, when one compares the overall frequency response of a directional microphone with the frequency response of the front axis, then the overall frequency response is very narrow, and many difficulties may be caused by untraceable spurious noise. With tape recorders, it is abundantly clear that not only will there be tape modulation noise, but that non-linear distortion will be at least one digit worse than that caused by other equipment. In the case of record reproduction, there are the problems of low high frequency limit of the cartridge (at present only 15 kHz) and crosstalk between the low and high frequencies.

Because of the characteristics of the limiter positioned before the modulator in the FM transmission chain, and because the transmission relay lines are independently synchronized, it is not actually possible to make a direct stereo transmission from conventional sources. What actually happens is that a 19 cm/s tape recording is used, which has previously been dubbed from the mother tape at 8 times normal speed at the local station. As a result, great discrepancies in sound quality and stability are possible, depending on the local station, because the local substation receives the signals from the parent station and the main local station, which, in effect, is transmission relay system which rebroadcasts the original transmission.

Speaker performance is affected by a number of factors linked to distortion; such as high frequency distortions of low frequency components because of the non-linearity of the diaphragm and drivers; substantial mid and high frequency distortion caused by unexpected oscillations of the diaphragm; and all types of modulation distortion. Generally speaking, if one looks at the various equipment used for sound pick-up, recording and reproduction from the point of view of design, then this equipment has to trace a signal which is varying greatly in amplitude over short periods of time. It is of course impossible to design a speaker sufficiently accurate to faithfully reproduce the original sound source air vibrations, and this is the reason why the speaker is the source of many elements which degrade the original signal.

To continue on the subject of speakers, it is first of all impossible to realize an actual method of cancelling the oscillation of the high frequency band on playback because of the pistonic motion of the diaphragm. In addition to this it is also extremely difficult to design adequate damping for vibrations generated in the frame and cabinet. Obviously all these are major contributory factors to the loss of sound quality, and all designers and engineers fully realize the enormous effort which has to be put into the development of materials, the analysis of optimum dimensions and so on.

To summarize, the areas which require the most attention to improve musical quality in the future, from the point of view of dynamic range, distortion (in its broadest sense), and evaluation of musical quality, are tape recorders and speakers. Besides this, some improvements should be made in the system of transmission for FM broadcasts. But fundamentally, looking back at the history, it is the tape recorder, the item of equipment which provided the motive force for the development and prosperity of the audio industry, which must be revolutionized to meet demands for improvement of sound quality over the coming decades. (The content of this section on FM broadcasting refers to customary practice in Japan.)

THE INTRODUCTION OF PCM AND THE REVOLUTIONIZING OF THE TAPE RECORDER

Sony took a new look at revolutionizing the tape recorder, from the standpoint of practical use, paying particular attention to the record and reproduction heads, and recording tape, as well as looking at the quality of the signal and subjective musical testing. With regard to the actual tape path, with particular reference to the capstan and pinch roller, a number of factors affect the level of modulation noise and wow and flutter: If the following arrangements are used, that is, direct coupling of the capstan and the motor; a closed loop system with the motor pulley and two capstan fly wheels forming a triangular arrangement linked by a belt, then the stability of the tape path is greatly increased, and both modulation noise and wow and flutter are reduced. At present, wow and flutter of 0.02% wrms at 38 cm/s, and of 0.04% wrms at 19 cm/s is obtainable from conventional tape recorders. A great deal of effort has been put into the improvement of both tape and heads. Oxide (2 $Fe_2 0_3$) particles have been reduced even further in size (an improvement of between 3 and 6 dB has been achieved by reducing the diameter of the acicular crystals from 0.8 μm to between 0.2 and 0.4 μm), and have also been packed much closer together without loss of coercivity. Improvements have also been made by the introduction of alloy tapes and evaporation techniques to achieve a wider dynamic range and a better high frequency performance. New types of alloy have been used to make more durable high performance heads, and the Hall effect has been re-adapted for head technology, to gain improvements in low frequency performance and to lengthen head life. Improvements have also been effected by means of compression/expansion systems such as Dolby, Noisex, dBX, etc., so that the sound quality available from conventional tape recorders has slowly but surely improved.

The performance of a conventional 38 cm/s 2 track stereo tape recorder using 6 mm wide tape is shown in the graphs for Figs. 1-3 through 1-6. Should improved characteristics be necessary, leaving aside the question of price, the tape speed may be increased and the tape breadth increased. However, at the present stage of development, it is impossible to carry out any further major improvements in the performance of conventional tape recorders due to physical limitations which cannot be overcome.

One of the ways to overcome the problem of having effectively reached the limits of conventional tape recorder development is to use the pulse code modulation (PCM) system of digitizing the

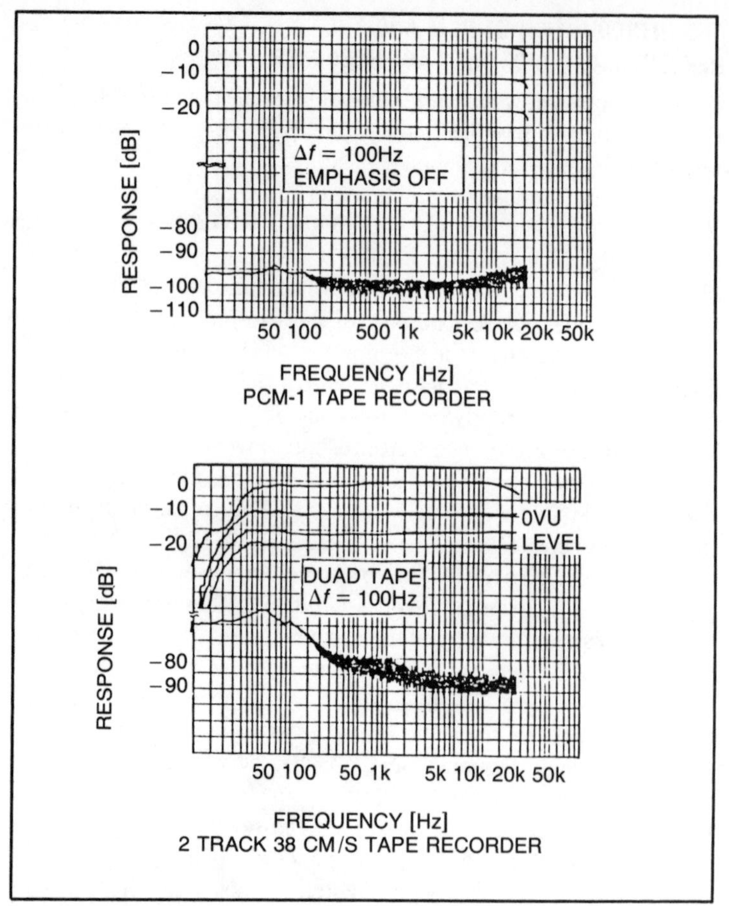

Fig. 1-3. Dynamic range.

original source signal for record and reproduction. The first group to become actively involved in practical application of digital technology as a way of achieving total fidelity to the original source was the NHK Technical Research Institute.

The equipment designed by NHK used a one-inch 2 head helical scan VTR as the recording medium, and the digital part employed a sampling frequency of 30 kHz and was a 12 bit companded system (five polygonal line quantization). The first public demonstration was in May 1967. At that time, the prototype was mounted in a vertical racking system (Fig. 1-7), and approximately 3,500,000, yen was necessary for parts of this prototype (the selling price might be probably at between 20 and 30 million yen). At that time, the impression of most of the people who heard it was that the clarity and excellence of the sound produced by the digital equipment just could not be matched by any conventional tape recorder. However, from a musical point of view the impact of live sound could still not be captured by the digital equipment. So, from that point of view, this particular PCM prototype had not fully succeeded. This prototype was not the ideal unit with which to assess the capabilities of digital recording equipment, because it employed microphones, preamplifiers, mixers, etc., and all the conventional attitudes to tape recording, rather than being a totally new departure into what was, and is, a totally new technology.

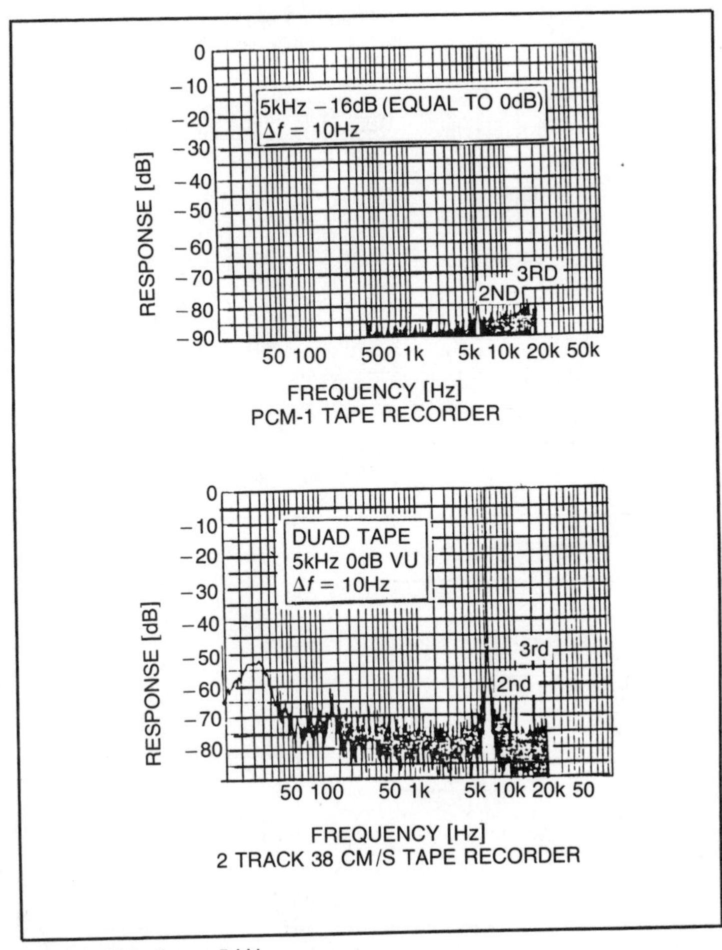

Fig. 1-4. Distortion at 5 kHz.

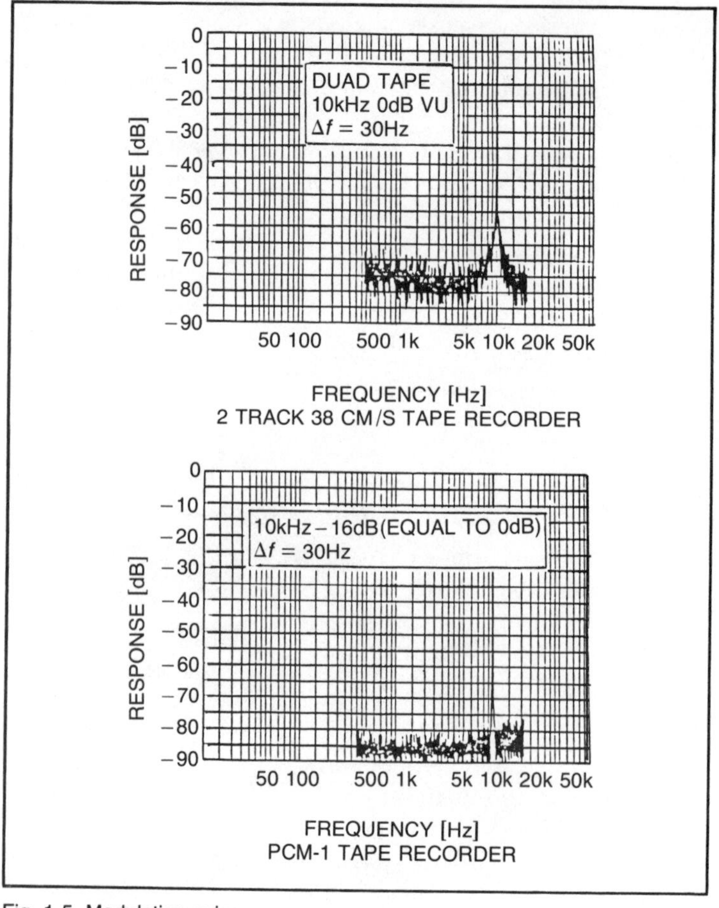

Fig. 1-5. Modulation noise.

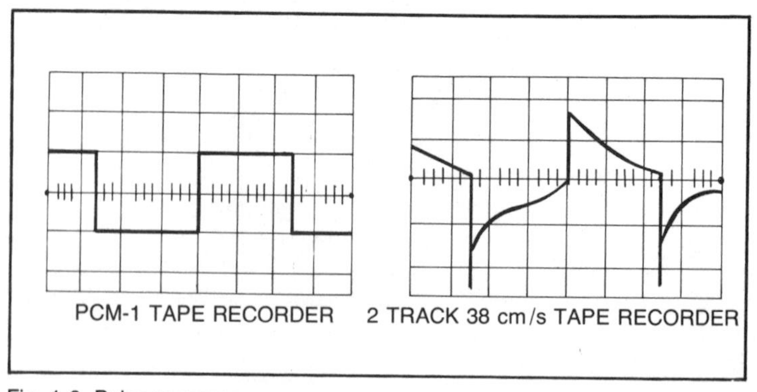

Fig. 1-6. Pulse response.

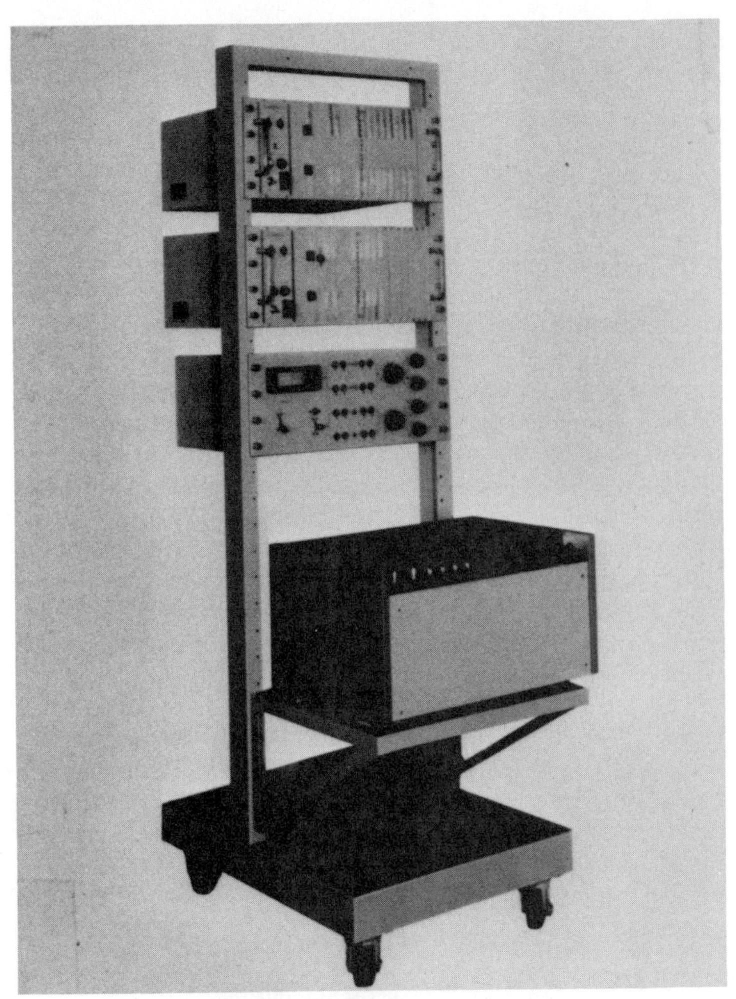

Fig. 1-7. The first PCM stereo tape recorder.

However, after this demonstration at least, there were no longer any grounds for doubting the high sound quality achieved by PCM techniques. The engineers and music lovers who were present at this first public PCM playback demonstration had no idea when this equipment would be commercially available, and many of these people had only the vaguest conception of the effect which PCM recording systems would have on the audio industry. In fact, it would be no exaggeration to say that owing to the difficulty of editing, the weight, size and difficulty in operation, as well as the high price, not to mention the necessity of using the highest quality ancillary equipment (another source of high costs), it was at that time very difficult to imagine that any meaningful progress could be made.

At the same time, Nippon Columbia had started developing a system for producing records by the direct cutting method. The logic behind this was due to a belief that the major cause of limitations in sound quality was the analog tape recorder. In direct cutting, the lacquer master is cut without using master and mother tapes in the production process: the live source signal is fed directly to the cutting head after being mixed. This sounds quite simple in principle, but is actually extremely difficult in practice. First of all, all the required musical and technical personnel, the performers, the mixing and cutting engineers, have to be assembled together in the same place at the same time. Then the whole piece to be recorded must be performed right through from beginning to end with no mistakes, because the live source is being fed directly to the cutting head. It was at this point that they decided to try to develop PCM equipment because, besides the advantages in improved sound quality, it had the merits of solving the problems of time and venue posed by direct cutting. PCM recording meant that the process after the making of the master tape could be completed at leisure. The development prototype was loosely based on the PCM equipment originally created by NHK; a 4 head VTR with two inch tape was used as a recording medium, with a sampling rate of 47.25 kHz for the 13 bit linear PCM adaptor by 1969[(5)]). This machine was the starting point for the PCM recording systems which are at present marketed by Sony, after a long process of investigation and adaptation.

THE PRESENT STATE OF PCM TAPE RECORDING

The introduction of PCM tape recorders was planned by one section of the recording industry, and because this equipment was to be introduced as a secondary source, it was necessary to develop a simple method of editing, and to obtain international standardization for sampling frequencies and signal format.

The system uses helical scan VTRs and has gone hand in hand with the development of video technology. The PCM system using unmodified VTRs for record coupled with a digital adaptor to digitize the source signal was developed by Sony in 1977. In the following year, the PCM-1600 digital processor coupled with a broadcast standard VTR, and the PCM-1 coupled with a domestic

VTR were marketed (see Figs. 1-3 through 1-6)[6]. In April, 1978, the use of 44.056 kHz as a sampling frequency (the one used in the above mentioned models) was accepted by the A.E.S. (Audio Engineering Society). At the same time, a committee for the standardization of matters relating to PCM audio adaptors using domestic VTRs was established in Japan by 12 major electronics companies. In May 1978 they reached agreement on the EIAJ (Electronics Industry Association of Japan) standard, and now all Japanese companies are pursuing research into commercial PCM equipment based on this standard.

In the 1970s the age of the video disc began, with three different systems being pursued: the optical system, where the video signal is laid down as a series of fine grooves on a sort of record, and is read off by a laser beam; the capacitance system, which uses changes in electrostatic capacitance to plot the video signal; and the electrical system, which uses a transducer. Engineers then began to think that since the bandwidth needed to record a video signal on a videodisc was more than the needed to record a digitized sound signal, the same system could be used for PCM/VTR recorded material. Thus the Digital Audio Disc (DAD) was developed, using the same technology as the optical video disc: in September 1977, Sony, Mitsubishi and Hitachi demonstrated their DAD systems at the Audio Fair.

1978 was used as an opportunity to announce their respective systems, rather optical, capacitance or electrical, by the various manufacturers. Everyone had announced their own particular system, but because everyone knew that the new disc systems would eventually become widely used by the consumer, it was absolutely vital to reach some kind of agreement on standardization. To further this, the DAD (digital audio disc) convention was held in September 1978. All together, 35 manufacturers from all over the world met to try and further international standardization.

To summarize, digital audio has passed through the developmental stage, to that of sales, and we have seen the rapid advances, and the moves towards standardization. It is already possible to translate PCM audio signals, given sufficient bandwidth, and experiments have been carried out concerning satellite transmission. At the 1978 CES (Consumer Electronics Show) held in the U.S.A., an unusual display was mounted. The most famous names among the American speaker manufacturers demonstrated their latest products using a PCM-1 digital audio processor and a domestic VTR as the sound source. Compared with the situation only a few years ago, when the sound quality available from tape recorders was regarded as being of relatively low standard, the testing of speakers using a PCM tape recorder marks a total reversal of the previous situation. The audio industry has made a major step towards true fidelity to the original sound source through the total redevelopment of the recording medium, which used to cause most degradation of the original signal as it passed through the recording chain.

References

1. Nakajima, Matsuoka, Hayashi, Ito, Fujita and Kurakake; "Special Developments in Stereo Transmission", NHK Publications, 10, 3 (1976-3).

2. Nakajima, Heitaro: "Problems in Sound Reproduction", Onkyoshi, 27, p.398 (1971).

3. Nakajima, Heitaro: *Audio Engineering* Jikkyo Publishing (1973-6).

4. Hayashi, Kenji: "Stereo Recording Equipment", NHK Publications, 12,11 pp.12-17 (1969).

5. Anazawa, Kiyoaki: "A PCM Recorder using a 4 Head VTR", Television Society Recording Society, 11-4 (1975-3).

6. Nakajima, H., T. Doi, T. Tsuchiya, A. Iga and I. Ajimine: "A New PCM Audio System as an Adaptor for VTR", AES 60th Conv., No. E11 (1978-5).

Chapter 2

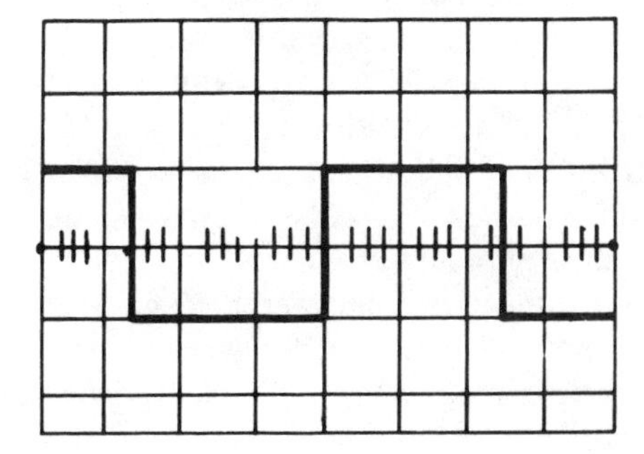

The Revolution of Audio Technology

Digital audio technology is beginning to gradually effect changes on convential audio thinking, fusing as it does two essentially unrelated fields of research and development: digital technology as used in computers and in satellite communication; and audio technology which pursues the art of attempting to reproduce reality (until now, by analog techniques.)

The type of audio equipment used in the domestic setting, as well as in studio recording, record production, and broadcasting, will undoubtedly be changed profoundly by the appearance of digital audio. In the following pages, you will find an outline of the basic principles involved in PCM, which is an essential background for learning about digital audio, and a brief description of the equipment which has so far been developed. Essentially, this chapter is a preview of the coming revolution in audio: based on the economy of VLSI, digital technology will soon become as much a part of modern life as the electronic calculator or the digital watch.

A NEW PHASE IN THE COMMUNICATION OF INFORMATION—PCM

During the course of history, the long struggle to perfect means of transmitting, storing, and organizing information has had enormous effects on human life and the structure of society. Every advance, from the printing press to the computer, has forced us to adapt ourselves and our lives to a new world of information.

Looking back over the past, the use of beacon fires or drums for communication, the invention of writing for making records, and the development of calculation devices, such as the abacus, led to development and consolidation of a structured society. Communications technology develops at an exponential rate, and especially since the nineteenth century with the inventions of the Morse code in 1837, the telephone by Bell in 1876, the recorder by Edison in 1877, and wireless telegraphy by Marconi in 1895, life for everyone has undergone many profound changes. If we move to the twentieth century and look at the period between 1940 and 1950, this was the time when the most decisive strides were made in the field of communications technology. One of the most revolutionary inventions in this period was the first computer: ENIAC (Electronic Numerical Integrator and Computer), developed at University of Pennsylvania in 1943 with the cooperation of the U.S. Army.

During the same decade, in 1948 to be precise, Shockley invented the transistor, and Shannon published his communications theories [1], which are the foundation upon which the PCM system is based.

These three events accelerated progress in development of communications (in the widest sense), because their mutual influence caused a myriad of multiplicative effects. Advances in the field of semiconductors, starting with the transistor, led eventually to the development of all types of digital IC and LSI. This in turn enabled the realization of PCM communication, linking all parts of the world by telephone lines, and satellite communication. Looking back over this process of development, we have progressed from the 30 ton ENIAC computer, using 18,000 valves, to the microprocessor, which sits comfortably in the palm of a hand. Conversely, advances in computer technology have stimulated the development of high density LSIs.

However, it was not until very recently that these new technologies, and in particular a close relationship of diverse technologies, could be applied to the audio field. This was because it is an area where highly specialized types of information, i.e., the musical signal, are handled.

The marriage of these diverse technologies will have a profound effect on the audio world. Or rather, the introduction of PCM technology, built upon semiconductor development, and advances in computer technology, will have a deep and lasting impact.

To put it simply, it is now possible to produce tape recorders and record players capable of a standard of reproduction, which would, only a few years ago, have been impossible. In the following pages, there is an outline of PCM, and of digital audio in general.

BACKGROUND INFORMATION ON PCM[2]

Since time immemorial, the most basic and most common form of communication has been the spoken word. At present, we are living in the type of society which demands a high level of communication. In fact, we seem to be demanding an ever-increasing level of communication, which has to be dealt with ever more accurately, and speedily. As a result, people are demanding quicker methods of transmission, and better methods of transcription.

As shown in Fig. 2-1, the world is linked together for communications purposes by orbiting satellites and undersea cables by wireless communication and microwave link. In addition, research into the use of fibroptics has been carried out increasingly over recent years. However, even in the field of person-to-person communication, let alone the high level communications media mentioned above, there still remain a number of rather troublesome problems which have to be ironed out satisfactorily. As shown in Fig. 2-2, whatever type of communication medium used, the message transmitted is subject to interference of some kind. Even in a normal conversation, excessive background noise may make it difficult to communicate accurately. In fact, interference is the biggest problem in the accurate transmission of information. Ever since radio waves were first used as a means of communication, the

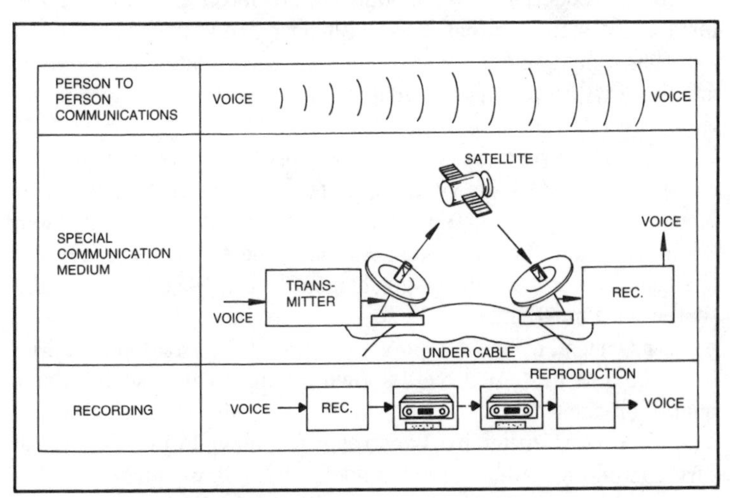

Fig. 2-1. Vocal communication systems "Transmission of Information."

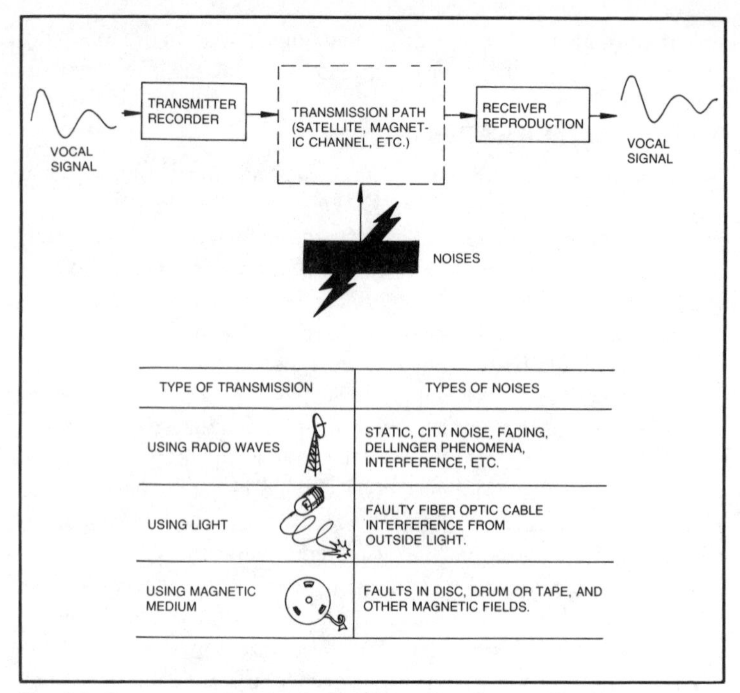

TYPE OF TRANSMISSION	TYPES OF NOISES
USING RADIO WAVES	STATIC, CITY NOISE, FADING, DELLINGER PHENOMENA, INTERFERENCE, ETC.
USING LIGHT	FAULTY FIBER OPTIC CABLE INTERFERENCE FROM OUTSIDE LIGHT.
USING MAGNETIC MEDIUM	FAULTS IN DISC, DRUM OR TAPE, AND OTHER MAGNETIC FIELDS.

Fig. 2-2. Common communication problems "Interference."

study of modulation systems has been the major item on the agenda for many researchers. Different modulation systems may bring a significant reduction in the amount of interference by altering the form of the signal so that it can be transmitted effectively.

The oldest, and the most commonly used modulation systems are the AM (Amplitude Modulation) and FM (Frequency Modulation) systems.

The AM system is most commonly used for commercial broadcasting systems. As shown in Fig. 2-3 (A), a fixed frequency carrier sinuous wave is modulated by the source signal for transmission. After reception, the varying amplitude of the carrier is recovered, and used to reproduce the original signal. In Fig. 2-3 (B) one can see that it is almost impossible to totally remove the interference patterns caused by the complex structure of the envelope in AM.

The FM system is well-known, owing to its use in higher quality broadcast systems, and was first introduced about 40 years ago, after its invention by Armstrong. As shown in Fig. 2-3 (B), in FM a fixed amplitude carrier sinuous wave is modulated by the signal to be transmitted. However, if the received carrier wave is limited sufficiently, as shown in Fig. 2-3 (B), it is possible to remove almost all interference patterns.

Thus, both AM and FM systems have several common features. They are both analog systems, because they deal with continuously varying quantities. In both cases, it is necessary to reconstruct the original transmitted signal upon reception, from a continuous signal which has been altered in some way (in the case of AM, the amplitude of the envelope, and in FM, changes in frequency.)

In everyday life, most people use analog qualities to discriminate between objects: for example, the shape or color of an object. However, such a method of description can be vague or possibly lead to inaccuracies. The most accurate way of identifying abstract principles is by digital analysis, which refers only to "presence" or "absence". For specifying some desired object or principle, digital analysis is by far the most accurate method.

At present, the PCM systems used in the communications field are, for the most part, modulation systems which operate by transmitting sound signals which have been converted to an "on", "off" format. As shown in Fig. 2-4 (A), modulation is a uniform process, with the "on", "off" components represented respectively by "1" and "0".

The basic process is as follows:

☐ The original musical signal is sampled by a fixed frequency, and the value of the sampled level is ascertained. The technical term

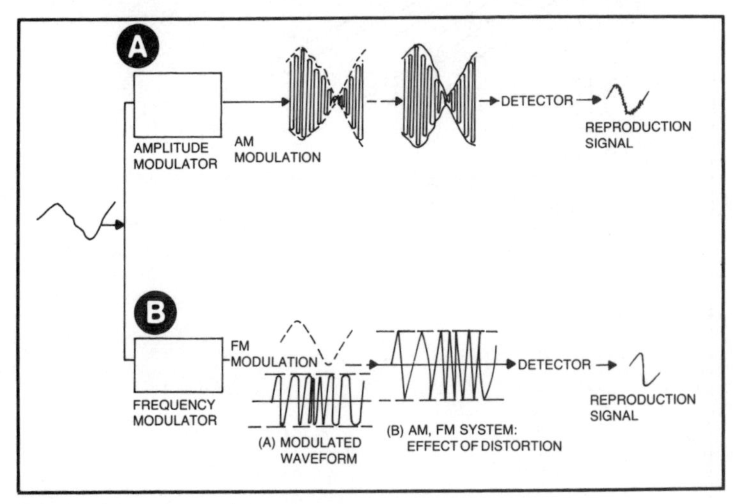

Fig. 2-3. Present modulation systems.

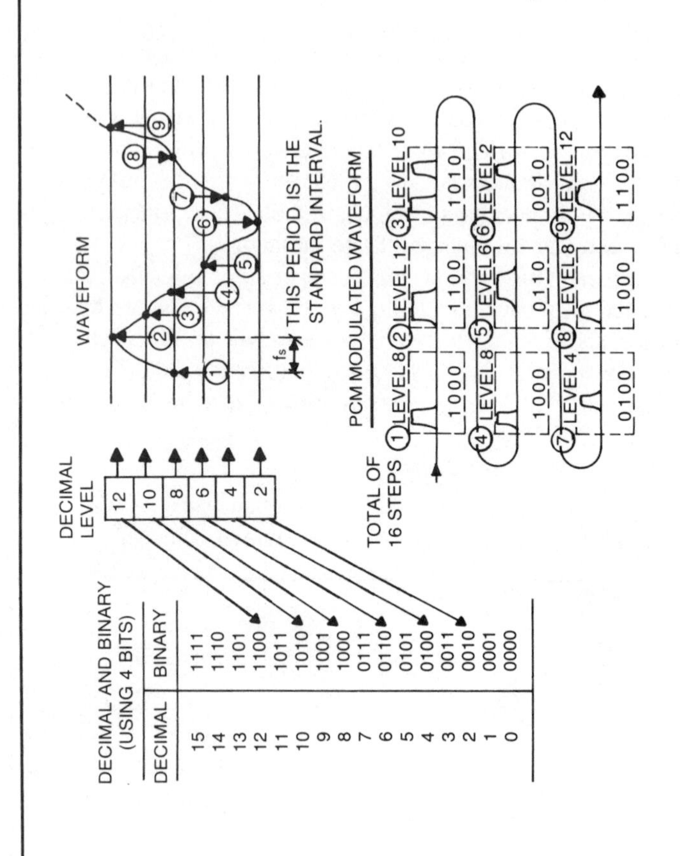
DECIMAL AND BINARY
(USING 4 BITS)
DECIMAL
BINARY
15 1111
14 1110
13 1101
12 1100
11 1011
10 1010
9 1001
8 1000
7 0111
6 0110
5 0101
4 0100
3 0011
2 0010
1 0001
0 0000
DECIMAL LEVEL
12
10
8
6
4
2
WAVEFORM
f_s
THIS PERIOD IS THE STANDARD INTERVAL.
TOTAL OF 16 STEPS
PCM MODULATED WAVEFORM
1 LEVEL 8
1000
2 LEVEL 12
1100
3 LEVEL 10
1010
4 LEVEL 8
1000
5 LEVEL 6
0110
6 LEVEL 2
0010
7 LEVEL 4
0100
8 LEVEL 8
1000
9 LEVEL 12
1100

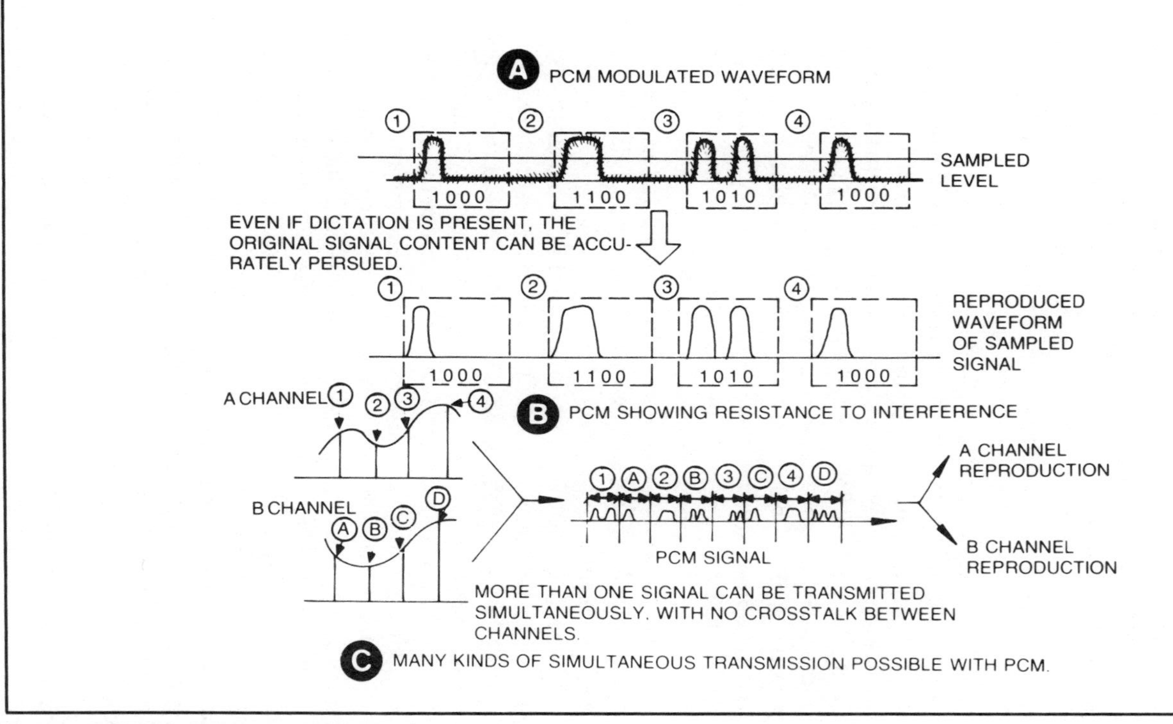

Fig. 2-4. PCM (Pulse Code Modulation) systems.

for this process is "sampling". If the fixed frequency used to sample the original waveform is chosen correctly, then the original sample can be recovered accurately by calculation using the sampled level.

□ The value of each sampled level is then converted into binary notation. The resolution of the system used (for example, 14 bit or 16 bit) determines the method of conversion into binary. The configuration of bits used are known as the quantization bits, and it is accepted that 14 or 16 bit systems are most commonly used.

□ The PCM modulation process is complete when digital values representing each original sampled level are collated, taking the form of strings of "1"s and "0"s.

A PCM signal encoded in the way outlined above has three major advantages. The first of these, as shown in Fig. 2-4 (B), is that a PCM signal is basically immune to the traditional bugbear of distortion. Even if the original signal is subject to distortion, sampling at a fixed level allows reproduction of the original in as near perfect a way as possible. However, even with a PCM system there are certain kinds of distortion which interfere with the accurate retrieval of the encoded signal. There are ways to overcome this problem; as explained in Chapter 6, the use of complex error correction codes enables the system to nullify the effects of interference with the digital signal. Thus, a PCM modulated system is, in one way or another, able to function without signal degradation caused by distortion.

The second major advantage inherent in PCM modulated systems is shown in Fig. 2-4 (C). PCM signals can be multiplexed; that is, many different signals can be mounted on one PCM signal. This is known as a time sharing multiplex transmission system. Even when numerous encoded signals are multiplexed in this way, there is absolutely no crosstalk between the different signals.

The third major advantage is the afore-mentioned quantization system. The number of quantization bits used determines the musical quality (signal-to-noise, S/N) at the time of reproduction. If a large number of quantization bits are used, then a PCM modulated system can obtain astoundingly high S/N ratios, which simply cannot be attained by other modulation systems. If one looks at a 14 bit system, for example, the theoretical S/N ratio for reproduction of a musical signal is in excess of 85 dB. This point will be further explained in the next chapter.

THE INCORPORATION OF PCM SYSTEMS INTO THE AUDIO CHAIN

PCM modulation systems have led to enormous improvements

in communications technology, and are beginning to demonstrate the same startling innovations in a field somewhat closer to everyday life: in home Hi-Fi systems. In order to fully appreciate the revolutionary improvements possible with PCM, it is necessary to examine the type of system which is, at present, being used both at home and in a studio environment.

The raison d'être of home Hi-Fi systems is to provide relaxation and enjoyment through the ability to reproduce high quality musical or sound programs from various sources. In many homes, the total audio system will be fairly comprehensive, as shown in Fig. 2-5.

Many people, especially concert-goers, demand an ever increasing standard of excellence from their Hi-Fi system. The enthusiast who wants to enjoy music will probably own a tuner, a tape deck, a record player, an amplifier, and, of course, a pair of speakers. The high quality sound source is supplied by FM radio broadcasts or from records, but these depend as much on manufacture as on studio recording techniques.

In this audio-chain from studio to home, there is one item which does not quite measure up to the technical excellence of the other pieces of equipment. And that is the tape recorder.

Both at home and at the broadcasting station the tape recorder is used as a "filing system" for important pieces of music. At the recording studio it is used in every stage of production from picking up the live sound to producing the record. (Editing, with dubbing from one tape recorder to another, is nowadays almost standard practice in a studio.) The deleterious effects on the musical signal can be gauged from the information that each record/reproduction cycle generates 0.5% distortion. Therefore, the first major step towards introducing PCM systems into the audio field was as a replacement for the conventional studio tape recorder. This meant that the distortion introduced and compounded by the conventional tape recorder before the source signal was broadcast, or released as a record, simply no longer occurred.

As shown in Fig. 2-6 (A), the main reason why conventional analog tape recorders cause such a deterioration of the original signal is firstly that the magnetic material on the tape actually contains distortion components before anything is actually recorded; secondly, the medium itself is non-linear, that is, it is not capable of recording and reproducing a signal with total accuracy. Distortion is, therefore, built-in to the very heart of every analog tape recorder.

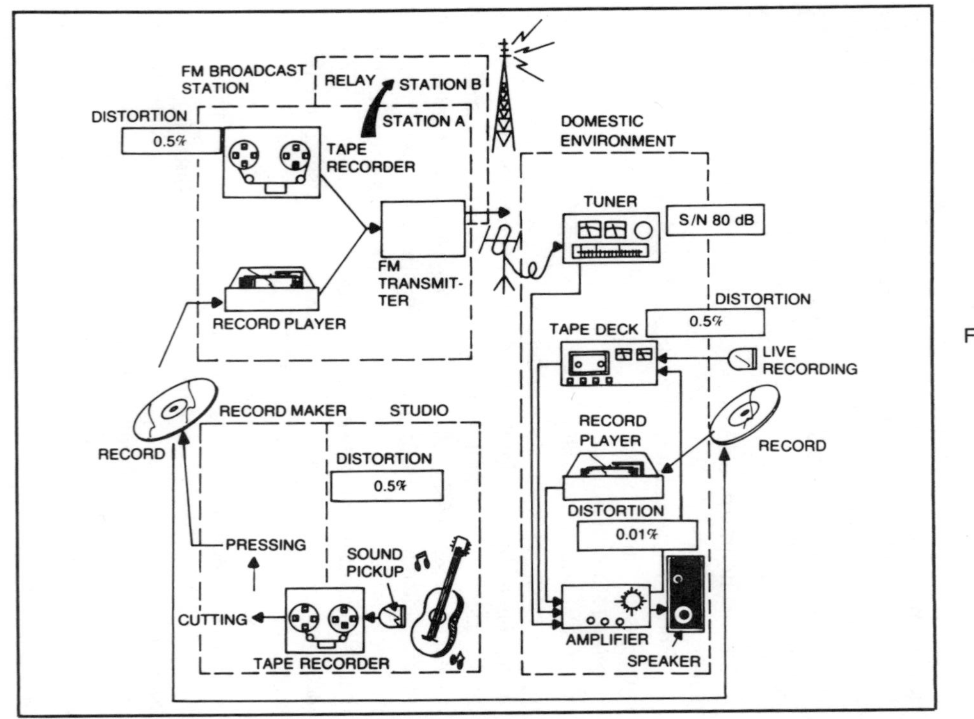

Fig. 2-5. Conventional audio chain.

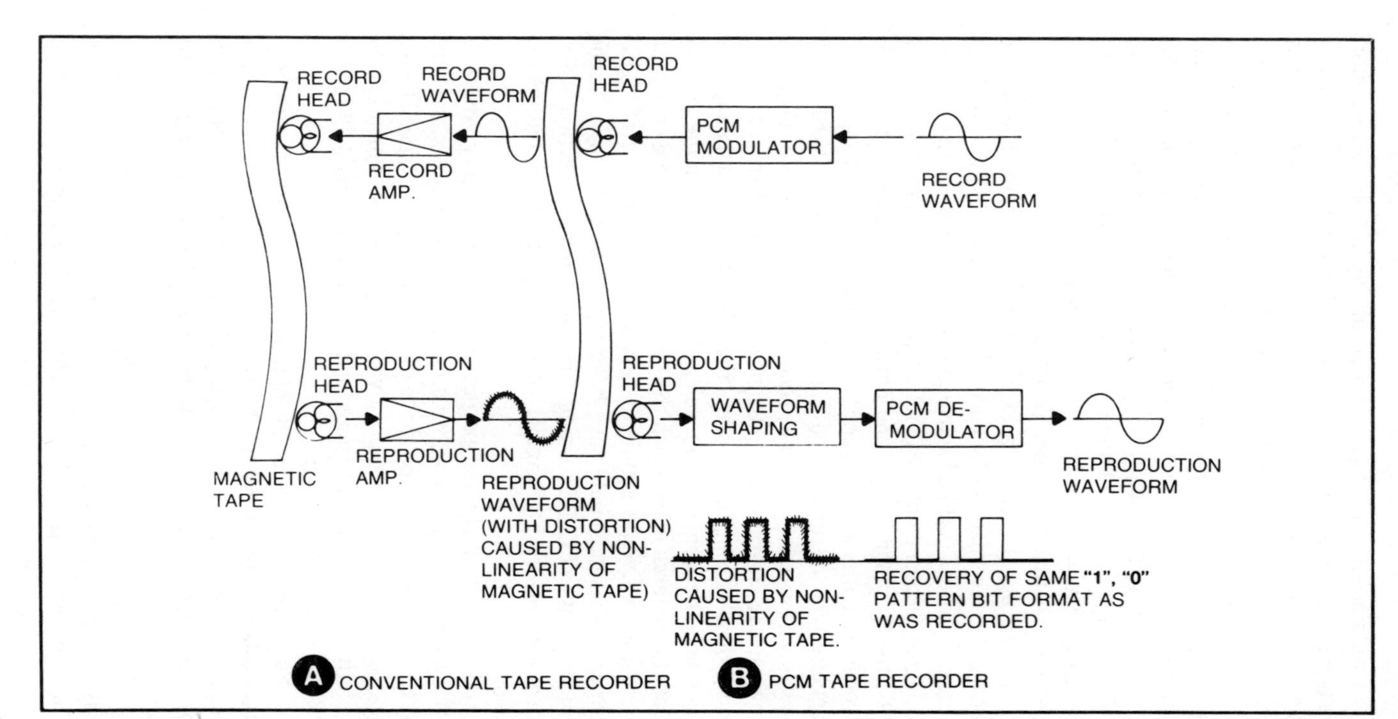

Fig. 2-6. Tape recorder using PCM system.

Ever since the development of ac biasing in 1940, every effort has been made to improve the sound quality available from analog tape recorders. Great improvements have been made in many areas; For instance, magnetic tape formulations, heads, mechanics, circuitry, and so on. In fact, modern analog tape recorders have only a passing resemblance to their forebears of only forty years ago. However, the analog tape recorder has reached the uttermost limits of development; it is simply not possible to refine it any further in a search for improved sound quality. In fact, the law of diminishing returns has set in, slight improvements require a disproportionately massive investment of time, money and manpower.

Figure 2-6 (B) shows a PCM recording system. The source signal to be recorded is transformed into a series of "1"s and "0"s, using the process outlined above, by PCM modulation. In this case, the signal recorded onto magnetic tape merely represents "on" or "off", the "1"s and "0"s of the binary code. Of course, this PCM encoded signal is also subject to the shortcomings of tape itself: magnetic tape distortion and non-linearity. However, the signal itself is very simple and can be recovered by a waveform shaper as shown in the diagram. Circumstances may arise where part of the digital signal is completely lost, if there is a dropout on the tape, for example. (The term drop out refers to a deficiency in the layer of magnetic material upon which the desired signal, analog or digital, is recorded.) Even a complete, albeit temporary, loss of digital information which should have been recorded on the tape, can be remedied with a PCM recording. The use of error correction codes, which can restore the original signal completely, are outlined in Chapter 6.

So, to all intents and purposes the musical signal, which is retrieved from the digital information actually recorded on the tape, has been rendered immune to the effects of magnetic tape. The net result is a recording of unbelievably high quality, which can be digitally rerecorded many times, and still maintain its fidelity to the original source.

Furthermore, the principle of time sharing multiplex transmission systems (shown in Fig. 2-4 (C) may be applied to PCM tape recorders. This means that a multi-channel PCM recorder which does not suffer from cross-talk between channels is a perfectly feasible proposition.

A domestic PCM tape recording system would, at present, consist of two items, a PCM processor and a Video tape recorder. (The latter would be needed to record the digital signal; a VTR can handle a much wider bandwidth than an ordinary tape recorder.) Figures 1-3 and 1-4 show the great performance differences between a PCM recording system and an analog 38 cm/s 2 track tape recorder.

However, the application of PCM technology to Hi-Fi is not by any means limited to tape recorders. The LP record can be digitized into a Digital Audio Disc (DAD), and many peripheral items such as mixers, reverberators and so on, have digital equivalents. following these additions, one can expect that, in the very near future, PCM broadcasts, PCM microphones and even PCM speakers will join the digital family. In fact, the advent of this wide range of digital equipment is not as far in the future as many people imagine.

THE APPEARANCE OF DIGITAL AUDIO EQUIPMENT[(3)-(10)]

The process of technology advance is a very curious one. Even more curious is the type of break-through which occurs when an existing technology has reached its practical limits when pursued in a traditional manner. The well-worn cliché, "Necessity is the mother of invention" covers the situation well enough to bear repeating. Technological innovation invariably occurs at the point where existing technology is causing a bottleneck, stultifying development. To put it concretely, in the audio field, it was absolutely necessary to provide something to supercede the analog tape recorder. However, as soon as the first, most pressing, technological gap has been plugged in this way, the second becomes more urgent, then another and another and another appears. A technological breakthrough resembles a stone thrown into a pond; the ripples caused by the stone affect an ever increasing area of water, long after the stone has sunk.

The following pages constitute a brief introduction to numerous items of digital audio equipment developed over the last few years. These PCM systems constitute the advance guard of digital equipment which will soon revolutionize the world of Hi-Fi reproduction.

Digital Audio Processor

It is a simple matter to talk glibly about PCM tape recorders and their putative effect, but it is a much more onerous matter to produce a marketable product. Before a theoretical concept can be made fact, many preconditions must be fulflled.

In the case of digital audio processors, the first steps of basic theoretical research had to be successfully completed: principles current in communications research had to be re-thought for audio purposes, and the technology to enable the design and creation of suitable equipment had to be established. But after these first basic steps came equally important considerations: that equipment is used by people, that it must be convenient to operate, that the price must be reasonable.

Before designing a PCM tape recorder, one would, first of all, need to consider the various basic sections into which such a machine might usefully be divided. There has to be a section for converting the source signal into a digital PCM signal, as well as the reverse, a decoder for the encoded PCM signal, and of course, there has to be a recording medium, using some kind of magnetic tape for record and reproduction of the PCM encoded signal.

The time period occupied by one bit in the stream of bits composing a PCM encoded signal is shown in Fig. 2-7, and is determined by the sampling frequency and the number of quantization bits, as previously mentioned. If we assume that a sampling frequency of 50 kHz is chosen, (sampling period 20 μs), and that a 16 bit quantization system is used, then the time period occupied by one bit when making a two channel recording will be 0.6 μs. In order to ensure the success of the recording, detection bits for the error correction system will also have to be included. As a result, it is necessary to employ a record/reproduction system which has a bandwidth of between 1 and 2 MHz.

Bearing in mind the parameters needed for the actual recording system, the most suitable practical recorder is a video tape recorder. As shown in Fig. 2-8, the VTR was specifically designed for recording TV pictures, that is, a video signal. To successfully record a video signal, a bandwidth of several Megahertz (MHz) is necessary, and it is a happy coincidence that this makes it eminently suitable for recording a PCM signal also.

Over the last few years, VTRs have been used more and more widely in many different fields. Rotary head VTRs were first introduced into the broadcast industry, and the U-matic format is widely used for many commercial applications. The U-matic is a general purpose industrial machine of the helical scan type. Nowadays, domestic VTRs are becoming popular, whether Betamax or VHS. The domestic VTR is now a mass-produced, competitively priced consumer item. In fact, it would be quite convenient to use it for digital audio purposes, in addition to its use as a home video recorder, because the domestic VTR is readily available, simple to operate and very convenient. (At the present time PCM processors

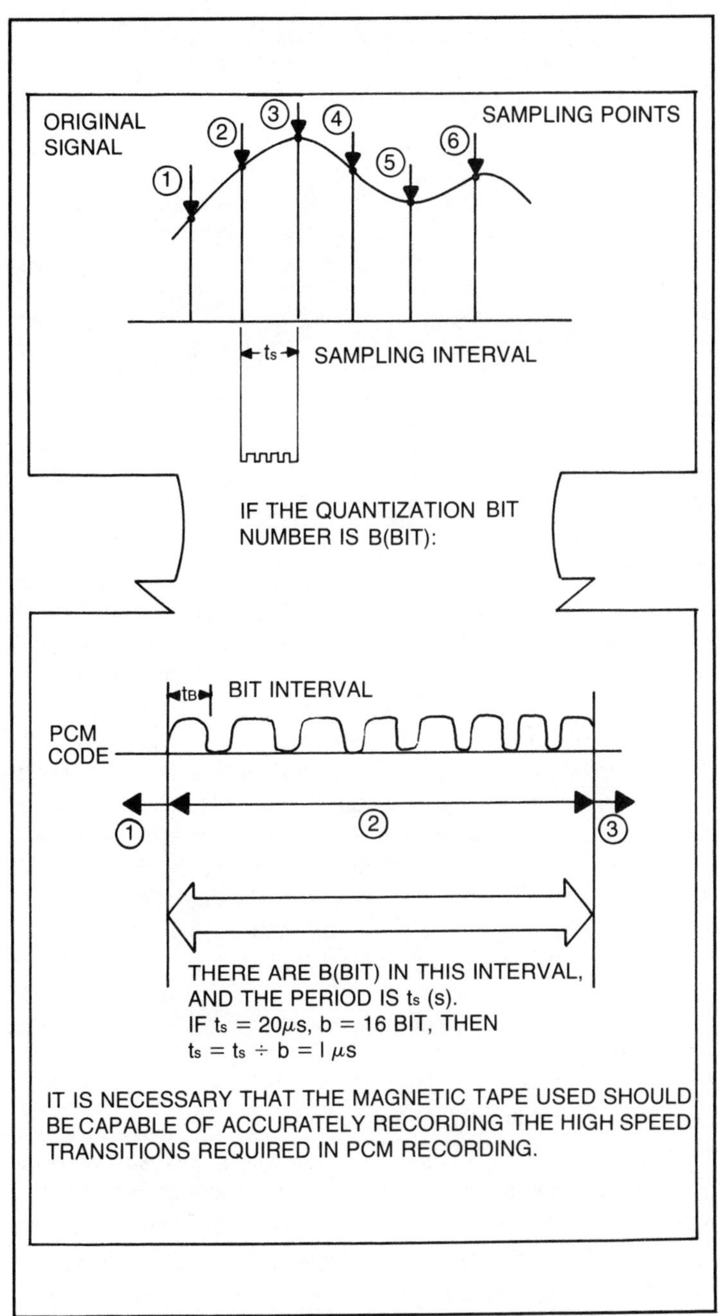

Fig. 2-7. The necessity of a wide bandwidth in PCM recording.

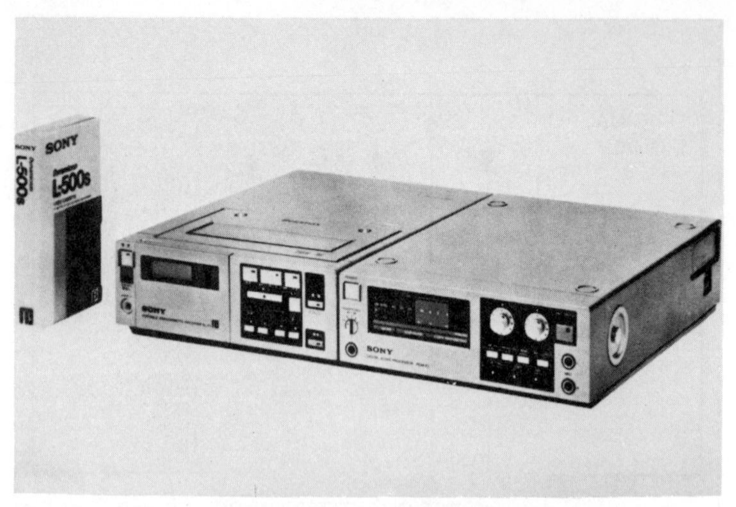

Fig. 2-8. VTR used as PCM recording medium (with digital audio processor).

are generally only suitable for use with EIA/NTSC standard VTRs, EIA/NTSC being the TV system used in Japan and the USA. The TV system used in most of Europe is CCIR/PAL or SECAM so a triple standard VTR would be needed in Europe is one wanted to watch video cassettes as well as listen to digital audio cassettes.)

The suitability of the VTR as a recording medium meant that the first PCM tape recorders could be developed as a two-unit system: a VTR and a digital audio processor. (The latter can be connected directly to an analog Hi-Fi system for actual reproduction.) Since a handy recording medium was readily available, development work could be concentrated on the digital audio processor itself. The first two channel domestic and industrial PCM processors are shown in Figs. 2-9 and 2-10. In fact, it would also be possible to construct a multi-channel recorder, which used a VTR as the recorder. This would be based on the time sharing multiplex PCM outlined in Fig. 2-4 (C).

Stationary Head Digital Tape Recorders

The most important piece of equipment in most studios is the multichannel tape recorder; different performers are recorded on different channels, so that the studio engineer can create the correct "mix" of sound before editing and dubbing. The smallest number of channels used is generally 4, the largest 32.

The digital tape recorder would be ideal for multi-channel use

Fig. 2-9. EIAJ standard digital audio processor (PCM-10).

because dubbing (re-recording of the same piece) can be carried out more or less indefinitely, so long as the dubbing is carried out digitally. On an analog tape recorder, however, the distortion increases with each dub. The digital tape recorder is also immune to cross-talk between channels, which can cause problems on an analog tape recorder. Therefore, a professional standard recorder designed for use in a studio needs the basic capacity to record a

Fig. 2-10. Industrial digital audio processor (PCM-100).

number of different channels simultaneously. There are also other necessary prerequisites allowing preparation of all the conventional types of software production and sophisticated editing.

It would be very difficult to satisfy studio standard requirements using a digital audio processor combined with a VTR. The reasons for this are covered in detail in Chapter 4. For a studio, a fixed head digital tape recorder would be the answer (track format and head shown in Fig. 2-12).

However, the construction of a stationary head digital tape recorder poses a number of special problems. The most important of these concerns the type of magnetic tape and the heads used. The head-to-tape speed of a helical scan VTR, as used with a digital audio processor, is very high, around 10 meters per second. However, on a stationary head recorder, the maximum speed possible would be around 76 centimeters per second. That means that information would have to be packed much more closely on the tape when using a stationary head recorder; in other words, it would have to be capable of much higher recording densities. As a result of this, a great deal of research had to be carried out into new types of modulation systems recording, and special heads capable of han-

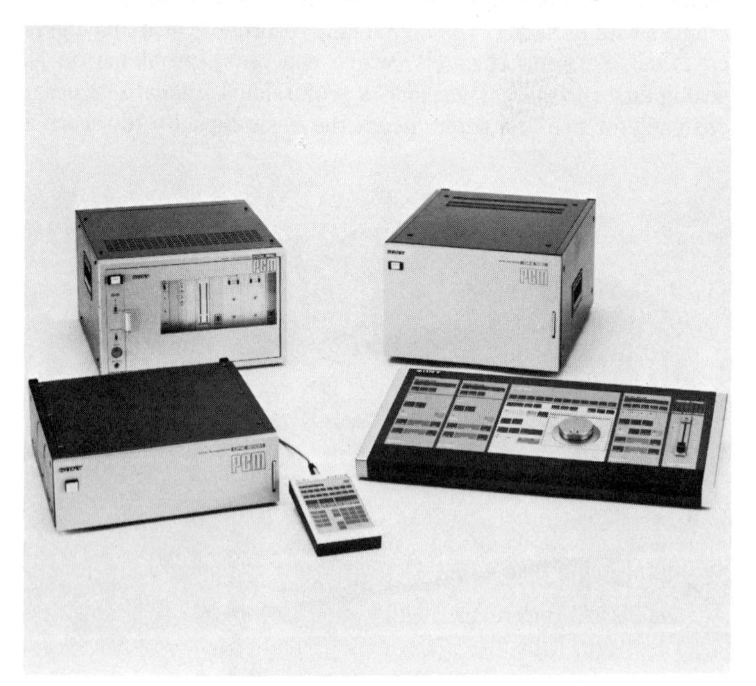

Fig. 2-11. Sony Compact Disc recording and mastering system.

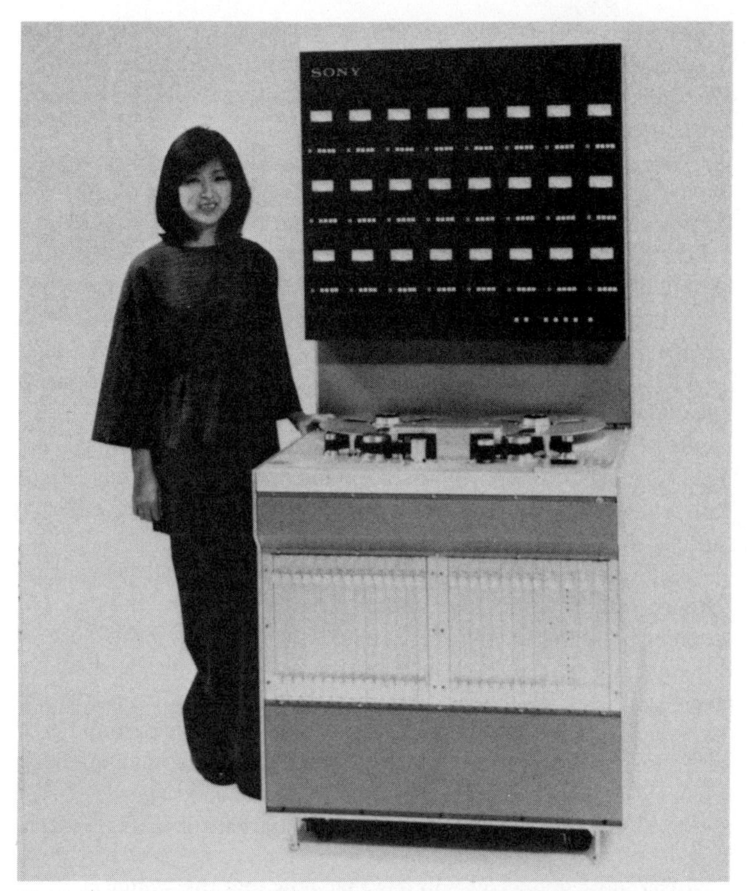

Fig. 2-12. Stationary head multi-channel PCM tape recorder (PCM-3224, Sony).

dling high density recording had to be developed. Finally, extremely strong error correcting codes had to be invented which were capable of handling the so-called "artificial" errors which are generated in the PCM signal during the process of electronic editing. The digital multi-channel recorder shown in Fig. 2-12 was finally developed after all the problems outlined above had been resolved. This machine has already been used by many studios and has met with great success.

The Digital Audio Disc (DAD)

The 33 1/3 rpm record was developed in the U.S.A. in 1948, and ever since, it has been the most popular source of recorded music. At present, it is the most widely available and also the richest source of prerecorded music. In comparison with recording tape, the conventional record is much easier to produce, and therefore, cheaper. This combination of good quality and reasonable price is the major reason for its success.

Because digital master tape recorders can now be used to make the initial master recording, which will later be used for LP production, studios will now begin to amass master recordings of unbelievably high quality. On the other hand, the public now also have access to digital technology in the form of a VTR plus digital audio processor. Thus, many people will become accustomed to the excellent quality available from digital recording, and by comparison, standard LP sound quality will begin to seem inferior. The next major development will be the digital audio Compact Disc (CD), containing PCM encoded music. These types of discs and players have just been introduced to consumers by a number of companies. Originally, there were three different prototypes of DAD, which differ in the method used to read the PCM signal off the disc.

One of these DAD systems, the optical system, where the DAD record does not come into physical contact with the pick-up assembly, is shown in Fig. 2-13. Besides the optical system, there is the capacitance system which operates by reading capacitance changes caused by unevenness (the PCM signal) on the record surface. The third type is the mechanical system, which uses a DAD with the same high performance as the other systems, but employs a stylus assembly to actually read the PCM encoded signal. These three different systems constitute the mainstream of DAD research at present.

The most important point is that now that the Compact Disc has become standardized as *the* digital audio disc system, the benefits to all will be enormous. As the CD is produced, it will improve the sound quality available from disc considerably, plus bring a wider range of music to the general public. Most importantly, all discs and players will be fully compatible with each other. Finally, the development of high density recording technology will allow much longer playing times: up to 74 minutes on one side of a disc (see Fig. 2-14).

Peripheral Equipment for Digital Audio

It is possible to make a recording with sound quality extremely close to the original source, when using a PCM tape recorder. In addition, digital tape recorders do not "color" the recording, a

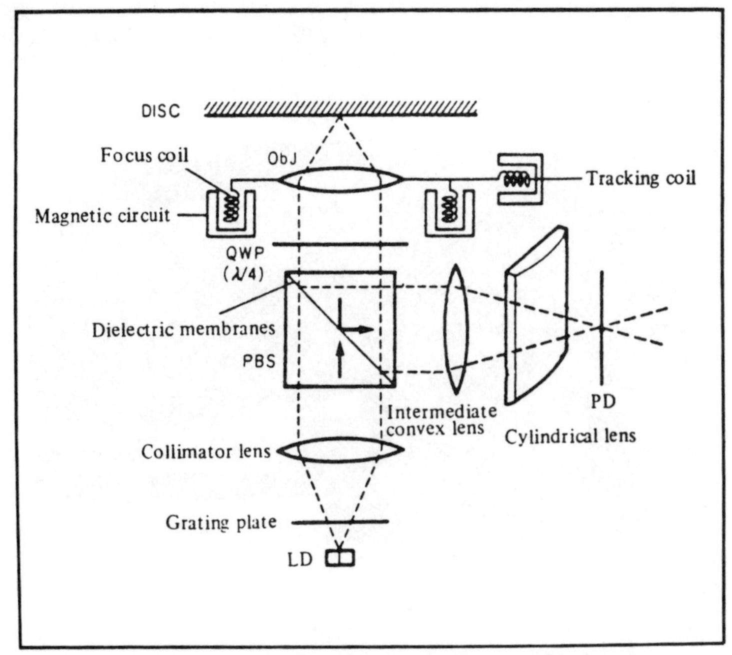

Fig. 2-13. The optical pick-up CD (digital audio Compact Disc) system.

failing inherent in conventional analog tape recorders. Finally, and possibly most important, a digital tape recorder offers scope for much greater freedom and flexibility during the editing process.

Below, there are brief explanations of peripheral equipment which would be used in a studio or a broadcasting station as part of a digital system for the production of software. All this equipment processes the signal fed to it digitally, as the types of digital tape recorder previously described.

Digital Mixer. A digital mixer which processed the signal fed to it digitally would prevent any deterioration in sound quality, and would allow the greatest freedom, within the limits of the PCM signal, for the production of software. The design and construction of a digital mixer is an extremely demanding task, and to date a prototype with only a few channels has been produced (see Fig. 2-15). However, it seems probable that a multi-channel version suitable for use in studios and broadcasting stations will soon be produced.

Digital Editing Console. One of the major problems associated with using a VTR based recording system was the difficulty

Fig. 2-14. Sony digital audio Compact Disc system (CDP-101).

in editing. The signal was recorded onto a VTR cassette, which meant that cutting and splicing the tape, as one would when editing conventional analog tape, or a tape from a stationary head digital recorder, was totally impossible. Therefore, one of the most pressing problems was designing an electronic editing console which would allow one to digitally edit tapes produced by a VTR based digital audio processor. The first prototype electronic digital editing console which has an accuracy of about 100 μsec is shown in Fig. 2-16.

Digital Reverberator. A digital reverberator is based on a totally different concept from conventional reverb units, which mostly use a spring or a steel plate to achieve the desired effect. If the addition of reverb is carried out on a digital signal, the same excellent effect is achieved as by using a specially designed reverb chamber. But in addition, the reverb effects available from a digital reverberation unit cover an extremely wide range, and are precisely variable. There are no moving parts, so there is no signal degradation at all (Fig. 2-17).

Sampling Frequency Conversion, Quantization Processing. The equipment shown in Fig. 2-18 is a sampling frequency

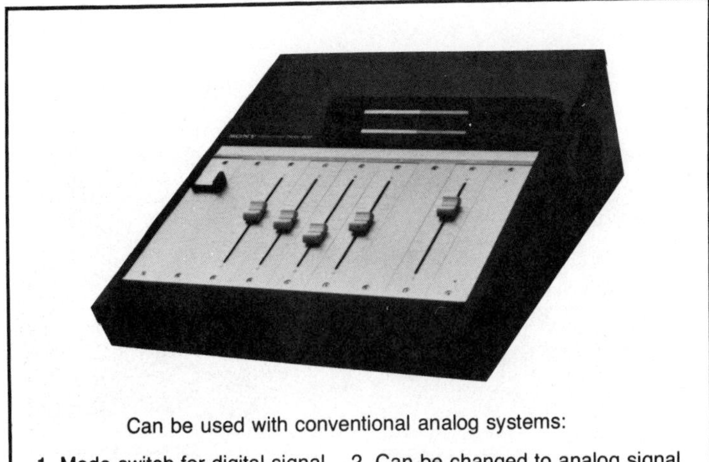

Fig. 2-15. Industrial digital audio mixer (DMX-500, Sony).

conversion unit, which would be used to connect together two pieces of digital recording equipment which did not use the same sampling frequency. Fig. 2-19 shows a quantization processor, for use with two pieces of equipment using different quantization bit numbers. These two pieces of equipment, the frequency converter

Fig. 2-16. Digital audio editing console (DEC-1000, Sony).

Fig. 2-17. Industrial digital reverberator (DRX-2000, Sony).

and the quantization processor, would allow free transfer of information between digital audio equipment of different standards. When building up a digital audio recording system, both would be indispensable.

THE CHANGING FACE OF HI-FI

The new developments in audio technology have been myriad over the last few years, and have provided subjects for discussion

Fig. 2-18. Digital sampling frequency converter (DSX87m, Sony).

Fig. 2-19. Digital quantization processor (DQP-6040, Sony).

enough and to spare. This chapter began with the story of how digital audio equipment grew out of PCM communications technology. On a small scale, this began with the microcomputer boom, and the application of digital technology to tape recorders and record players. On a large scale, there are FM broadcasts using PCM as a medium, and experimental satellite transmission. As with all fields of new technology, the complexity and the possible future effects are breathtaking.

Now, we are beginning to see the shape of things to come as the audio industry, which has always had the reputation of taking best advantage of the latest technology, begins to put this new digital revolution to positive use.

In this section, we shall look at the way that all this new technology, PCM, digital audio equipment, and the other new developments related to audio, will affect the recording studios, the broadcasting stations and domestic audio equipment. Changes there most certainly will be, and perhaps this is the time to examine them.

THE RECORD PRODUCTION SYSTEM

In a conventional analog recording system, as shown in Fig. 2-20 (A) there are a large number of analog tape recorders. The main drawback of this type of setup is that the final quality of the record depends, to a large extent, upon the initial quality of the master tape.

As a result, some studios decided to return to the direct cutting system, with the intention of improving the quality of sound available from records. However, with the direct cutting system the slightest mistake on the part of the performers, the mixer or the cutting engineers spelt disaster. This process demands the utmost in terms of skill and technology to be successful.

As soon as digital tape recorders appeared, studios seized on the idea as a great advance on analog systems, and a less demanding alternative to direct cutting. The system shown in Fig. 2-20 (B) is widely used for the production of so-called "analog PCM" discs. (A digital tape recorder is used for mastering, but in other respects, disc production is identical to the conventional analog system.) This type of record production will no doubt become increasingly popular as multi-track digital tape recorders and peripheral digital studio equipment become more widely used.

Recently, with a standard digital audio CD format, the record production chain has become totally digital, as shown in Fig. 2-20 (C). Using this type of full digital system, especially keyboards and other electronic instruments, all of which would have a digital output, digital audio discs could be recorded directly. One can only imagine that the range of program material that could be recorded would far surpass the capabilities of conventional equipment, providing an aural feast for music lovers and audio fans.

High Quality Broadcast Systems

FM broadcasts are, at present, probably the most readily available source of high quality program material. As shown in the brief outline in Fig. 2-21 (A), a large number of conventional analog tape recorders are used in FM broadcasting for program production, just as in conventional analog record production. The result of this is that program quality is not quite as good as it could be. When programs produced centrally are distributed to subsidiary stations, then either analog tape, or special transmissions are used. The transfer process, however, generally means that the same program may have deteriorated slightly, giving a disparity of quality between the main-and sub-stations.

Since 1978, the FM broadcasting stations have expressed a great deal of interest in digital tape recorders, realizing the benefits it could bring almost as soon as it had been developed. Fig. 2-21 (B) shows an FM broadcast set-up using digital tape recorders to maintain high quality broadcasts.

In the future, high quality broadcasts will probably be made through completely digitized PCM systems, as shown in Fig. 2-21 (C). Once PCM broadcasting has become established, it will be possible to broadcast the live PCM signal direct to the family living room. As mentioned previously, a bandwidth of several megahertz would be needed, a much higher frequency than used at present in FM broadcasting. In fact, a frequency with a wavelength of less than one centimeter would have to be used. The most effective method for broadcasting a PCM signal, bearing in mind the high frequencies involved, and also from the point of view of areas which could be covered, would probably be via satellite. A great deal of research has been carried out in this field in recent years.

In 1971 at the WARC-ST meetings (World Administrative Radio Conference, world-wide meetings held to agree on wireless communications, with specific reference to satellites), frequency allocations to be used in satellite broadcasting were agreed. Since that time, 4 satellites have been put into orbit: ATS-6 for the U.S.A., CTS for Canada and the U.S.A.; plus satellites for the U.S.S.R. and Japan. More satellite launches are planned for the 1980s. Before PCM broadcasts can be made using these satellites, many problems will have to be ironed out in wireless communications. However, success in this field would gladden the hearts of many Hi-Fi fans.

Home Audio Systems

Home electronics is a booming business, which does not stop at audio, but spills over into every area of life. Only recently, the home VTR was produced, and met instant success, and no doubt the home computer will appear very shortly as an adjunct to everyday life.

So far, we have discussed the effects of new technology on the record production chain and on broadcasting. The net result will be the general availability of much improved musical sources to the public. This will have a certain effect on the conventional Hi-Fi system shown in Fig. 2-22 (A), but eventually the advent of digital technology will have much more far-reaching effects.

At the moment, the simplest way to enjoy digital audio is as shown in Fig. 2-22 (B), where a PCM processor and VTR have been linked to a standard Hi-Fi system. This one change would provide a recording system which far surpasses the quality of a conventional professional 38 cm/s 2 track analog tape recorder. Live recording with a dynamic range in excess of 80 dB made with this type of system should satisfy even the most demanding enthusiast. How-

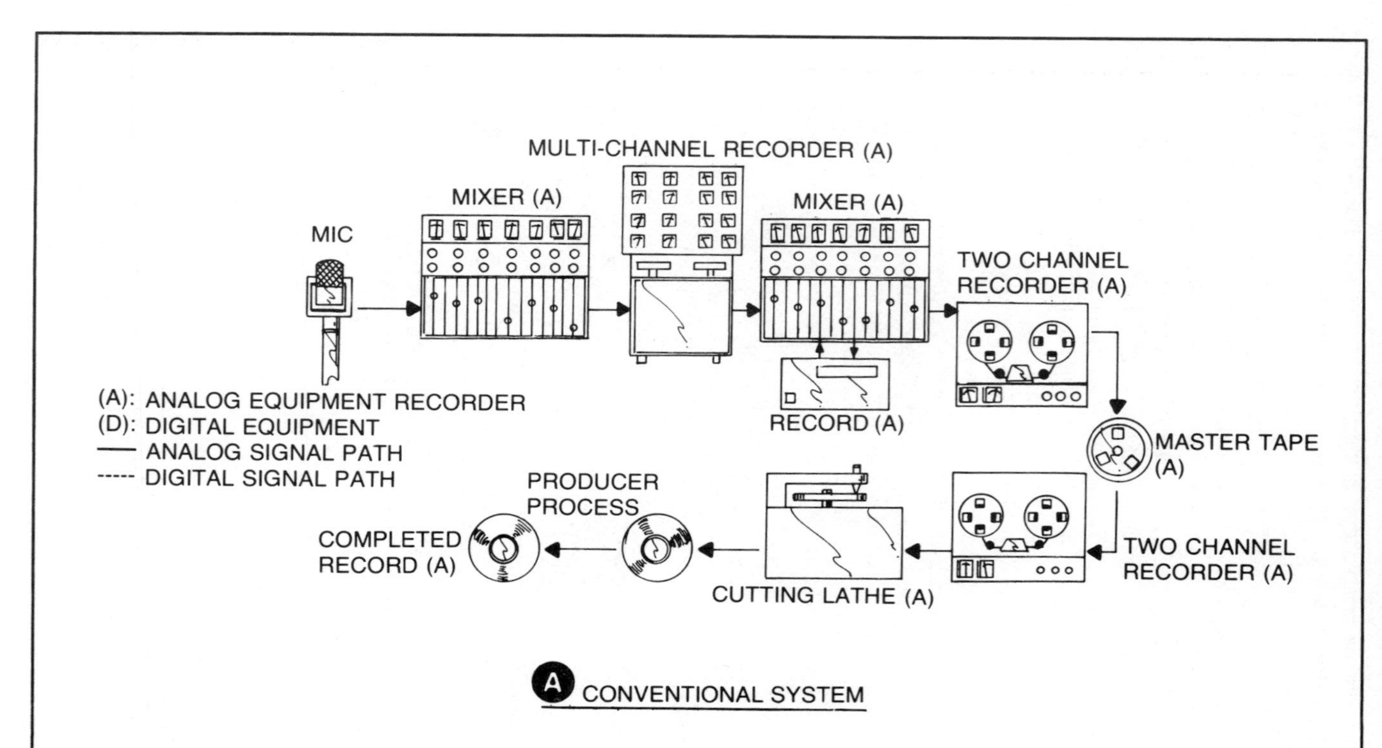

A CONVENTIONAL SYSTEM

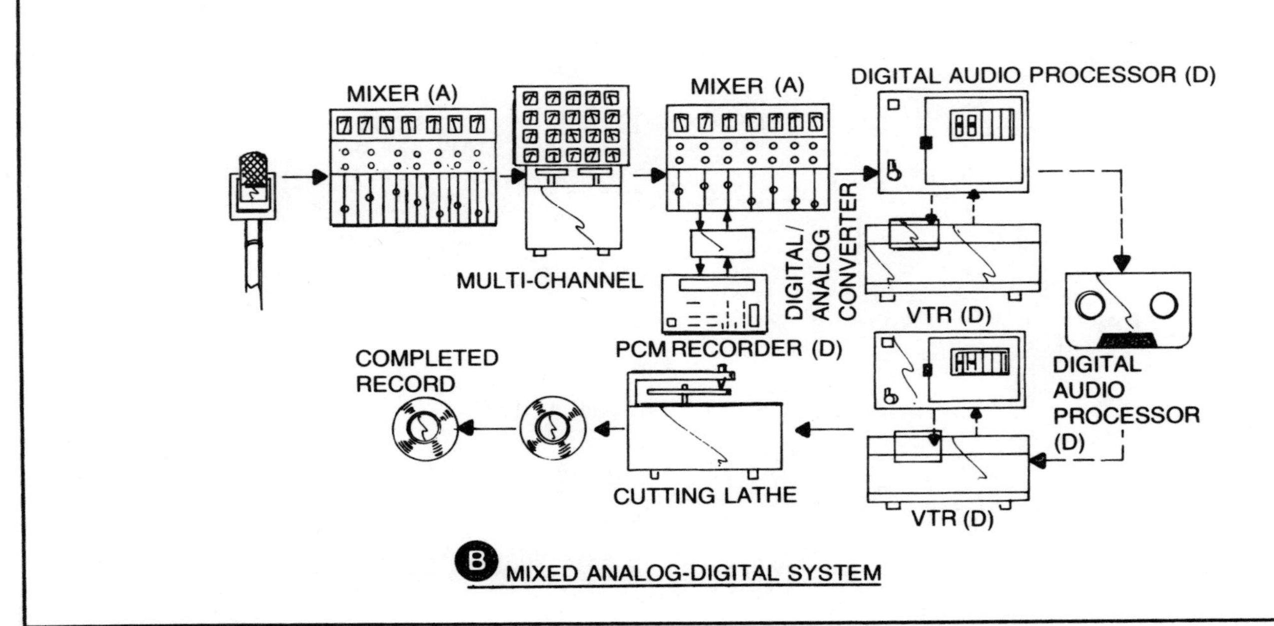

Fig. 2-20. The record production chain.

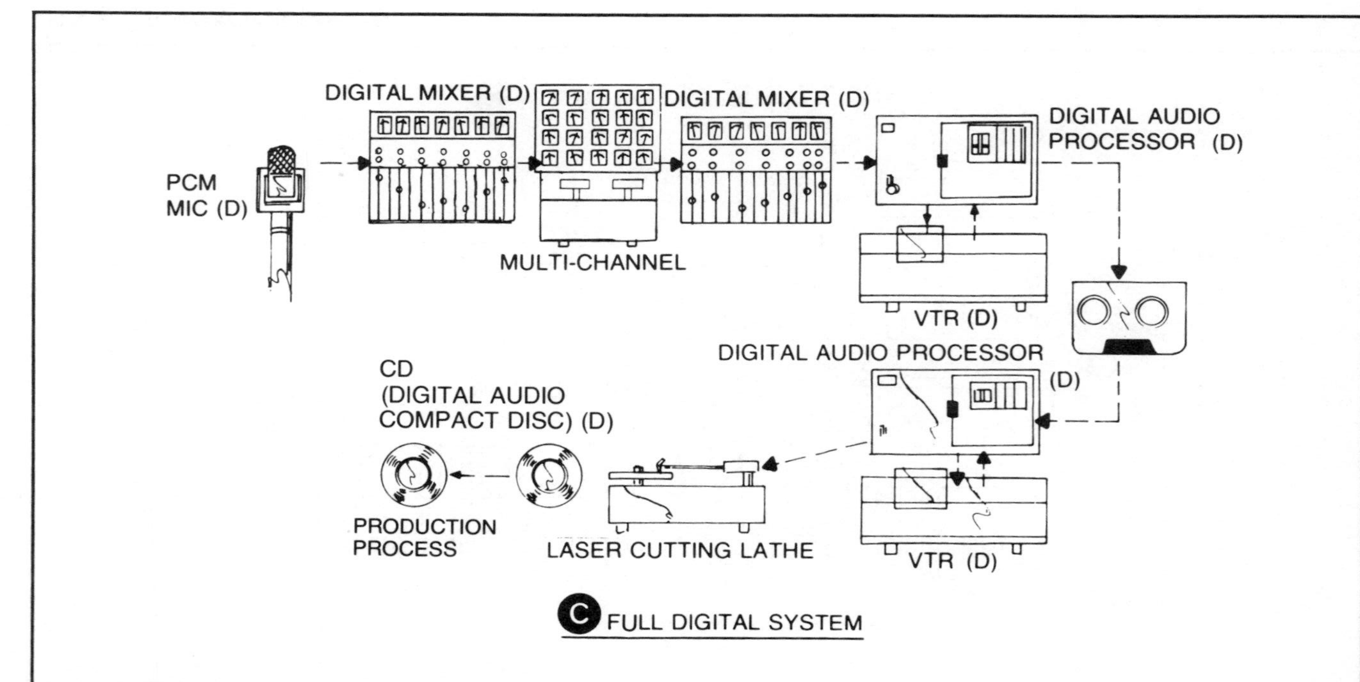

Fig. 2-20. The record production chain. (Continued from page 43).

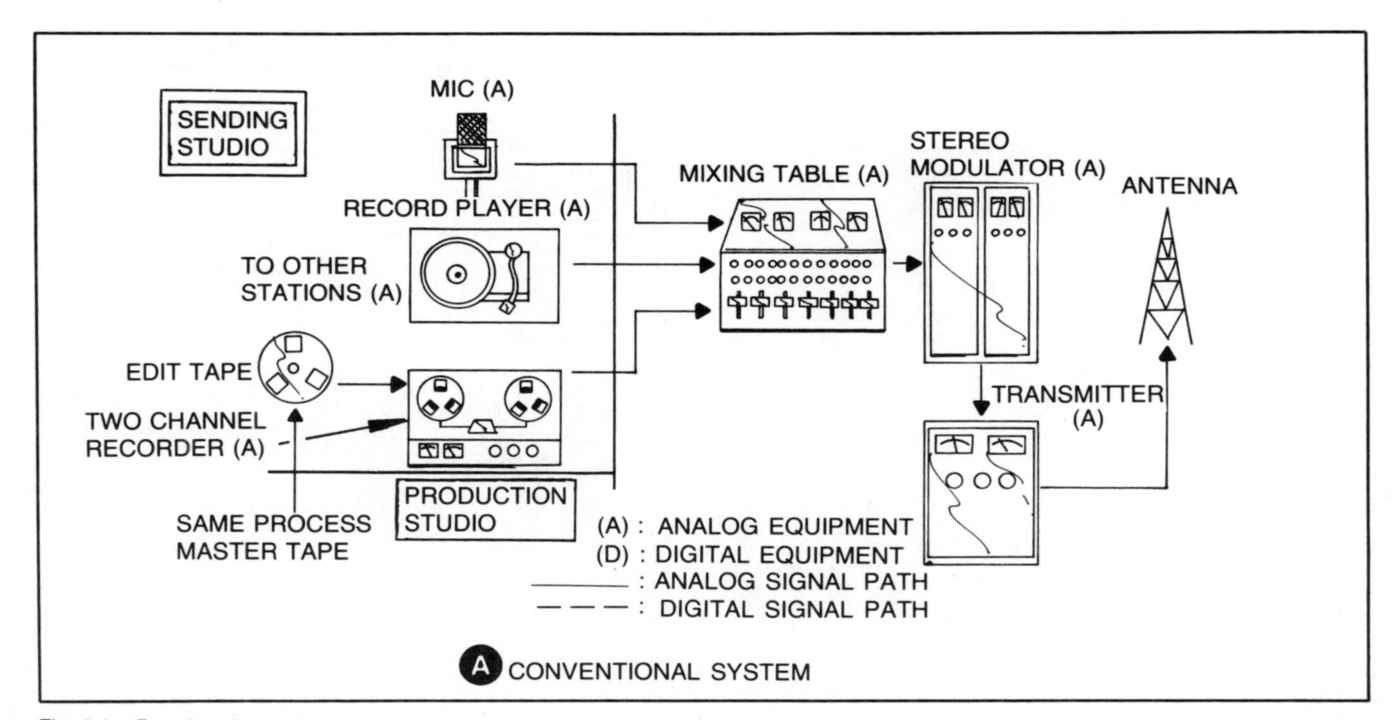

Fig. 2-21. Broadcasting systems.

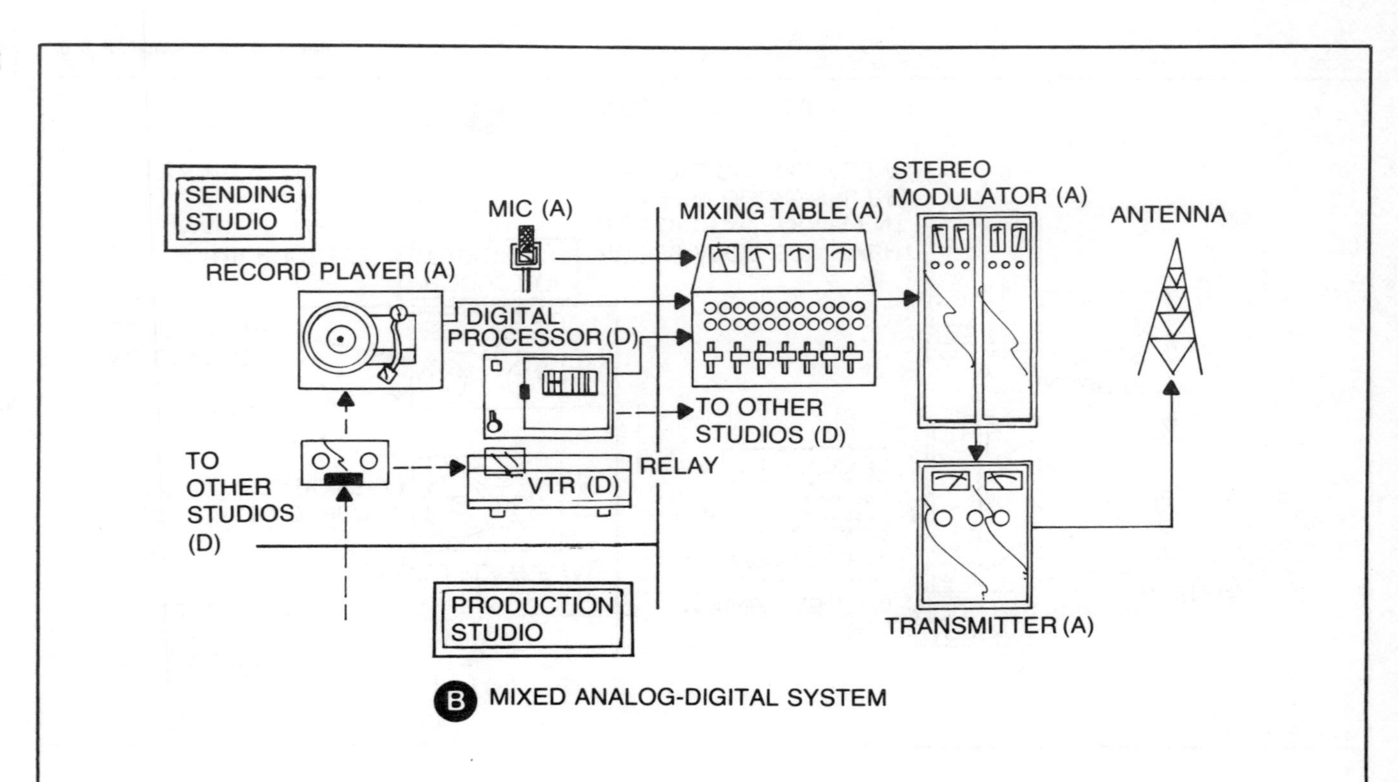

B MIXED ANALOG-DIGITAL SYSTEM

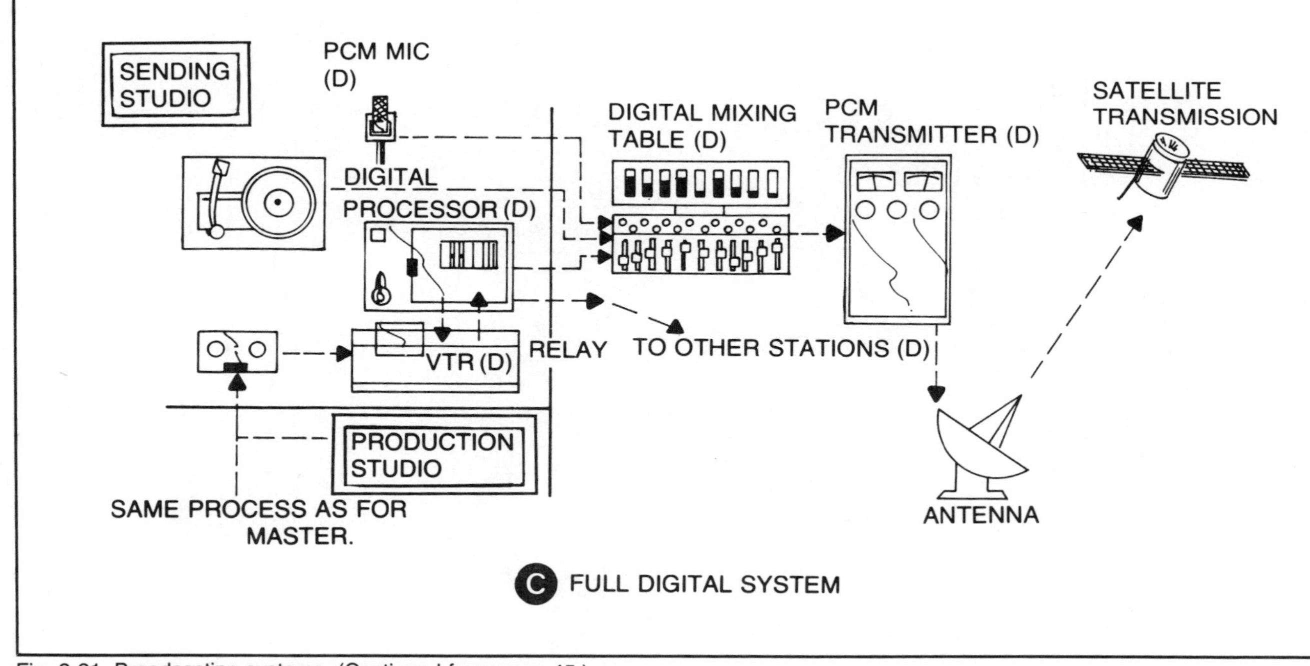

Fig. 2-21. Broadcasting systems. (Continued from page 45.)

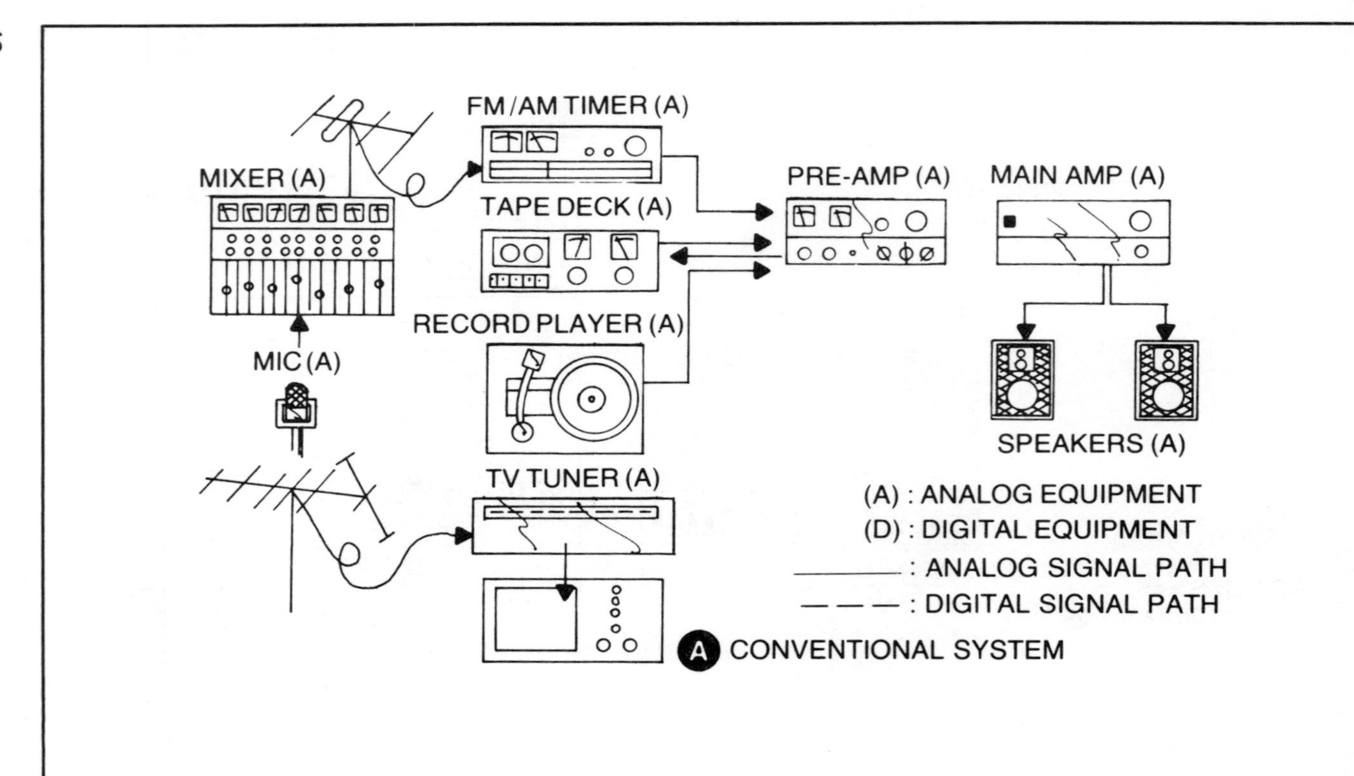

A CONVENTIONAL SYSTEM

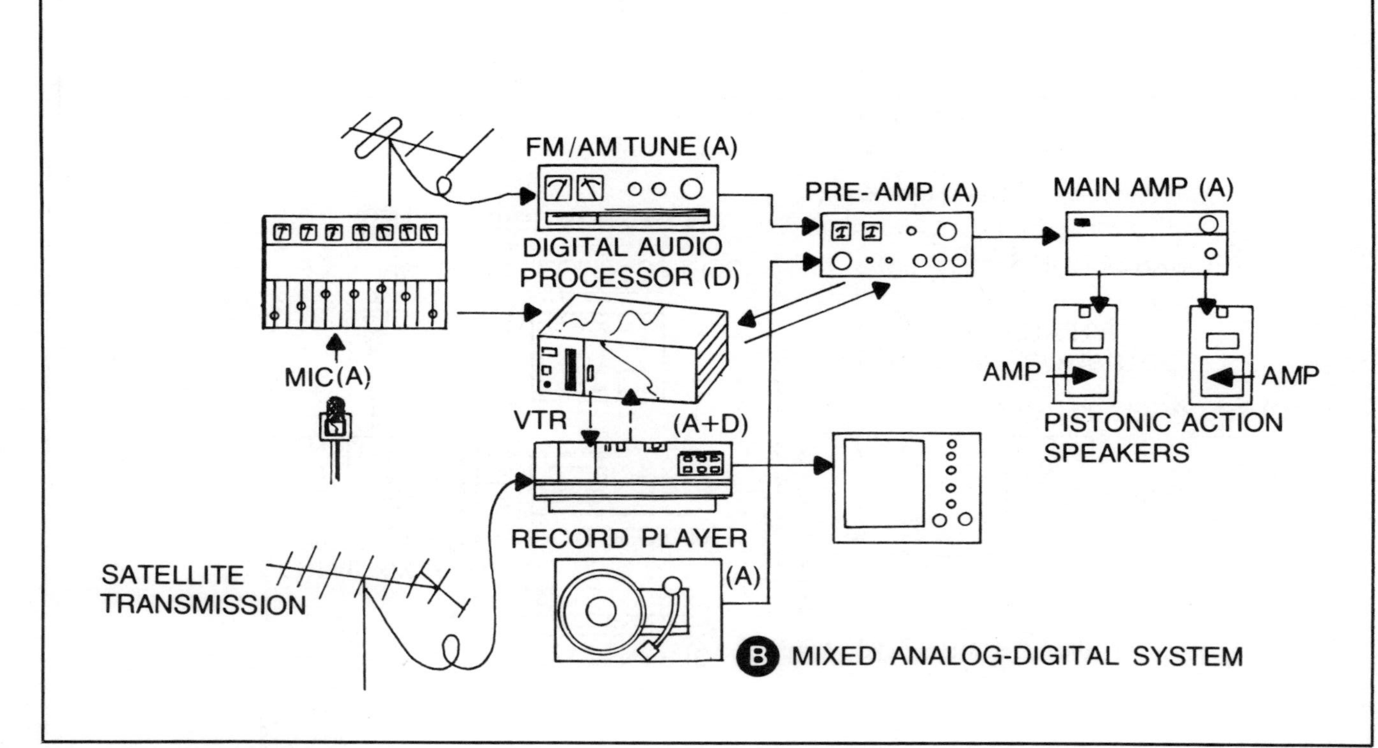

Fig. 2-22. Home audio hi-fi systems.

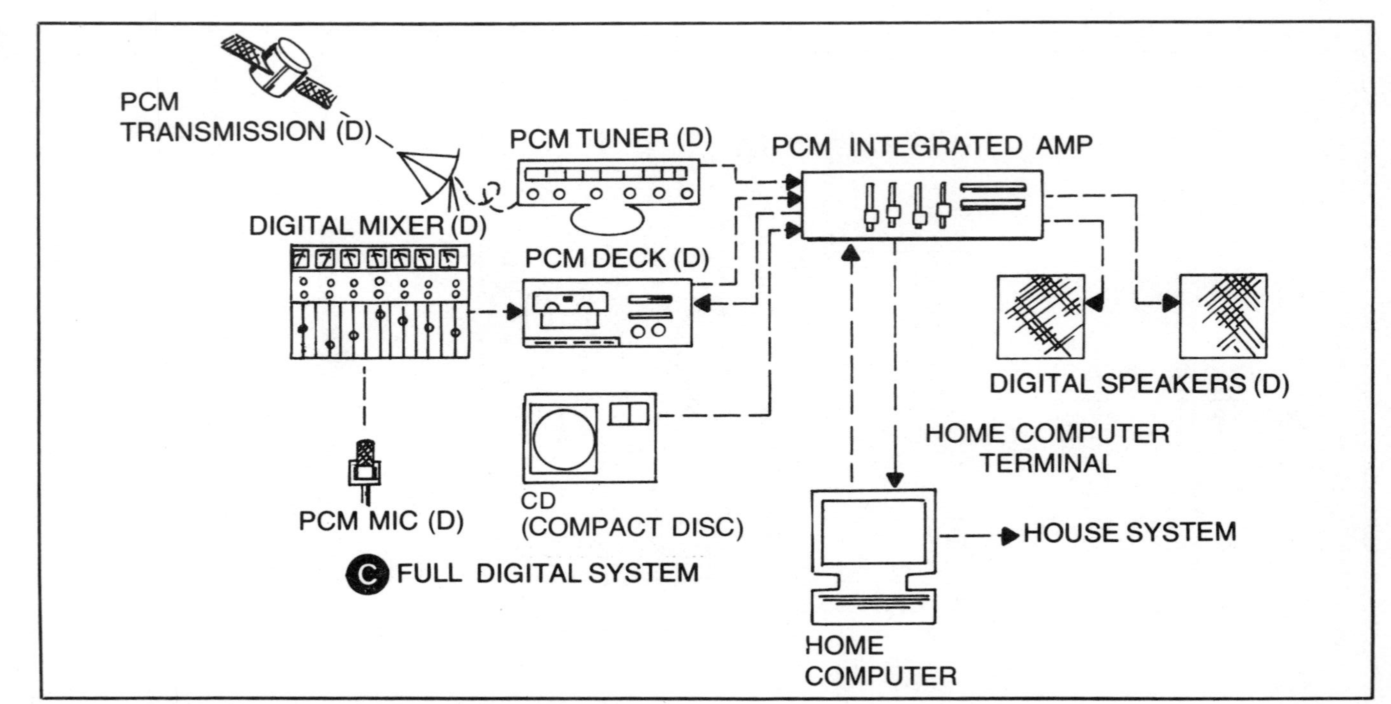

Fig. 2-22. Home audio hi-fi systems. (Continued from page 49).

ever in these circumstances, other musical sorces available would be conventional LP records and broadcasts. The owner of a digital recording system will soon start demanding similar digital quality from his other musical sources: that is, digital broadcasts and CD systems. These developments will take place in the very near future, and the home audio system will soon look more like the one shown in Fig. 2-22 (C); a full digital audio system.

References

1. Shannon, C.E.: "A Mathematical Theory of Communication", BSTJ, 27, pp. 379-423, pp. 623-656 (1948).

2. Kaneko: Electro-chemical series 69, PCM communication techniques, Sanpo.

3. Oba, Tsuchiya, Otsuki, Kazami: *Adaptors for high quality PCM record and playback,* Hoso-gijutsu (1977-9).

4. Iga, Oshima, Ogawa, Hashimoto, Masaoka, Yasuda, Yokota, Doi: "A Consumer PCM Audio Unit Connectable of Home Use Video Tape Recorder," presented at the Conference of the Acoustical Society of Japan. 3-2-9 (1977-10).

5. Ogawa, Ito, Doi: "A System of Optical Digital Audio Disc "IECE, EA 78-27 (1978-7).

6. Ajimine, Otsuki, Kazami, Anju, Okude: "A 2-channel/16 bit PCM Recorder for Professional Use (VTR Adapter Type)", IECE, EA 78-35 (1978-7).

7. Ito, Ogawa, Doi, Sekiguchi: "Advanced System of Digital Audio Disc", presented at the Conference of the Acoustical Society of Japan., 3-P-13 (1978-10).

8. Tsuchiya, Sonoda, Nakai, Ishida, Doi: "24 Channel Stationary Head Type PCM Magnetic Tape Recorder", presented at the Conference of the Acoustical Society of Japan. 3-P-17 (1978-10).

9. Naito Yokota, Yasuda, Odaka, Doi, Sekiguchi: Rotary Head PCM Deck presented at the Conference of the Acoustical Society of Japan. 3-P-18 (1978-10).

10. Tsuchiya, *Stationary head 24 channel PCM magnetic record and reproduction equipment."* Hoso Gijutsu (1978-10).

Chapter 3

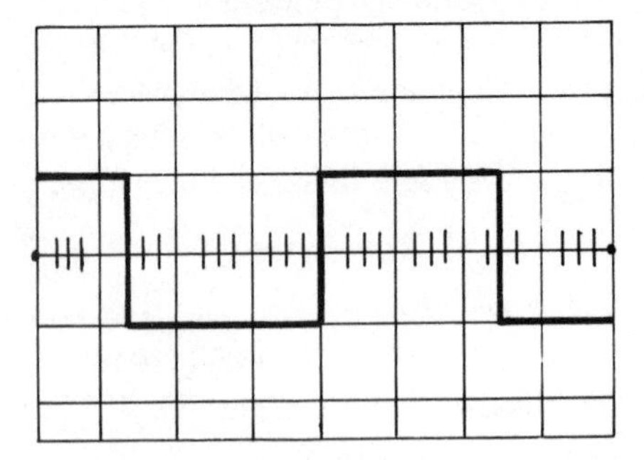

The Fundamental Principle of Digital (PCM) Recording and Reproduction

PCM (pulse code modulation) was invented by A.H. Reeves in 1939 [1], and was analyzed and developed as a modulation system from the point of view of communications theory by C.E. Shannon in 1948. Today, pulse code modulation is widely used in telephone systems and in satellite communications.

Many other modulation systems beside PCM have been developed for a variety of purposes over the years. AM (amplitude modulation) and FM (frequency modulation) systems are, of course, used in broadcasting. Other examples of modulation systems are: PPM (pulse position modulation), PAM (pulse amplitude modulation), and PNM (pulse number modulation).

In this chapter, we will examine the special features of PCM as compared with other modulation systems, and the advantages of using PCM for record and reproduction equipment for music. In addition, a technical explanation of this modulation system will be given.

A COMPARISON OF DIFFERENT MODULATION SYSTEMS

Modulation is a technique used for converting a particular signal into a different form: for example, the signal used for transmission or for record and reproduction has to be converted (modulated) into a form suitable for the transmission path or the recording medium. If we consider a music source, then we can appreciate that the highest possible signal-to-noise ratio (S/N) is necessary for broadcasting.

Therefore, a modulation system which has an extremely wide frequency response, but a low S/N ratio, would not be an ideal choice. Modulation is really a method of matching signal to transmission path in such a way that the signal is transmitted accurately and also efficiently.

The most widely used modulation systems are shown in a comparative form in Table 3-1, as we do not have sufficient space to carry out a detailed investigation into all types of modulation system. To carry out a full-scale comparison between different modulation systems, it would be necessary to investigate relative signal strength, the density distribution of distortion, and the depth of modulation.

In the case of AM, FM and PM, the signal is still continuous after the signal has been modulated (as shown in Fig. 3-1). These systems are classified as continuous wave parameter modulation. PAM, PWM, PPM, PNM, and PCM, on the other hand, are known as pulse parameter modulation systems because, as shown in Fig. 3-2, a pulse chain is generated after the modulation.

In the PAM, PWM and PPM systems, the information contained in the original analog signal is converted into pulses. These pulses correspond to the original analog signals in terms of analog values of pulse height, width or time position. PNM and PCM, on the other hand, convert the original signal into a number of discrete pulses or into a code. Some of these modulation systems are particularly suitable for equipment reproducing excellent record and reproduction of musical signals.

If we begin by looking at the recording medium only, then it is obvious that there is no such thing as an ideal recording medium. Whether one uses magnetic tape, or an optical or capasitive disc system, the S/N ratio, the linearity and the extent of temporal changes may be unsatisfactory in some way. Basically, there is not really a satisfactory way of recording and reproducing a high quality musical signal in its original form.

However, a modulation system cannot be affected by the characteristics of the recording medium, although, as shown in Table 3-1, the modulation method itself may affect the signal. Therefore, it is desirable to examine these effects on S/N ratio, linearity and temporal disturbance.

Looking at Table 3-1 it is self-evident that the systems with the

Table 3-1. A Comparison of Various Modulation Systems.

Modulation System		Frequency Bandwidth	Improvement of Signal to Noise Ratio	Improvement of Linearity	Improvement of Timing Jitter
AM	amplitude modulation	SMALL	no	no	no
FM	frequency modulation	MEDIUM	excellent	excellent	amplitude changes
PM	phase modulation	MEDIUM	good	good	amplitude changes
PAM	pulse amplitude modulation	MEDIUM	no	no	good
PWM	pulse width modulation	MEDIUM	good	excellent	amplitude changes
PPM	pulse position modulation	MEDIUM	good	excellent	amplitude changes
PNM	pulse number modulation	EXTREMELY LARGE	excellent	excellent	excellent
PCM	pulse code modulation	LARGE	excellent	excellent	excellent

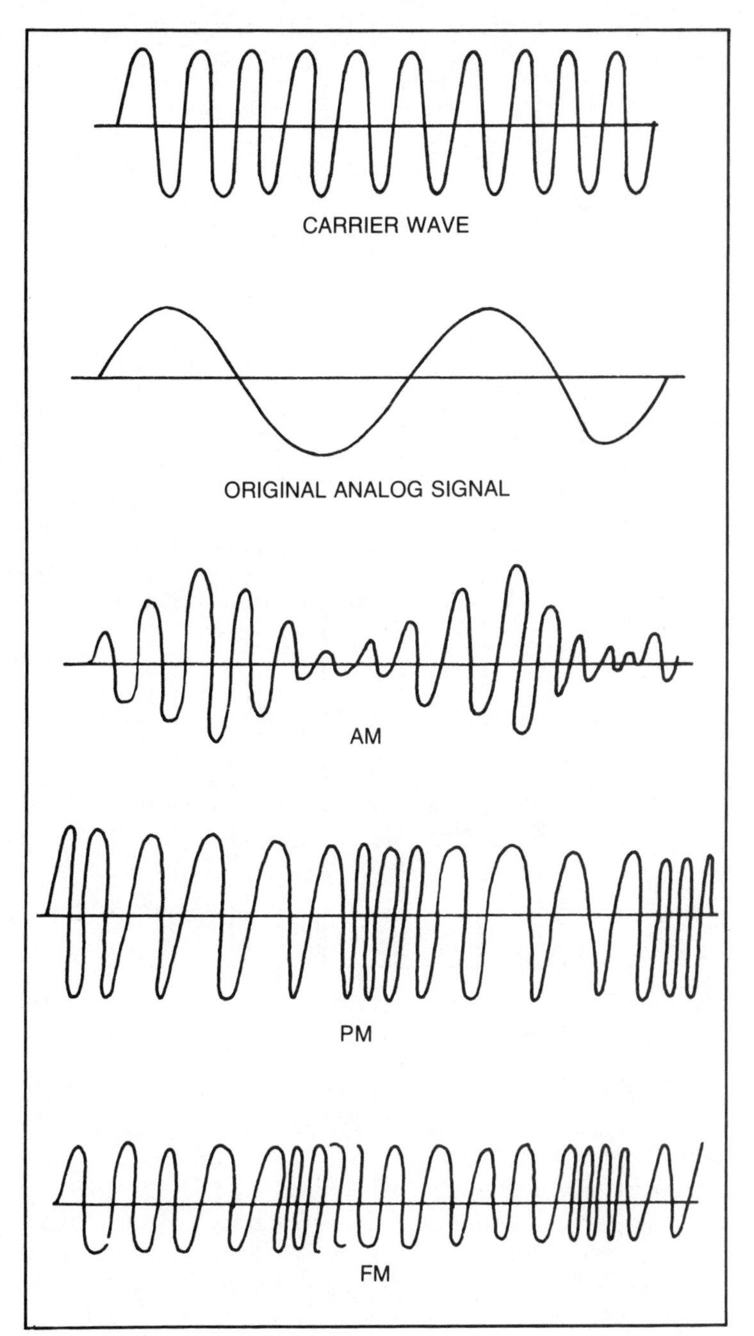

Fig. 3-1. Continuous wave modulation.

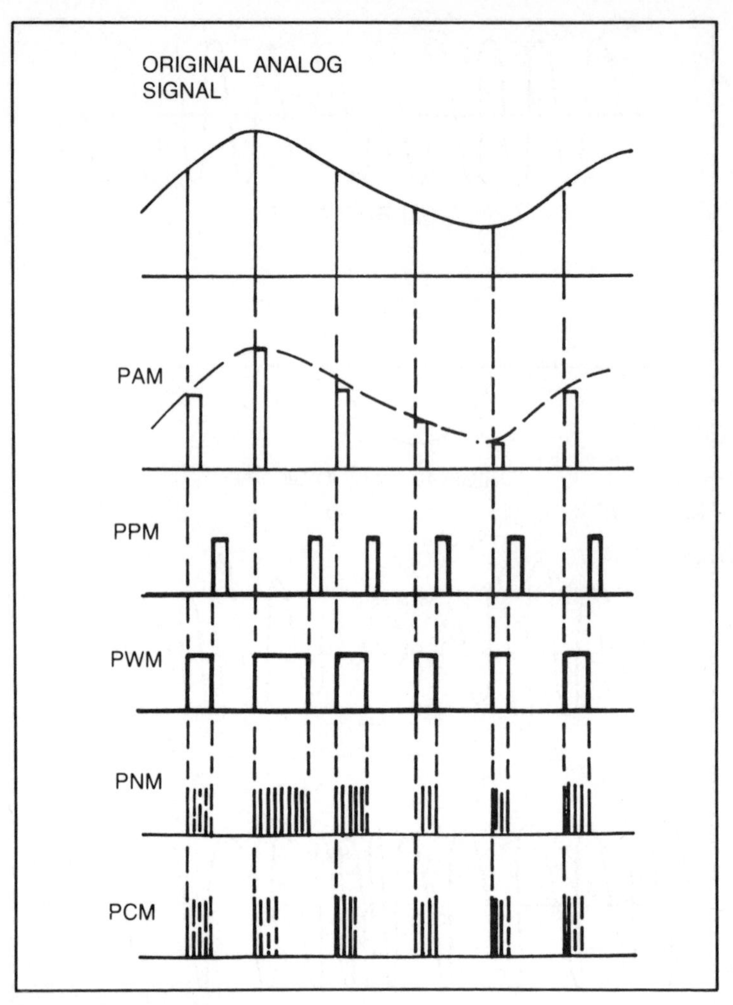

Fig. 3-2. Pulse modulation.

best performance from an audio point of view are PNM and PCM. However, the PNM system modulates the original signal as a number of pulses, and although resolution is high, it requires the use of an extremely large number of pulses. In short, it is a modulation system requiring an extremely wide bandwidth. PCM, on the other hand, can provide the same degree of effectiveness, but because it combines a pulse chain with a coding system, the bandwidth needed is much less than that required for PNM.

Again, although the effect of temporal change is a problem,

wide band width FM compares favorably with PCM. In fact, under certain conditions it can outperform PCM. The major drawback is the change in time and amplitude of the demodulated signal, as mentioned above. It is, however, possible to correct errors at the time of demodulation by recording a pilot signal derived from a frequency division process. Therefore, granted that the recording medium is capable of continuous recording, both wide bandwidth FM and PCM can be used effectively for record and reproduction of a high quality music signal.

Although we stated that the recording medium must be "capable of continuous recording," in actual fact, neither magnetic tape nor discs wholly fulfill this precondition. For example, a record reproduction system using magnetic tape and rotary heads is not truly "continuous" because the head switchover point causes a portion of the signal to be distorted. In the case of a stationary head recording system, or a disc, even though "continuous" recording is possible from the point of view of the actual construction of the machines, dropouts will occur on reproduction because of slight faults in the recording medium. Thus, none of these systems really satisfy the precondition of being "capable of continuous recording."

Whatever modulation system is used in an attempt to achieve a high quality musical signal, dropouts on reproduction are bound to be a problem. It follows, therefore, that whichever system is used, comprehensive error correction is necessary.

If the location of the error is known in advance, the signal which ought to be recorded at that position can be recorded using correct information elsewhere. On reproduction, the correct signal can be used at the appropriate time. However, it is often not feasible to predict dropout which may affect the signal, so in effect, the same signal is recorded in two different places on the tape. Thus, if there is a dropout on reproduction, the error can be corrected by using the correct signal from a different place on the tape. (See Chapter 6 for a detailed explanation of error correction).

This type of error correction requires special processing circuits as well as memory circuits. To maintain a high degree of accuracy during the error correction process, it is much better to use digital rather than analog circuits.

All in all, it is much more advantageous to use a modulation system where the signal is in a digital form after modulation, in short, a pulse code system. Therefore, although PCM needs a wide bandwidth, it is the most suitable modulation system to use in equipment designed for high quality sound reproduction.

AN OUTLINE OF THE PCM SYSTEM

In the PCM system, samples of the original continuous analog signal are taken at fixed intervals of time, and a parameter is generated which corresponds to the sampled value. Then, using only two alternative pulse values (1 and 0), a pulse chain of specific length is generated which indicates the sampled value.

There are two other modulation systems which closely resemble PCM : DPCM (Differential Pulse Code Modulation) and ΔM (Delta Modulation). As shown in Fig. 3-3, in PCM, each sampled value is converted into a pulse chain of specific length. In DPCM, the difference between two consecutive sampled values is converted into a pulse chain, again of specific length. Finally, in delta modulation, the differences between the two consecutive samples values are expressed not as a pulse chain of specific length, but as a

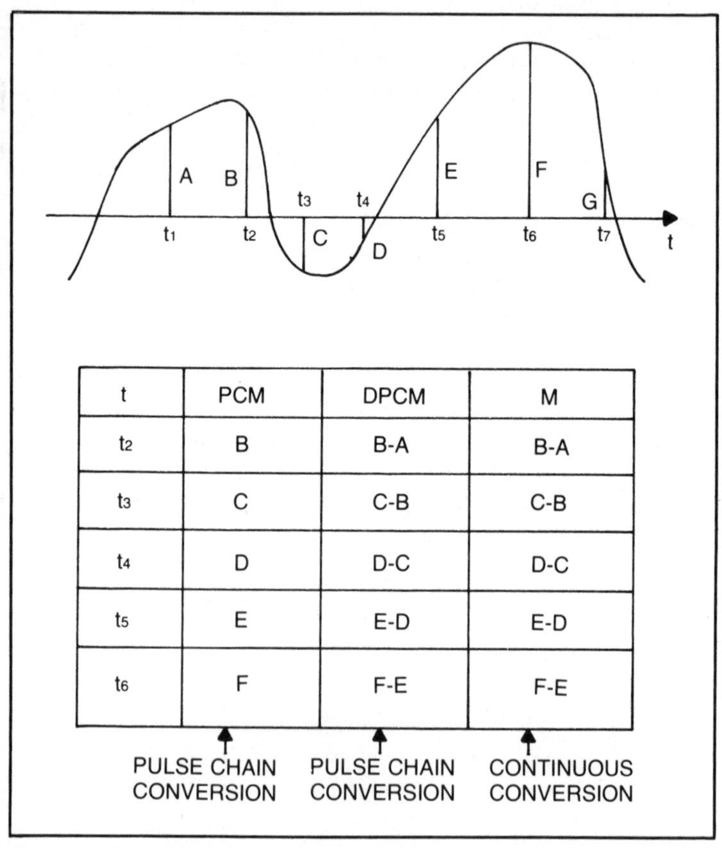

t	PCM	DPCM	M
t_2	B	B-A	B-A
t_3	C	C-B	C-B
t_4	D	D-C	D-C
t_5	E	E-D	E-D
t_6	F	F-E	F-E

Fig. 3-3. PCM, DPC, ΔM conversion systems.

Table 3-2. Binary and Decimal.

Decimal	Binary
0	0
1	1
2	10
3	11
4	100
5	101
6	110
7	111
8	1000
9	1001
10	1010
11	1011
12	1100
13	1101

continuous stream of pulses, which indicate only positive or negative differences of the signal. The pulse chains representing sampled values are expressed in binary notation for both PCM and DPCM.

Binary Notation

The number system that most people commonly use is called the decimal system, and is based on ten different numerals from 0 to 9. Numbers greater than 9 are indicated by combinations of the ten basic digits in different "columns" (i.e. "hundreds", "tens", and "units".)

However, when dealing with the PCM system, we want to use pulses to express many different sampled values. These pulses can only express "on" or "off": the presence or absence of a pulse signal. Unfortunately, we cannot use the ten different numerals of the decimal system to express these patterns of PCM pulses. Therefore, we have to use a number system which is based only on two different numerals: the binary system.

In the binary system, the two different numerals, used in many combinations, are 1 and 0. When we want to express numbers greater than 1, we have to use the other "columns" (i.e., "twos", "fours", "eights", etc.), just as we would in the decimal system when we want to express numbers greater than 9.

Table 3-2 gives the binary equivalents of some decimal numbers. As shown in Fig. 3-4, one decimal numeral is sometimes called

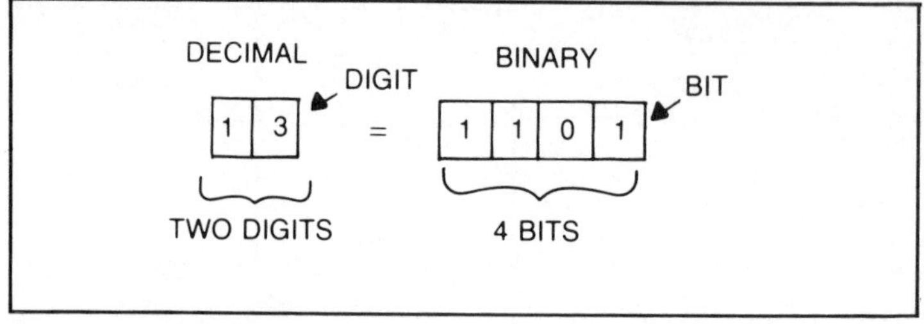

Fig. 3-4. Digits and bits.

a "decimal digit", and one binary numeral is called a "binary digit" or "bit".

The ordinary electronic calculator carries out all its functions using the binary system. However, people are accustomed to using the decimal system, so the numbers can be entered, and read off in decimal rather than binary. Thus, even though a calculator operates in binary, it is accessed via the decimal system. In pieces of equipment where this decimal-binary-decimal conversion is needed, a variant of the binary system known as BCD (Binary Coded Decimal) is used. This is a special type of binary where each decimal digit is represented by a group of 4 binary digits (a 4 bit word). Even though binary equivalents for ten different decimal digits have to be supplied, a 4 bit binary word is quite adequate. Figure 3-5 gives an example of BCD.

If we now consider the suitability of using the binary system to represent the sampled values, the first problem is that the samples are taken from an audio waveform and are, therefore, negative as well as positive values. A normal analog audio waveform is equally divided into a positive going and also a negative going half. Therefore, the binary system we have described thus far is not suitable for representing this type of signal, because we have only covered positive numbers. Next we must examine the expression of negative or minus numbers.

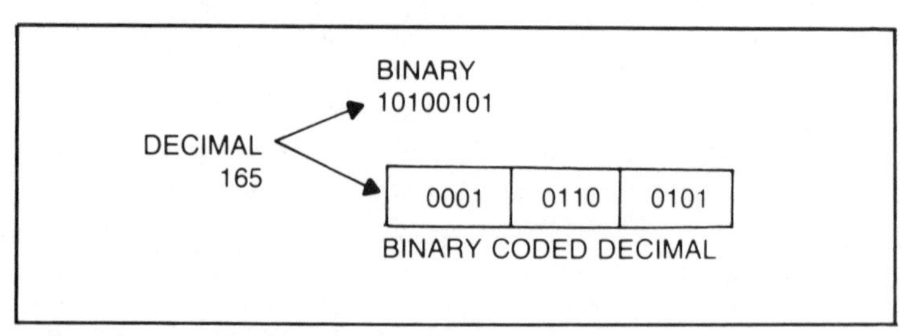

Fig. 3-5. Binary coded decimal.

There are, in actual fact, various types of binary which can be used to express positive and negative numbers. Table 3-3 gives examples of 4 bit binary often used in PCM. To distinguish between the type of binary we talked about earlier and the systems shown in the table, we shall call the former "natural binary".

The left-hand column gives examples of "offset binary"; this is a system where the negative numbers are created by offsetting natural binary. This system is often used in PCM equipment because conversion from natural to offset binary is a simple process.

In "2s complement" there is a simple transition between positive and negative numbers, and it is very commonly used in calculations because addition and subtraction can be carried out quite simply. Also 2s complement is safe for digital audio equipments, because it will terminate audio signal at some unusual situations where the bits generated for a pulse chain are either all 1s or all 0s.

In sign and magnitude binary, depending on their magnitude, values of positive and negative samples are expressed in natural binary, and the leftmost bit (MSB = most significant bit) is used to indicate whether the number is positive or negative.

By using one of these binary systems, the correct code can be assigned to both positive and negative sampled values. There is, however, a limit to the number of codes which can be used to express sampled values.

For example, a code constructed from 3 bits would give a total of 8 different expressions; one constructed from 4 bits would give a

Table 3-3. Various Types of Binary.

	Offset Binary	2's Complement	Sign and Magnitude Binary
+7	1111	0111	0111
+6	1110	0110	0110
+5	1101	0101	0101
+4	1100	0100	0100
+3	1011	0011	0011
+2	1010	0010	0010
+1	1001	0001	0001
+0	1000	0000	0000
−0	(1000)	(0000)	1000
−1	0111	1111	1001
−2	0110	1110	1010
−3	0101	1101	1011
−4	0100	1100	1100
−5	0011	1011	1101
−6	0010	1010	1110
−7	0001	1001	1111
−8	0000	1000	

total of 16 expressions. However, the sampled values represent the original analog signal, and the number of possible values is infinite. Unfortunately, we will always have a finite number of bits to express an infinite number of values, and therefore, a totally exact correlation is impossible.

The sampled value is achieved as follows: the analog signal is sampled at short, fixed intervals, and the sample is taken at the center of this interval. The sample is then rounded off to the nearest available number. The precise amount of rounding up or down needed is determined by the number of bits used. As a result, we end up with a series of values which have been taken at fixed sampling intervals. In Fig. 3-6, there is an example showing an analog signal being sampled at fixed intervals, using a 3 bit code.

Therefore, sampling is a method of converting a continuous signal into a series of non-continuous values. The process of assigning finite values to the sampled levels is known as quantizing.

Sampling

If we consider a continuous signal, which has been sampled, then only the sampled values are recorded or played back. This raises an interesting theoretical problem: namely, whether or not the information from the original signal which fell between samples can be accurately reproduced on playback. In actual fact, so long as the original input analog signal has been bandwidth limited, then a

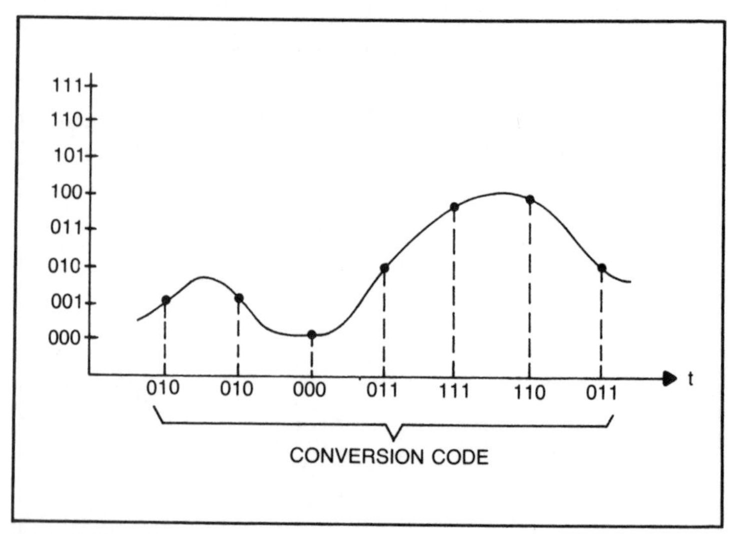

Fig. 3-6. Encoding with 3 bits.

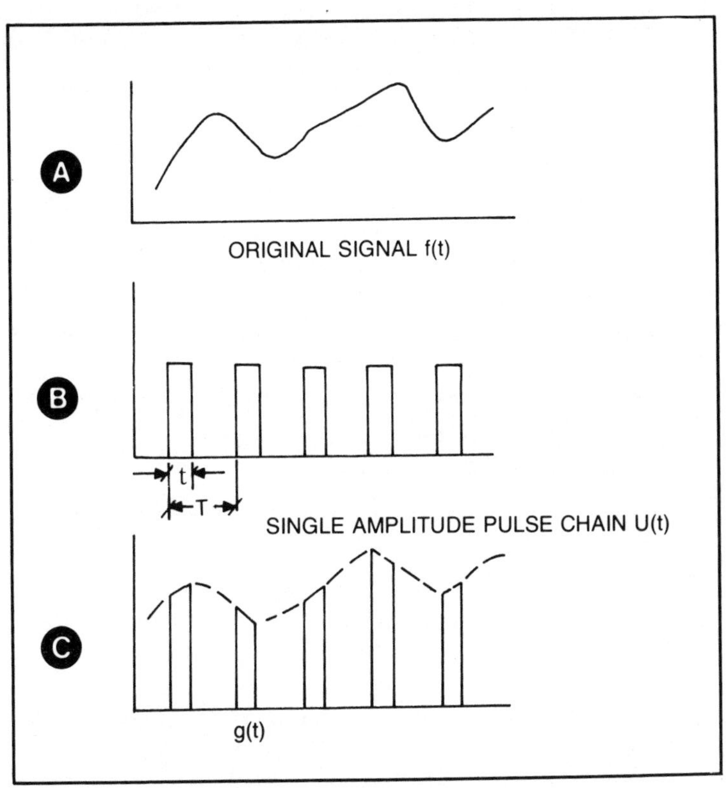

Fig. 3-7. Extraction of signal from pulse chain.

totally accurate reconstruction of the analog signal can be achieved for record and reproduction by using only the sampled values. This can be proved by a rather important sampling theorem used in digital audio.

The sampling theorem in question runs as follows: "The signal waveform f(t), which has been bandwidth limited, can be accurately reproduced under conditions where the intermediate values are unknown, provided that the values $f(t_1)$, $f(t_2)$, $f(t_3)$,......, $f(t_i)$ are known at points t_1, t_2, t_3,......, t_i."

We can explain this theorem in the following manner: First of all, the original waveform f(t), shown in Fig. 3-7(A), is sampled at the interval T (Fig. 3-7(B)), with a unit pulse width of t. The pulse chain known as U(t) is composed of unit pulses with a width of t. As a result, the signal g(t) is obtained:

$$g(t) = f(t) \times U(t) \qquad \textbf{Equation 3-1}$$

By progressive expansion, we can show that for the pulse chain U(t):

$$U(t) = k \sum_{n=-\infty}^{\infty} \frac{\sin nk\pi}{nk\pi} \cos n\omega st \qquad \textbf{Equation 3-2}$$

Here $K = t/T$, $\omega s = 2\pi fs = 2\pi/T$ (fs = sampling frequency). From equation (3-1) and (3-2) we see that:

$$g(t) = f(t) \times k\sum_{n=-\infty}^{\infty} \frac{\sin nk\pi}{nk\pi} \cos n\omega st \qquad \textbf{Equation 3-3}$$

When analyzing the signal component level represented by the sampled value, we should regard g′ (t) as having been obtained from very small width of the pulse chain g(t).
Here, because $\lim_{x \to 0} (\sin x/x) = 1$:

$$g'(t) = f(t) \times k \sum_{n=-\infty}^{\infty} \cos n\omega st$$

$$= f(t) \times k + 2f(t) \times k \sum_{n=1}^{\infty} \cos n\omega st \qquad \textbf{Equation 3-4}$$

It is clear from the first term above that the original continuous signal f(t) can be perfectly preserved, and the second term shows a modulated signal of f(t) by U(t). Figure 3-8 shows sampling with a suitable frequency.

One important point which must be remembered is that the original signal f(t) must be less than half the sampling frequency f(s). Or to put it another way, frequencies above f(s)/2 cannot be included. The expression "...signal... which has been bandwidth limited..." from the sampling theorem quoted earlier, refers to this phenomenon. Only frequency components below f(s)/2 can be used in the sampling process. If f(t) contains frequencies above f(s)/2 and is sampled, then there is a frequency overlap between the basic waveform shown in the first part of Equation 3-4, and the modulated signal shown in the second part. Upon demodulation, it then becomes impossible to retrieve only the basic signal f(t). This state of affairs is shown in Fig. 3-9, and is one possible source of distortion of the basic signal when using digital equipment. This distortion

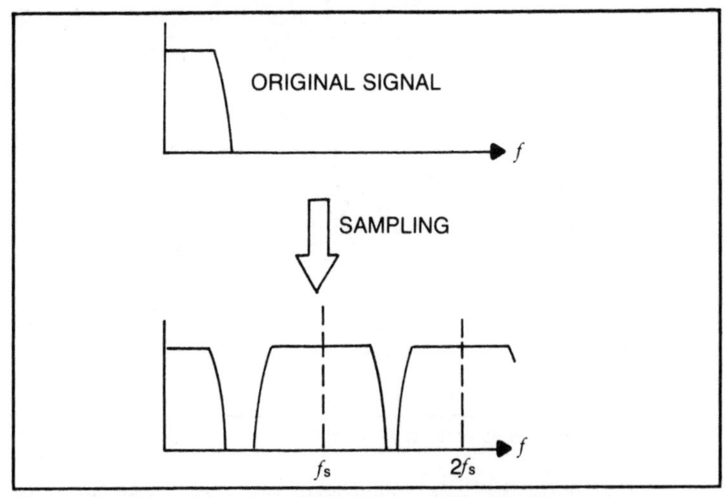

Fig. 3-8. Changes in frequency spectra caused by sampling.

phenomenon is known as **aliassing.** It resembles jamming on wireless receivers.

If we look at sampling theory from another point of view, we can say that there is only one signal which the frequency components are below fs/2 where the sampled values have been taken at time intervals of 1/fs.

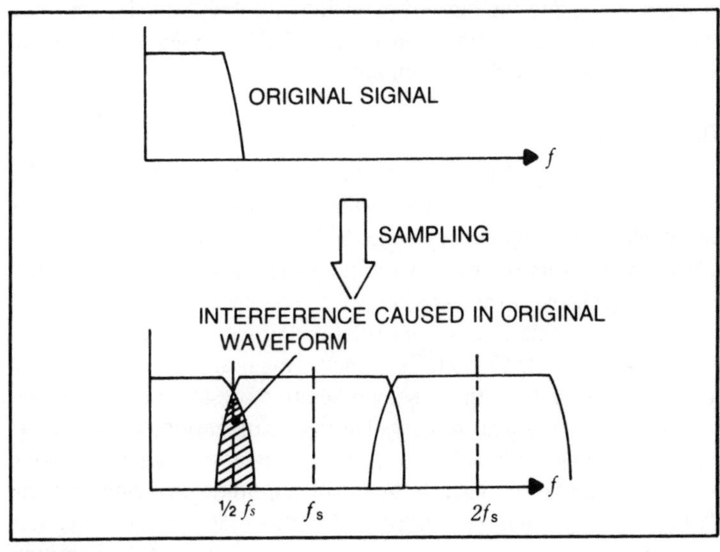

Fig. 3-9. Aliassing.

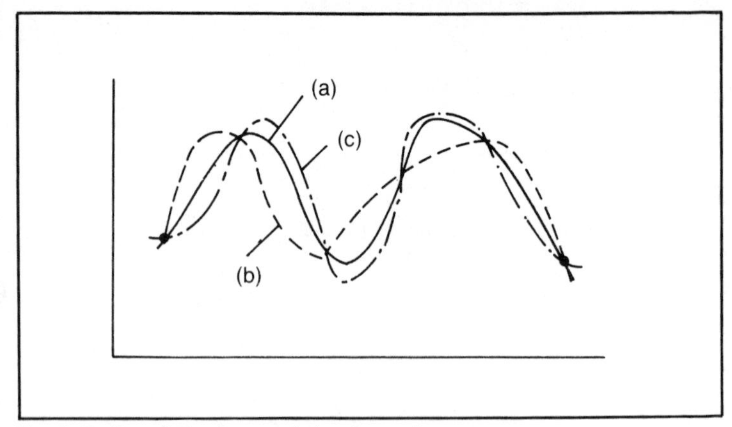

Fig. 3-10. Sampled values and signal waveforms.

In Fig. 3-10, the signal waveform (a) is the only one with frequency components below fs/2, following the sampled values taken at the points marked with a solid dot. Then if we carefully consider waveforms (b) and (c), which also conform to the pattern of solid dots (sampled values), it is obvious that these two waveforms contain frequencies exceeding fs/2.

In PCM equipment, because of the effects of the sampling theorem, the sampling frequency must be set at a level which is twice the maximum anticipated audio input signal. The upper limit of human hearing for high frequencies is below 20 kHz, and thus sampling frequencies between 40 and 50 kHz are generally chosen for use in digital audio equipment.

Quantization

After the signal has been sampled, it is then quantized, and the modulation process is complete when the quantized sample values are converted into a binary code. Since quantization and encoding are carried out more or less simultaneously, we shall consider both these processes in this section.

Provided that the correct relationship between sampling frequency and maximum input frequency is maintained, as explained in the previous section, there should be no information losses and no deterioration in signal quality. During quantization, a continuous amplitude analog signal is processed to produce a signal consisting of a finite number of discrete levels of amplitude. As shown in Fig. 3-11, the inputs and outputs resemble those of a signal which has been passed through a stepping transducer. If the input and output

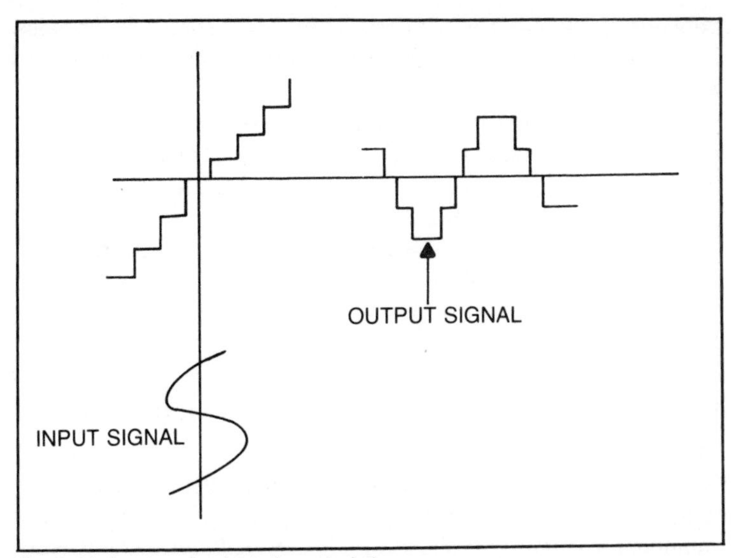

Fig. 3-11. Input and output characteristics.

signals are compared, a slight discrepancy is apparent, as shown in Fig. 3-12.

This can be a source of distortion, but the error can be reduced, because for every increase in the number of steps used, there is a corresponding reduction in the height of each step. The discrepancy

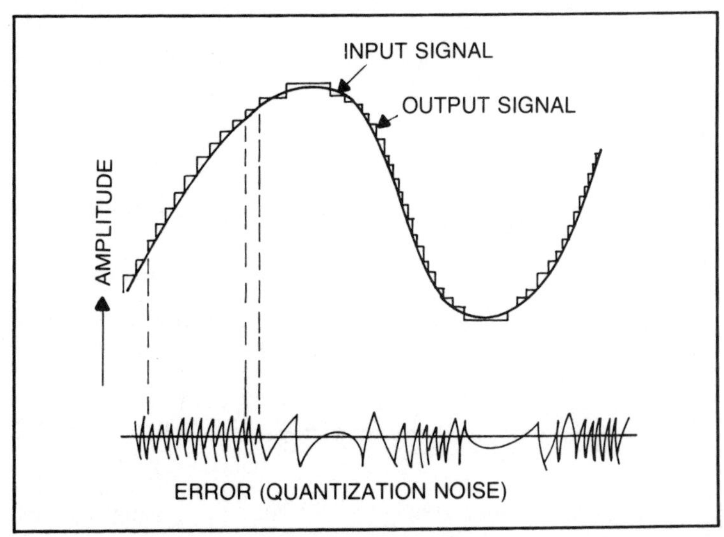

Fig. 3-12. Comparison of input and output signals.

between the input and output signals can produce quantization noise if the steps used are too large. We shall now consider the level of quantization noise.

When quantizing one sampled value, assuming that M bits are used, and that the number of steps (or the amount of quantization) is N, then:

$$N = 2M \qquad \textbf{Equation 3-5}$$

When the amplitude of O to V is used, then one quantization step will be Eo, and:

$$Eo = \frac{V}{N-1} \qquad \textbf{Equation 3-6}$$

Therefore, the peak-to-peak (p-p) amplitude of the quantization noise becomes Eo, and the probability of generating a distorted amplitude from an input signal where the amplitude is sufficiently large, is uniform within ± Eo/2.

Thus, the quantization noise power N_Q can be calculated as follows:

$$N_Q = \frac{2}{E_0} \int_0^{E_0/2} x^2 dx = \frac{1}{12} = E_0^2 \qquad \textbf{Equation 3-7}$$

Working on the hypothesis that the input signal is a sine wave with a p-p amplitude of V, then signal power S is:

$$S = \frac{1}{2\pi} \int_0^{2\pi} \left(\frac{V}{2} \sin x \right)^2 dx = \frac{1}{8} V^2 \qquad \textbf{Equation 3-8}$$

The relationship between signal power and quantization noise power is as follows, based on Equations 3-6, 3-7, 3-8:

$$\frac{S}{N_Q} = \frac{V^2/8}{E_o^2/12} = \frac{3}{2} N^2 \qquad \textbf{Equation 3-9}$$

The dynamic range D may be calculated thus, using Equation 3-5:

$$D = 10 \log_{10}\left(\frac{S}{N_Q}\right) = 10 \log_{10} \frac{3}{2} \times 2^{2M} = 6 \times M + 1.75 \text{ (dB)}$$

Equation 3-10

The dynamic range on, for example, equipment where sampled values are quantized using 16 bits, will be approximately 98 dB. Strictly speaking, the quantization noise power is spread at random across a wide frequency bandwidth, and the digital signal is passed through a filter at the time of demodulation so that only the original signal is retrieved. As a result, the dynamic range actually achieved may vary slightly from the theoretical result shown in Equation 3-10.

It is obvious from the above explanation that the quantization process will produce quantization noise, and that this will be a major problem to be solved during the design process. From a practical point of view, an increase in the number of quantization bits used will reduce this objectionable noise to a level where it cannot be perceived.

However, there is a limit to the efficacy of increasing the number of quantization bits, because as the amplitude of the input signal decreases, the correlation between quantization noise and signal becomes more marked. In other words, it would not be perceived as distortion but as higher order distortion component of the input signal.

In the case of a PCM recorder, this situation poses a problem from the point of view of audible quality. If white noise is added to the input signal being quantized, the generation of higher order distortion is controlled. This is because the middle of the quantization step is reproduced at the time of demodulation by the effect similar to PWM by noise. The process, however, does not mask higher order distortion by inclusion of noise in the input signal, but it does basically control the generation of higher order distortion.

The type of noise which is added to the signal for this purpose is known as dither, and from the point of view of a listener, the effect is satisfactory if the distortion amplitude is suitably large compared to the quantization steps.

Non-Linear Quantization (Companded Systems)

In the quantization system described above, the code corresponds to amplitude steps of a fixed width. This is known as linear quantization. There is another quantization system which is non-linear: the companded system. The quantization steps are of different width, depending upon the amplitude value presented.

The input and output of a non-linear quantized system is shown in Fig. 3-13 (in relation to audio use, the non-linear system most commonly used is the companded system). Compared with linear quantization, very high amplitude input signals can be handled. To explain this more fully, when quantizing an input signal with the same maximum value on both linear and non-linear systems with the same bit number, at low signal levels, a companded system will have less quantization noise.

However, when quantizing signals with a large amplitude, the quantization noise generated by a companded (non-linear) system, tends to increase. The quantization noise generated in a companded system will change depending on the level of the input signal.

From the point of view of a person listening to a digital system, even though a high signal level will increase the incidence of quantization noise, this will be masked, and be imperceptible. A companded system may, therefore, be highly effective in a PCM system for handling a musical signal. Companded systems are, for the most part, widely used in telecommunications systems, and partially in the PCM tape recorders.

However, when using a companded system in a high quality PCM unit designed for audio use, it should be noted that when a very low frequency input signal (dc-20 Hz) is fed to the machine, the changes in level of quantization noise may be distinctly audible, whereas the signal is not. Furthermore, unless the ratios between changes in quantization steps have been chosen very carefully, cross-modulation may also become objectionable.

Modulation Systems Used for the Recording Medium

In a PCM tape recorder, the analog signal is converted into a signal consisting of two information values, either 0 or 1, and these are recorded onto a suitable medium.

If we consider equipment using magnetic tape as the recording medium, the recording is made in the form of a magnetic imprint after the tape has passed through a magnetic field generated by the recording head. During reproduction, the residual magnetic flux on the tape (the recorded signal) is "read" by the reproduction head. The magnetic heads used differ slightly for record and reproduction, but are both constructed from a circular magnetic core with a coil wound around. For both record and reproduction, relative movement between the magnetic tape and the head is necessary.

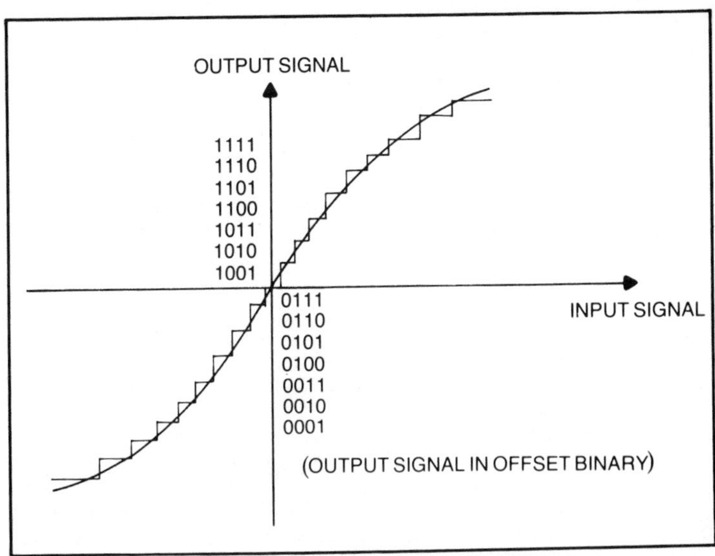

Fig. 3-13. Input and output characteristics of a non-linear quantization system.

The reproduction of a dc signal using this type of record/reproduction equipment is not possible. A digital signal is composed of 0s and 1s, corresponding to the polarity of the pulses, and reproduction would not be possible whenever a continuous string of 0s or 1s is generated.

During reproduction, the bit period must also be correctly distinguished, because upon this depends the accurate reproduction of the information contained in the recorded digital information. In normal circumstances it would not be possible to distinguish between bit periods when the signal is composed of long strings of 0s, and 1s, and this would be particularly necessary when a synchronous signal is required.

Luckily, all these problems can be solved by using various types of modulation. Thus, the recording medium is not required to record digital information corresponding to pulse timing in equipment used for digital audio recording. (PCM itself is, of course, also a type of modulation system.) Figure 3-14 shows examples of the signal waveforms of modulation systems in common use.

NRZ (Non-Return to Zero). NRZ is the most simple and basic form of modulation: the 0s and 1s represent pulse polarity or level. It is most commonly used in digital circuitry for PCM equipment or electronic calculators for the transmission of information or for control functions. It is probably more appropriate to refer to it as

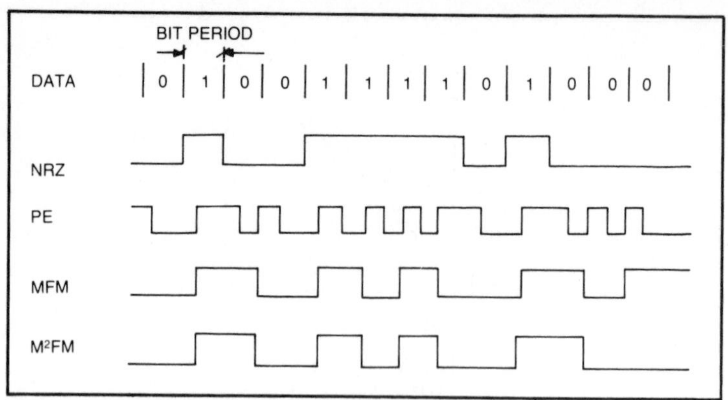

Fig. 3-14. Modulation waveforms.

a basic signal waveform than as a modulation waveform. It contains frequency components down to dc, and it is difficult to synchronize bit cycles using only an NRZ signal. It is, therefore, not suiable for magnetic recording.

PE (Phase Encoding). Using this modulation system, positive and negative pulse transitions are used to represent the 1s and 0s, and when a continuous stream of 1s and 0s occurs, a redundancy transition is added at the bit boundaries. The pulse period is around 1~½ bits, so no dc is included. Because a transition occurs more than once per bit, this system can easily maintain bit sync within the signal.

MFM (Modified Frequency Modulation). This system performs a transition for 1s, but not for 0s. If a long string of 0s occurs, then transitions are performed at the bit boundary. MFM is sometimes called Miller modulation or DM (delay modulation), and can be used very effectively to maintain bit sync.

M²FM (Modified Modified Frequency Modulation). This system resembles MFM in so far as transitions are performed for 1s and not for 0s. When a string of continuous 0s is generated, once again, a transition is carried out at the bit boundary. However, when a transition occurs at the previous bit boundary, the next transition is not carried out. This system is also suitable for maintaining bit sync.

4/5 Rate. This is a modulation system where one redundancy bit is added for each four bits of data, thus giving a five bit code. After modulation, whatever permutation of information is handled, the maximum number of 0s can never be greater than three. In a 4 bit data system, 16 data expressions can be handled ($2^4=16$), whereas,

Table 3-4. 4/5 Rate Modulation.

Data	Code
0000	11001
0001	11011
0010	10010
0011	10011
0100	11101
0101	10101
0110	10110
0111	10111
1000	11010
1001	01001
1010	01010
1011	01011
1100	11110
1101	01101
1110	01110
1111	01111

in modulated form of a 5 bit system 32 expressions (2^5=32) can be used, and it is possible to select the arrangement mentioned above.

The conversions are shown in Table 3-4, and during the record process, the 1s and 0s are represented by magnetic transitions. At least one magnetic transition is carried out during 3 bits, and bit sync can be easily maintained.

3PM (Three Position Modulation[2]. In 3PM, 3 data bits are converted into a 6 bit code; the conversions are shown in Table 3-5. After conversion, between each 1 and the succeeding 1, there are at least two 0s. However, after studying the conversions shown in the table, we can see that from these figures alone our coding conditions

Data	Code
000	000010
001	000100
010	010000
011	010010
100	001000
101	100000
110	100010
111	100100

Table 3-5. 3PM Modulation.

are not satisfied. This is because, according to the table, the fifth bit of one converted code may be 1, and the first bit of the following code may also be 1. Under these conditions, the 5th bit of the preceding code and the first bit of the following code are converted to 0, and the 6th bit of the preceeding code is converted to 1. This system resembles 4/5 rate in that, after conversion, 1s and 0s correspond to magnetic transitions and no-transition. The special feature of this modulation system is that the minimum distance between transition is 3/2 of original bit cell. It is possible to use this system where a signal requiring bit sync is necessary.

THE STRUCTURE OF PCM RECORD/REPRODUCTION CIRCUITRY

PCM record/reproduction equipment uses either magnetic tape or disc as the recording medium. The actual circuits used in PCM equipment will vary slightly depending on the recording medium used and on the recording format. The circuits for the correction of dropout and those used for processing the serial or parallel digital data may also vary slightly. However, in general, the actual pulse code modulation and demodulation will be basically the same in any machine.

We will now examine the role of these circuits in two channel PCM record/reproduction equipment. Figure 3-15 shows a block diagram showing the record and reproduction chains separately.

Line Amplifier

The line amplifier appears at the beginning of the record chain, and at the end of the reproduction chain, and is used for adjusting signal level and impedance. It is also used to ensure that the frequency response and phase characteristics are kept flat within the audible frequencies. It is exactly the same circuit as that commonly used in analog audio equipments.

Dither Generator Circuit

This circuit generates white noise to randomize the higher order distortion caused by quantization of low signal levels. A zener diode is used as a noise source, and the output is amplified to the correct uniform level. As an alternative using a shift register the M series may be generated and used.

Because the dynamic range of the digital equipment is determined by the level of dither, noise with equal dither amplitude must be made less objectionable; the frequency components of the noise

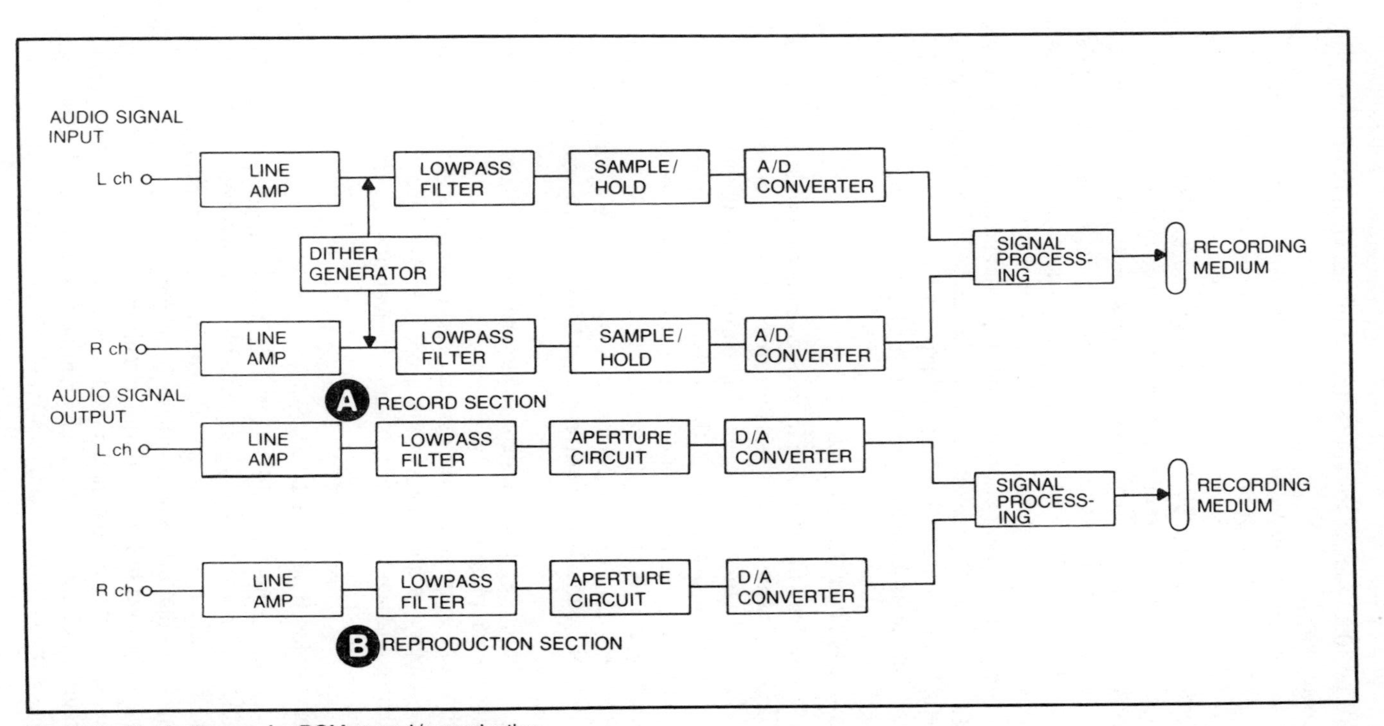

Fig. 3-15. Block diagram for PCM record/reproduction.

are not flat, and so are concentrated in the vicinity of particular frequencies.

The same effect can be achieved by using a single frequency sine wave instead of noise. Here, the frequency chosen must be outside the audible frequencies, but must also be less than half the sampling frequency (so that no aliassing occurs). Using this type of sine wave does not reduce the dynamic range of the audio frequencies, and is also an effective countermeasure to quantization noise.

However, in practical terms, the upper levels of the audio frequencies and the level of half the sampling frequency can be very close, so it is not generally feasible to insert the requisite sine wave between them. Another drawback is that, should linearity be bad at later stages in the circuit, there will be cross-modulation with the actual audio signal, achieving a result opposite to that intended. No such similar problems occur with dither circuitry using white noise.

Low-Pass Filter (Record Chain)

Low-pass filters appear in both record and reproduction chains. Filters with similar characteristics are commonly used in the record and reproduction sections, but the purposes are different.

The low-pass filter in the record section is used to protect against aliassing, mentioned earlier in this chapter. When the input audio signal includes frequencies higher than half the sampling frequency, Equation 3-4 comes into play, and they are made to fall within the audio bandwidth, the area below half the sampling frequency.

Therefore, this filter has a sharp cut-off for frequencies above half the sampling frequency. In effect, the signal is limited to fall within the same area as the dynamic range, which is determined by the number of quantization bits used. It is desirable that the frequency characteristics of the audio frequencies should be kept as flat as possible.

Sample and Hold Circuit

The sample and hold circuit, as its name suggests, is used for sampling the continually changing audio input signal, and then for holding the sampled value until the conversion processes in the A/D (analog to digital) converter have been completed.

As shown in Fig. 3-16, this circuit consists of an input buffer amp, an analog switch, and a hold capacitor. The sampling period occurs when the analog switch is in the ON position. The voltage across the capacitor changes according to the input signal, so that

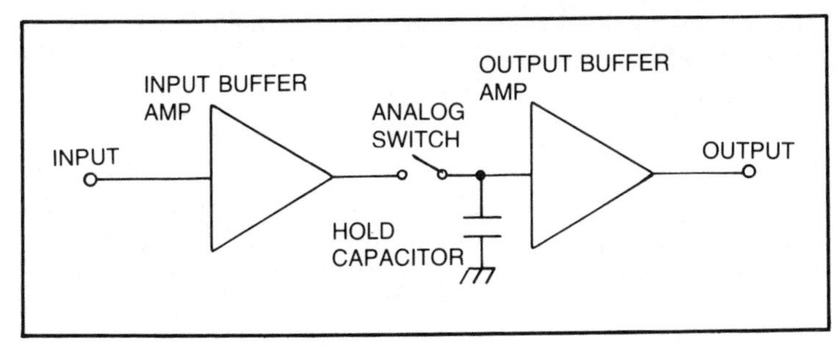

Fig. 3-16. Sample-and-hold circuit.

the actual input and output signals are the same. As soon as the switch returns to the OFF position, the voltage across the hold capacitor is maintained, so that the output voltage is then effectively the previous voltage, the one before the switch moved to the OFF position.

The various inputs and outputs of the sample and hold circuit, and their relationship to the signal controlling the analog switch are shown in Fig. 3-17. We can see that the input signal is sampled at intervals of t_1, t_2, t_3, and that these sampled values are held until the beginning of the next sample.

A/D Converter

The A/D converter changes the sampled values obtained from the analog signal into binary code, in other words, into a digital signal. The A/D converters most commonly used in audio equip-

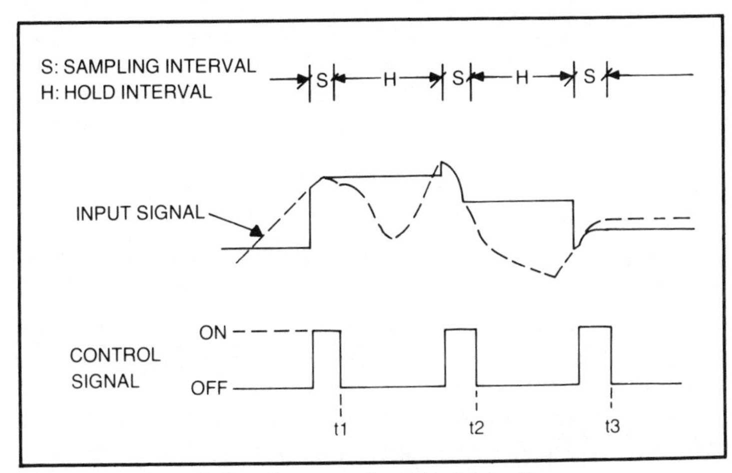

Fig. 3-17. Signal waveform from sample-and-hold circuit.

ment for PCM record/reproduction have between 12 and 16 bits. Using Equation 3-10 we can easily calculate the dynamic range theoretically available (strictly speaking, as mentioned before, other factors must also be taken into account.) Provided that the A/D converter speed is high enough, it is perfectly possible to use just one converter to handle both right and left channels by time division multiplexing. There are many different types of A/D converters, and the operation and performance of A/Ds is covered in Chapter 7.

Digital Signal Processing Circuitry (Record Chain)

After passing through the A/D converter, the digital signal enters the digital signal processing circuitry of the record chain. It is at this point that the redundancy bits, which are used in reproduction for detecting errors caused by tape dropout, and subsequently for error correction, are added. In this circuitry, time compression of the digital signal using a digital memory occurs, so that the signal can be recorded on a helical scan VTR. These digital signal processing circuits vary very widely depending upon the recording medium and the recording format used.

Digital Signal Processing Circuitry (Reproduction Chain)

The digital signal processing circuitry in the reproduction section carries out the same operations as in the record chain, but in reverse. It carries out detection and correction of any errors or dropouts, and expands the digital signal which has been compressed during record.

The reproduction signal may have been affected by mechanical oscillation in the record medium, and thus have had timing errors introduced into it. To counter this, a stable synchronizing signal is added; this is derived from the digital memory and the master oscillator.

D/A Converter

After error detection and correction in the digital processing circuitry, the digital signal is converted back into an analog signal. Similar to A/D converters, the number of bits for the D/A converters is usually between 12 and 16. Compared to A/D converters, D/A converters have a much higher conversion speed, and it is the rule rather than the exception to use just one D/A and time division multiplexing. Once again, a great variety of D/A converters are available, and a fuller description will be found in Chapter 7.

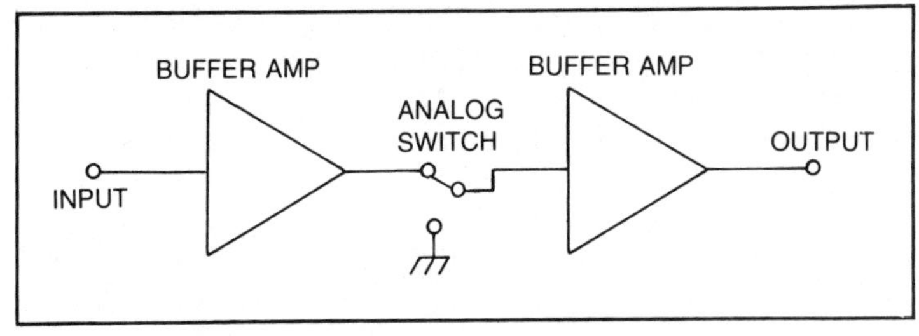

Fig. 3-18. Aperture circuit.

Aperture Circuitry

The output of the D/A is fed to an aperture circuit composed of an analog switch and a buffer amp, as shown in Fig. 3-18. There are two reasons for the necessity of this circuit.

First, a certain amount of time is necessary for the input digital signal to reach the correct value. The waveform in Fig. 3-19A shows that the digital signal requires a short time to elapse before it becomes fully stable, even though it enters the D/A during a sampled period. A control signal from the analog switch is then used, as shown in Fig. 3-19B, to read off a voltage from the stable area of the waveform. The output waveform of the aperture circuit is shown in Fig. 3-19C.

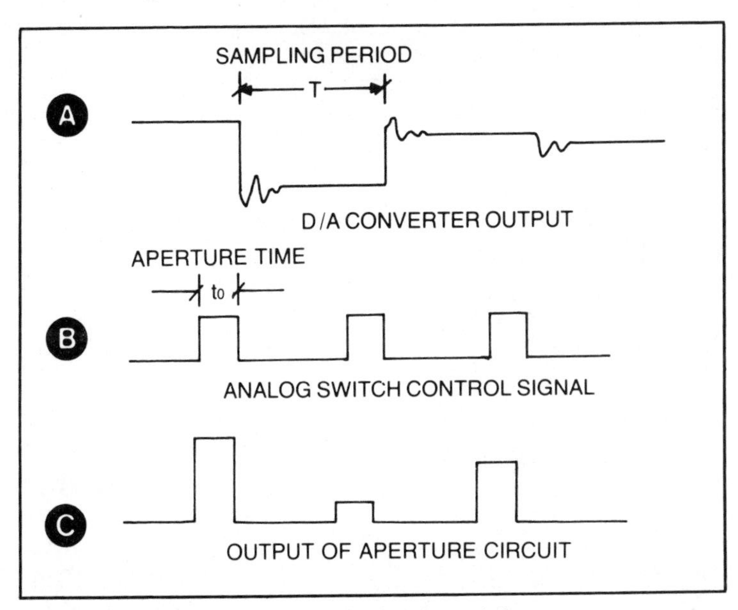

Fig. 3-19. Output of aperture circuit and analog switch.

The second reason for using this circuitry is so that frequency characteristics can be improved. The signal retrieved after passage through the aperture circuit is PAM. The height of the various pulses corresponds exactly to the sample values recorded (assuming, for the purposes of argument, that there is no quantization error.)

Assuming that the aperture time is t_0, then the PAM output wave M(t) will be equal to the sum of the pulse chains of width τ as shown in Equation 3-4, up to pulse width t_0:

$$M(t) = \frac{1}{T}\int_0^{t_0} f(t-\tau \times \sum_{n=-\infty}^{\infty} \cos n\omega_s (t-\tau)\, dt$$

Equation 3-11

Then, if the input signal f(t) is a sine wave, $f(t) = Av \cos \omega vt$, thus:

$$M(t) = \frac{Av}{T}\int_0^{t_0} \sum_{n=-\infty}^{\infty} \cos [(n\omega_s + \omega v)(t-\tau)]\, dt$$

$$= \frac{t_0 Av}{T} \sum_{n=-\infty}^{\infty} \frac{\sin \frac{t_0\pi}{T} \frac{n\omega_s + \omega v}{\omega s}}{\frac{t_0\pi}{T} \frac{n\omega s + \omega v}{\omega s}} \cos[(n\omega s+\omega v)(t-\frac{t_0}{2})$$

Equation 3-12

Then, M(t), when n=0 becomes the basic waveform for the reproduced analog signal.

$$m(+)n = 0 = \frac{t_0 Av}{T} \frac{\sin \frac{t_0}{2} \omega v}{\frac{t_0}{2} \omega v} \cos \omega v\, (t - \frac{t_0}{2})$$

Equation 3-13

The frequency characteristics of the basic waveform H(wv) are:

$$H(wv) = \frac{t_0}{T} \frac{\sin \frac{t_0}{2} \omega v}{\frac{t_0}{2} \omega v}$$ **Equation 3-14**

However, this assumes that $A_v = 1$.

If we assume that the time required for full stabilization of the D/A converter output voltage is 0, and does not pass through the aperture circuit, then the pulse width to is equal to the sampling period T, and the frequency characteristics are:

$$H(\omega v) \text{ to} = T = \frac{\sin \frac{T}{2} \omega v}{\frac{T}{2} \omega v}$$ **Equation 3-15**

When the signal frequency is Wv = 0, then H(Wv) will be 1, and when the maximum frequency is Wv = π/T, there is a reduction close to 4 dB:

$$H \frac{(\pi)}{T} \text{to} = T = \frac{\sin \frac{\pi}{2}}{\frac{\pi}{2}} = 0.64$$ **Equation 3-16**

However, when to is reduced to ¼ of the sampling period by the aperture circuit, the amplitude ratio between Wv = 0 and Wv = π/T becomes:

$$\frac{H(\frac{\pi}{T}) \text{ to} = T}{H(O) \text{ to} = T} = \frac{\sin \frac{\pi}{8}}{\frac{\pi}{8}} \div 0.97$$ **Equation 3-17**

and there is a reduction of only 0.22 dB.

Figure 3-20 shows the frequency characteristics when the aperture time to is changed in relation to the sampling period.

From this diagram, we can also appreciate the fact that the smaller we make to, the flatter the frequency characteristics will become. From Equation 3-14 we can see that:

$$H(0) = \frac{t_0}{T}$$ **Equation 3-18**

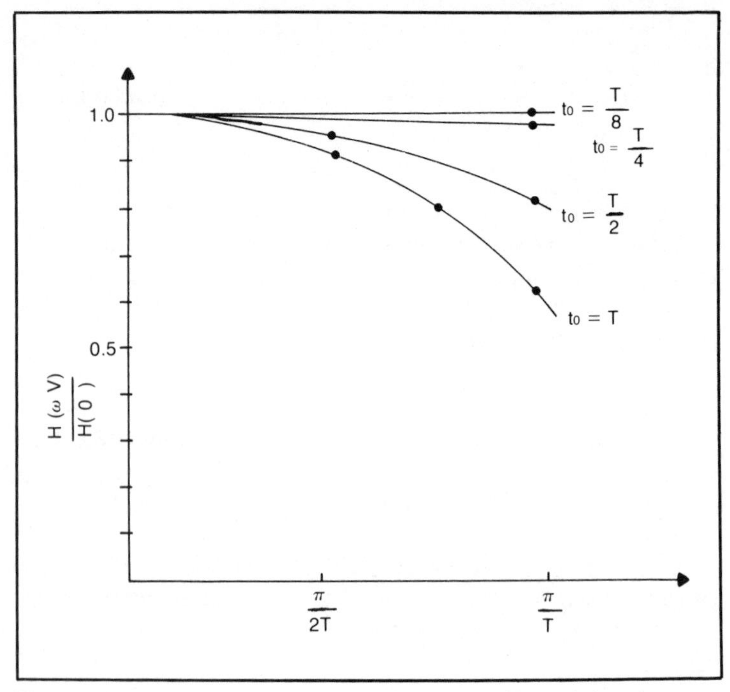

Fig. 3-20. Changes in frequency characteristics caused by variation of aperture time.

and so an overall fall in output level will occur.

However, if we make to too small, the S/N will become worse, so in general we would choose around to = T/4. These are, then , the major reasons for including aperture circuitry on the D/A converter output.

Low-Pass Filter (Reproduction Chain)

The output signal from the aperture circuit contains a high proportion of high frequency components, as shown in Equation 3-12, apart from the basic frequencies required. These extra high frequencies which are superfluous to the basic signal, are outside the audible range. It would, therefore, be perfectly possible to reproduce the desired audio signal without using this filter at all. However, in the circuits after this, the slightest amount of nonlinearity will cause cross-modulation, and will, therefore, cause degradation in what would otherwise have been a high quality audio signal. Therefore, it is desirable to include this low pass filter in the reproduction chain of high quality digital audio equipment.

We have now looked at the basic functions of the major sections of the block diagram, and a more detailed description of the basic structure and operation will be found in Chapter 7.

References

1. Reeves, A.H.: American patents 2272070 (1942-2).

2. Jacoby, G.V.: "A new look-ahead code for increased data density", IEEE Trans., Mag - 13, 5 (1977-9).

Chapter 4

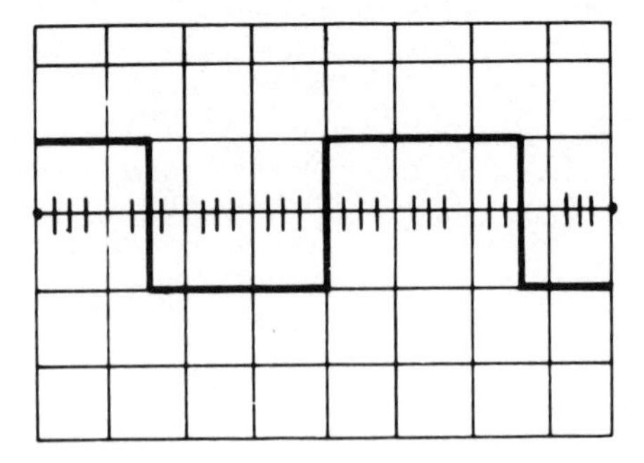

PCM Tape Recorders

The basic principles for converting a sound signal into electrical energy and recording this on tape are the same for both PCM tape recorders and conventional analog tape recorders. The three fundamental differences between digital and analog tape recorders are as follows:

□ The bandwidth required for a PCM tape recorder is more than 30 times that required for analog, necessitating improvements to increase recording density.

□ Bits rather than waveforms are recorded on PCM equipment, and it is possible both to record and disperse the data bits comprising one channel onto a number of tracks, or to record data from a number of channels onto one track. (Time sharing multiplexing : see Chapter 2).

□ Only two values, 1 and 0, are used in PCM recording, so linearity is not necessary.

Various techniques have been employed to increase recording density: in some cases a rotary head VTR (video tape recorder) is used. Stationary head tape recorders, capable of recording and dispersing one channel of data on a plurality of tracks, have also been developed. Conventional analog tape recorders require an ac bias to achieve the necessary linearity, but PCM tape recorders do not. The latter may be used by saturating the tape by a strong enough magnetic field.

In this chapter, we will investigate the relationship between the bandwidth required by PCM tape recorders and the minimum wavelength to be recorded on the tape. This will lead to a comparison between rotary and stationary head systems, and the different ways of solving the problem of recording density. Finally, we will look at the construction of various PCM tape recorders.

RECORDED WAVELENGTH AND HEAD-TO-TAPE SPEED

The following example gives an indication of how much bandwidth is required to accommodate a PCM signal. The bit transmission rate for 16 bits of information sampled at 50 kHz would be:

$$16 \times 50 = 800 \text{ kb/s}$$

If this were to be modulated using NRZ and MFM, then the maximum recorded frequency (f max) would be half the value 400 kHz, as explained in Chapter 3. Assuming that one channel were made up of two tracks, it would be acceptable to use half this figure, 200 kHz, for each track. It would also be necessary to allow for a certain amount of redundancy to accommodate error detection correction bits and synchronization bits.

Figure 4-1, indicates these relationships. The horizontal axis shows redundancy (the percentage of bits used for purposes other than information), and the vertical axis shows f max (maximum frequency to be recorded.) The value of f max will change depending on the modulation system and the sampling frequency, and thus four representative examples are given along this axis.

The star marks represent actual PCM tape recorders. However, the curves are calculated on the basis of 16 bits, so the examples for 13 and 14 bit machines have a slightly lower f max value.

Figure 4-2 shows the relationship between head-to-tape speed (V) and minimum recorded wavelength (λ min). With the recent advent of metal tape, the wavelengths recorded by conventional analog tape recorders have become shorter, and are now not very different from those recorded by PCM tape recorders. Successful recording of the bandwidth demanded by PCM tape recorders is due entirely to the increase in head-to-tape speed. Where the tape speed is 38 cm/s, one channel must be distributed over two tracks (PCM-3200). Where one channel is laid down on one track, the speed exceeds 1 m/s, and tape consumption gets very high. This is why a VTR with rotary heads is used: the high head-to-tape speeds

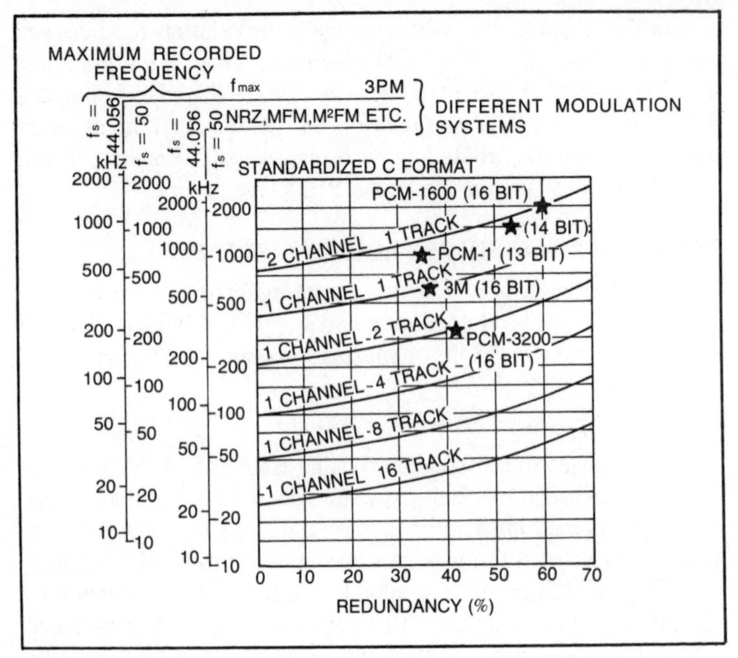

Fig. 4-1. Redundancy and maximum recorded frequency (16 bit quantization, sampling frequencies 44.056 kHz, 50 kHz).

mean margin in wavelength, and thus, a reduction in tape consumption.

ROTARY AND STATIONARY HEAD SYSTEMS

The differences between rotary and stationary head systems are shown in Fig. 4-3. The tape path for a typical stationary head array is shown in 4-3A. It is very similar to that of a conventional analog tape recorder, except that there is no erase held while there are two reproduction heads. The reason for this will be explained later.

Figure 4-3B shows the tape path for a rotary head VTR. The videotape extracted from the cassette is wrapped around the head by guides so that it covers an arc of more than 180°. The two heads are affixed to the head drum which revolves in the same direction of flow (the opposite direction to the tape), so that each head traces a path across the tape as shown in 4-2. It is clear that the relative head-to-tape speed is extremely high. Figure 4-3C shows the area around the head drum from the side. The guides are positioned so that the tape flows past the head drum diagonally, so that, as shown

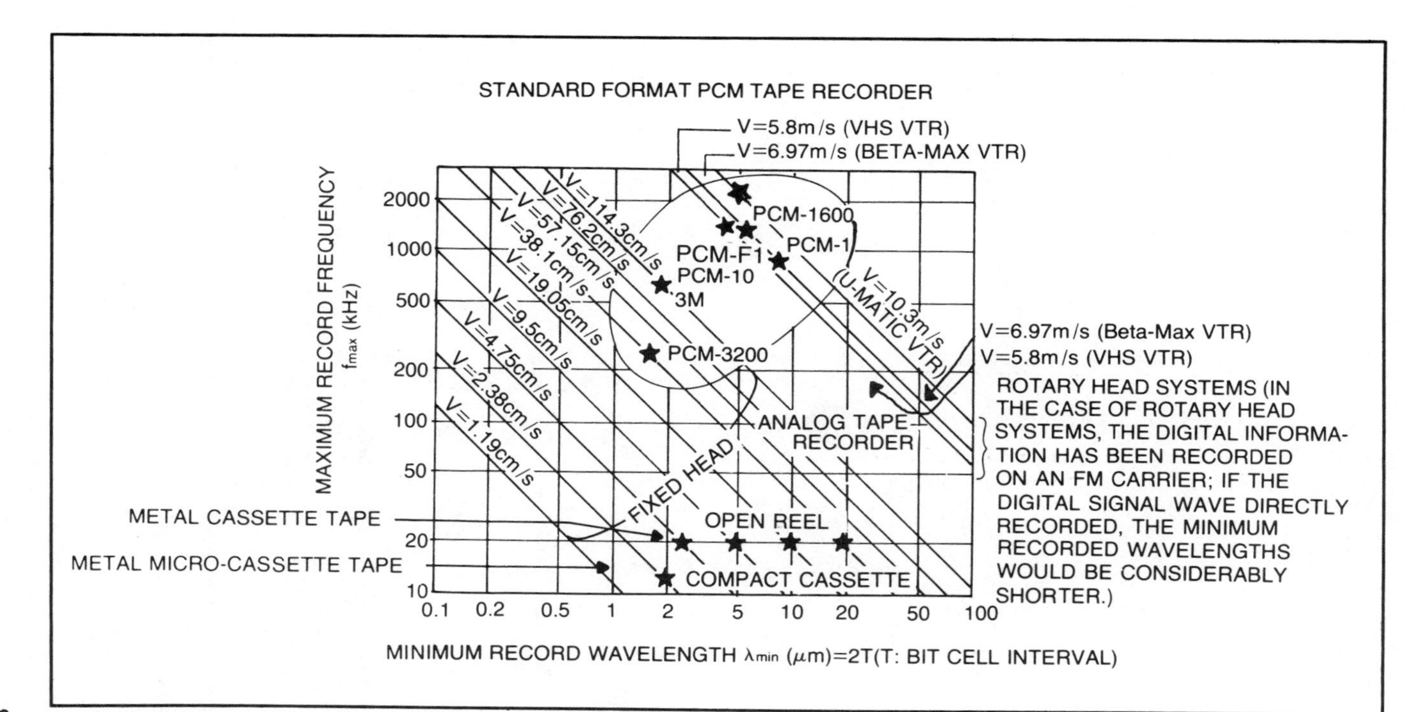

Fig. 4-2. Head-to-tape speed (V) and minimum record wavelength λ min.

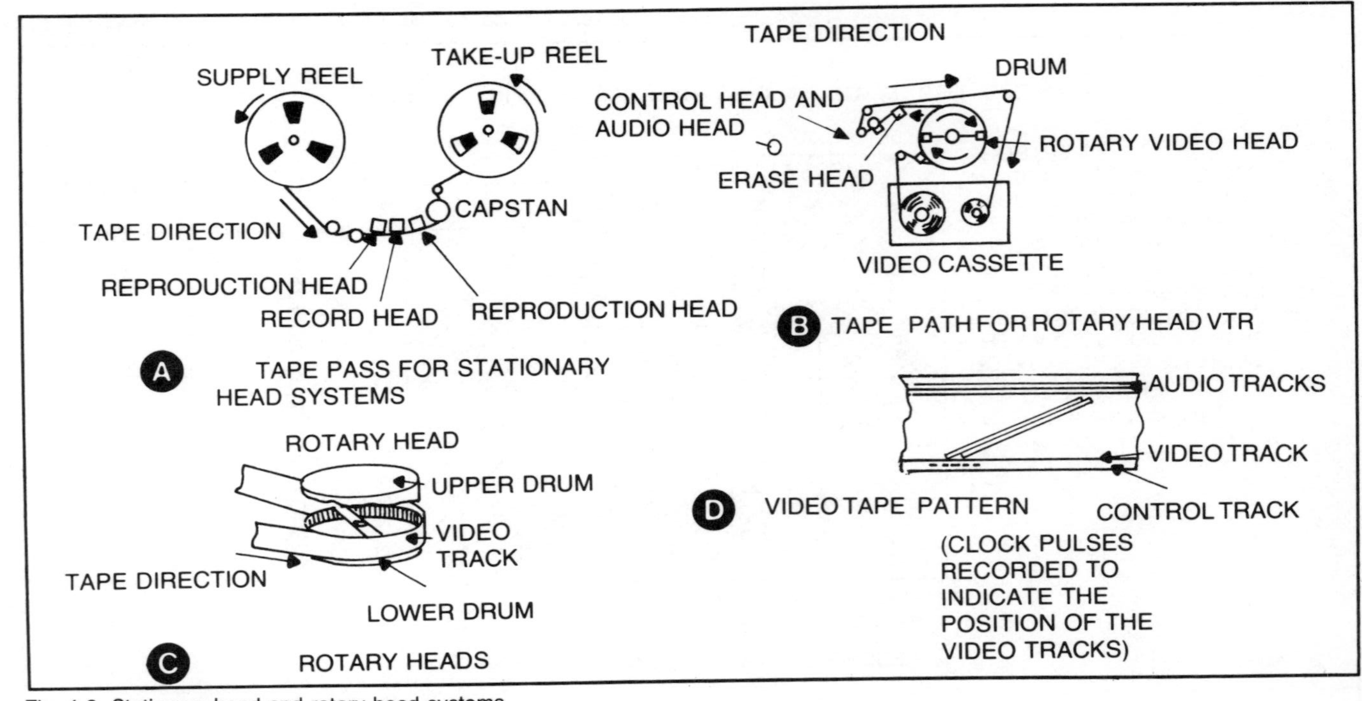

Fig. 4-3. Stationary head and rotary head systems.

in Fig. 4-3D, the video tracks are laid down diagonally across the tape. The rotary heads lay down alternate tracks on the tape. When pictures are recorded switching point from one head to the other does not appear in the picture as one picture is recorded on one video track. However, this causes a big problem when the VTR is being used in PCM tape recorders. The signal supplied to the heads cannot be recorded during the switching period, so in the case of a PCM recording, the data must be compressed along the time axis. During reproduction, the compressed signal is then re-expanded so that the original, continuous signal can be reproduced. This process will be explained in more detail later.

Table 4-1 shows a comparison between rotary and stationary head machines. The detailed implications of the various headings will become clear as we progress through later sections. However, generally speaking, the rotary head system is simpler, cheaper, and more convenient than a stationary head system, except for the compression and expansion of the signal along the time axis. A rotary head system also has the advantages of a high recording density, and low tape consumption. However, editing and synchronous recording pose some problems.

The professional multi-channel recorders (4 channels and above) requiring sophisticated editing capabilities will use the stationary head system. Two-channel machines for home use will, on the other hand, be based on the rotary head system. Falling between these two types is the professional two channel PCM

Table 4-1. A Comparison between Rotary and Stationary Head Recorders.

	Rotary Head Systems	Stationary Head Systems
1. Recording density	O	Δ
2. Tape consumption	O	Δ
3. Complexity of circuitry	O	Δ
4. Time axis compression	X	O
5. Sync	O	Δ
6. Price	O	Δ
7. Tape cut editing	Δ (X for cassette systems)	O
8. Synchronous recording	X	O
9. Rec./pb. of each channel	X	O
10. Punch-in, punch-out	X	O
11. Ease of use	O (particularly for cassette systems)	Δ

O: very good, Δ: a little inconvenient, X: very difficult.

recorder, and it is in this area that two systems are in effect competing with each other. We will no doubt see many more technological innovations, such as thin film heads, which will solve this problem by making stationary head systems more competitive.

ROTARY HEAD PCM TAPE RECORDER

Of the rotary head systems used in PCM recording most use not only the same mechanics as a VTR, but also use a simulated video waveform to carry the digital data. Figure 4-4 shows a PCM signal waveform from a PCM tape recorder (PCM-100). The signal includes horizontal and vertical sync pulse signals, and the signal recorded on the tape looks exactly the same as a real video signal. The PCM signal processing circuits can be regarded as transducers producing the following chain of signals: analog audio signal ⇄ PCM signal ⇄ pseudo-video signal. The arrow → indicates the recording chain, and ← the reproduction chain.

Basic Principles of the Rotary Head System

In the PCM processing, two channels of information are alternately recorded on one track, using the technique of time sharing multiplexing. This is shown in Fig. 4-5. The main point to be drawn from this diagram is that because the pseudo video signal must be provided with horizontal sync, vertical sync, a head switching period, and, of course, with redundancy bits for the PCM signal, the sampling period (τ' on the diagram) is shorter than the original sampling period (τ).

Around the vertical sync period there are between 17 and 18 TV lines where no signal is recorded, as shown in Fig. 4-4A. The input signal is of course, continuous, but has to be divided up into blocks (the simulated video signal) of uniform length. As a result, the signal has to be compressed into sections during recording. This process occurs along the time axis and is known as compression; the opposite process is carried out for reproduction, and is called expansion. The process as a whole is often called compansion or companding. Figure 4-6 shows the process of compression.

However, if the signal is to be compressed along the time axis, some kind of temporary store must be provided. This store is known as a buffer memory. Figure 4-7 explains the role of buffer memories in simple terms. Two "buckets" representing, respectively, the record and reproduction buffer memories are shown. Bucket No. 1 (corresponding to the record memory) is supplied with a uniform flow of water, at its base is a tap which can be used periodically to

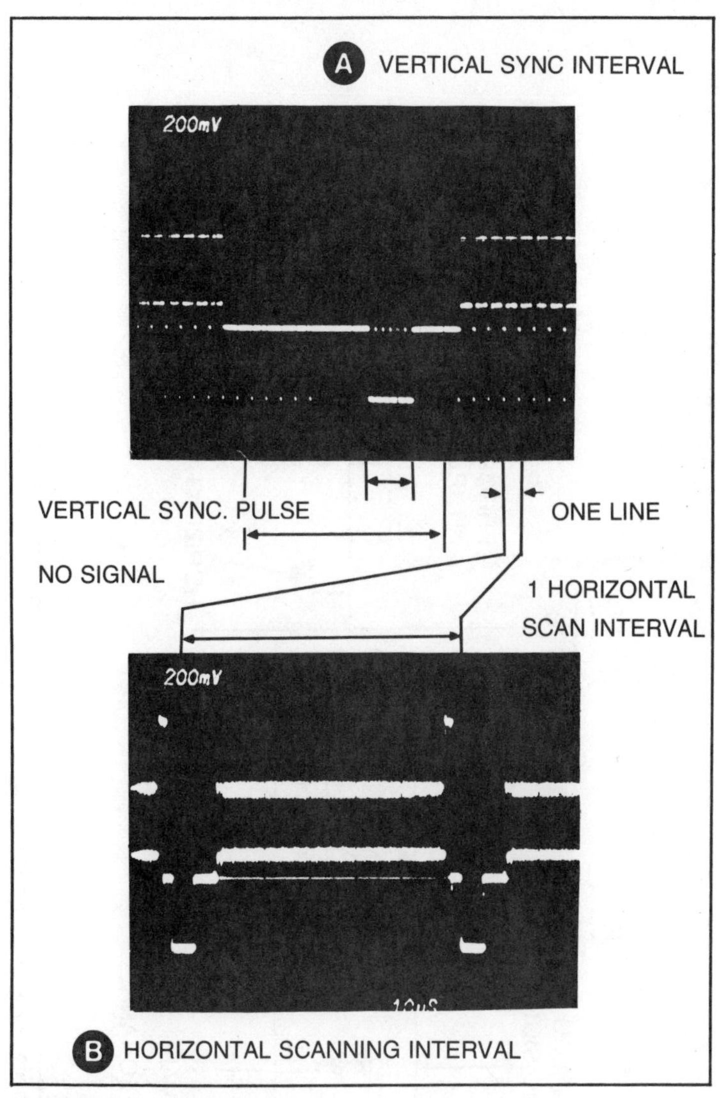

Fig. 4-4. Pseudo video signal.

cut off the outflow. (This corresponds to the no-signal period.) While the tap is shut off, the water still flows into the bucket, and the volume of water increases, so the tap must be opened again before the overflow point is reached. Bucket no. 2 corresponds to the reproduction buffer memory, and a continuous stream of water flows out of the hole in the side. (This flow corresponds to the music

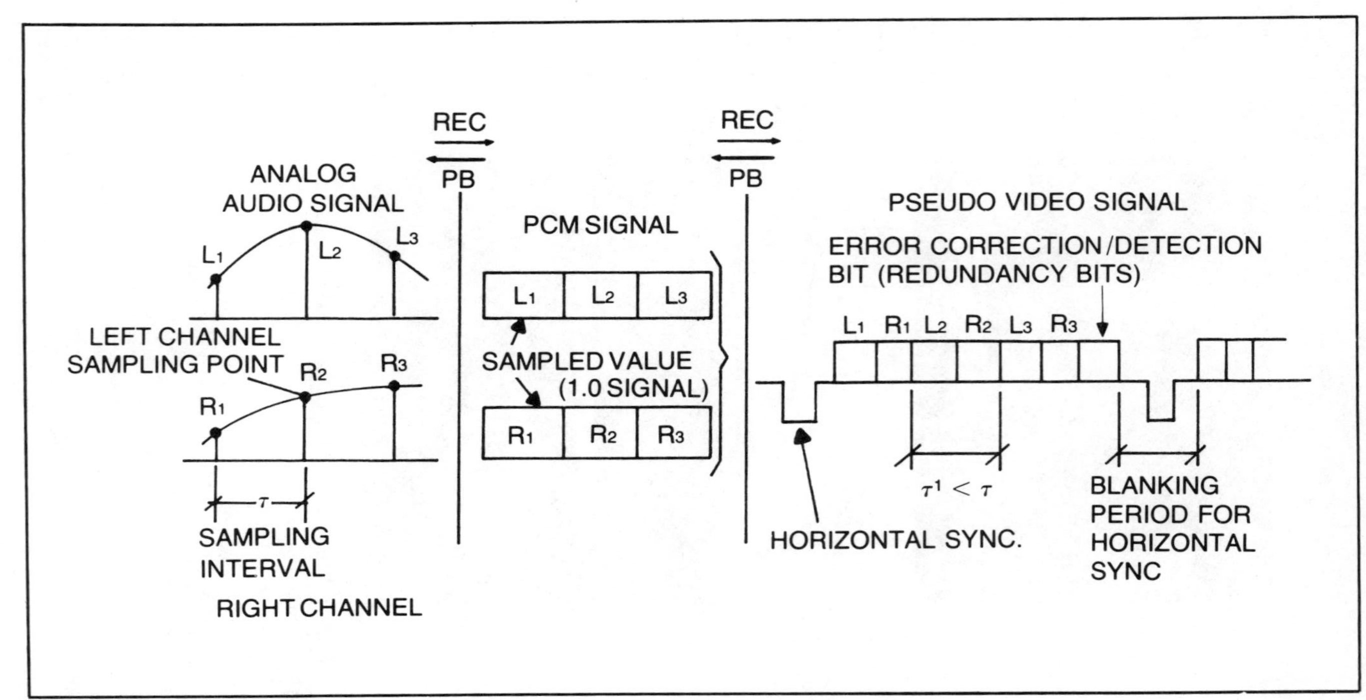

Fig. 4-5. How two channels are laid down as one track.

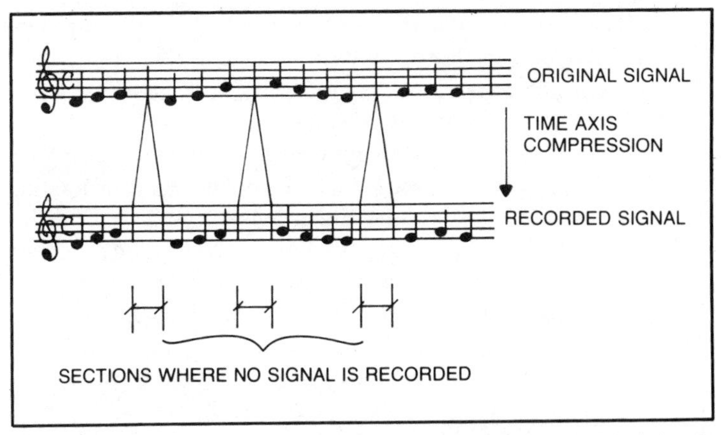

Fig. 4-6. Compression along the time axis during uniform flow (original signal).

signal being reproduced.) Of course, instead of buckets, the PCM recorder uses RAMs (Random Access Memories).

In this way, two channels of information can be written on one track via time sharing multiplexing. This and the ability to compand the signal along the time axis to insert periods of no-signal are unique to PCM, and cannot be imitated by an analog system.

Therefore, because an unmodified video signal is used to carry the PCM data, a certain amount of unnecessary information, such as

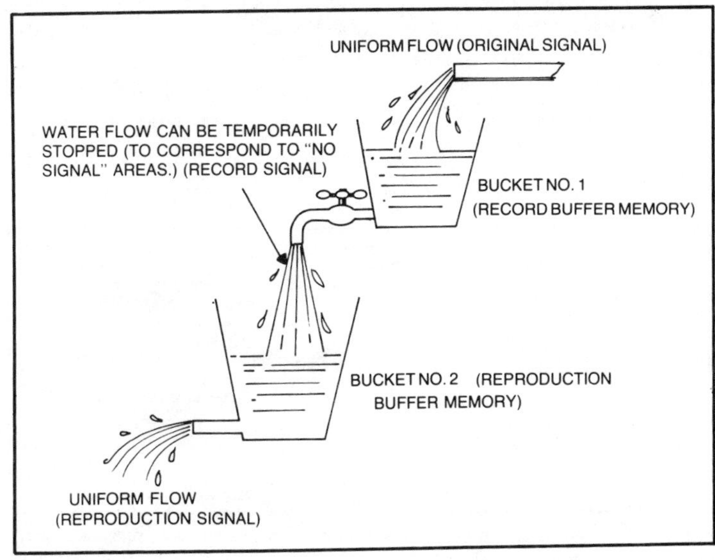

Fig. 4-7. Companion along the time axis using buffer memories.

horizontal and vertical sync. pulses are included. This is why companding must be used along the time axis.

There are benefits as well as drawbacks. As unmodified VTRs can be used as the recording medium, it becomes possible to visualize a PCM tape recorder made up of two separate items: VTR and PCM adaptor. In this case, it is self-evident that broadcasting companies and home video users will be able to use their VTRs for PCM audio as well as video. This, of course, makes sound economic sense to all video users who also require high quality audio.

Referring back to Table 4-1, where the benefits of a rotary head system are laid down, we see the headings "recording density" and "tape consumption". If we also look at Fig. 4-2, we can see that the recorded wavelength is greater on rotary head systems. The key to this lies in the track pitch. Track pitch refers to the distance between neighboring tracks recorded on tape (Fig. 4-8). Let us take an actual example. According to Fig. 4-2, when using a Betamax VTR the track width for a PCM-100 is 30 μm, and there is no guardband. Thus, the track pitch is also 30 μm. On the PCM-3200, however, the track pitch is 300 μm, while the guardband is 180 μm, giving a track pitch of 480 μm, the above figures result in a 16 fold difference in trade pitch between the two machines.

This leads us to consider why the track pitch on a rotary head machine can be so small. The reason is two-fold. First there is the problem of S/N. Generally, doubling the head-to-tape speed on a digital machine results in doubling of the S/N ratio. However, if the tape width is doubled, then the improvement in S/N ratio is no more than $\sqrt{2}$. As shown in Fig. 4-2, the head-to-tape speed for a PCM-100 is approximately 8 times that of the PCM-3200. Therefore, since the same S/N ratio is required for both machines, the tape would have to be 16 times wider on the PCM-3200.

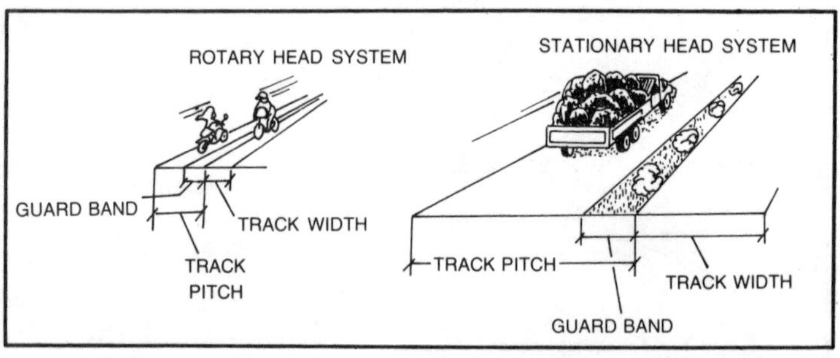

Fig. 4-8. The reason why recording density on rotary head systems is high is the narrow track pitch.

Another problem can occur while the tape is running. Variations in tape width of up to ±10 μm can occur during manufacture when the tape is being slit, and the quality of each end of the tape will not be as high as it is in the middle. As a result, the contact of head with tape can vary by tens of microns. With a stationary head system, the wide tracks and guardband will guarantee that these variations do not adversely affect the S/N ratio or the crosstalk figures.

To overcome these problems on a standard rotary head VTR, a control track (see Fig. 4-3D) is recorded on the tape, and various electronic devices are incorporated to ensure that the video tracks are positioned and re-traced accurately. In this way, tape variations are compensated for. Video tracks are laid down diagonally in the center of the tape, with audio tracks and control tracks conventionally recorded along the edges of the tape. Thus, any discrepancies on the upper and lower edges do not affect the signal. Horizontal variations (incorrect tape speed) are compensated for by the servo systems. Finally, all VTRs are provided with a tracking control, which gives complete interchangeability of tapes between different machines of the same standard. The mounting of heads on modern VTRs allows the tracing of tracks which are not straight, and it is also possible to use very narrow tracks.

The third heading in Table 4-1 refers to the amount of circuitry needed. In a rotary head VTR, two channels of information are recorded on one track, making the circuits relatively simple. However, if, for example, we compare the PCM-100 with the PCM-3200, the latter makes one channel into two tracks, so four tracks are needed to record two channels of information. This in turn means that most of the circuitry, record and reproduction heads, head amps, bit sync and frame sync circuits, jitter compensation, and so forth, has to be multiplied by four. Finally, the tape speed is slower, so the same length tape dropout will have a proportionately greater effect, necessitating more comprehensive correction circuitry.

Item number six in Table 4-1 refers to respective prices of stationary and rotary head machines, because differences in circuitry may contribute to price differences. On the one hand, a rotary head system is technologically extremely complex, including sophisticated mechanical parts. This is the major reason for their high cost. On the other hand, VTRs are produced in high volume from highly automated factories, so there is in fact not a great price discrepancy caused by mechanical parts. Item number five in Table

4-1 refers to synchronization: the video signal is used as it stands which makes sync for rotary head machines extremely simple. On stationary head machines, however, the only signals available are ones and zeroes, so the start of separate blocks of data must be indicated by a special pattern of ones and zeroes in order to locate the sync interval accurately. A rotary head machine, however, mounts PCM data on a standard video signal, and this pseudo-video signal is of course supplied with horizontal and vertical sync. Thus gaps between data blocks are very easy to identify (Fig. 4-9).

The question that must immediately arise is, why aren't sync periods included for stationary head systems as they are for rotary head systems. The answer to this lies in the difference between "digital direct recording" and digital signal-FM modulated recording", which will be explained later on.

In this type of high density recording, "saturation recording" is the technique used. The word "saturation" is often applied to amplifiers and transistors and has the same kind of meaning when referring to magnetic particles on a tape. In the same way as a transistor switching circuit is capable of indicating only two values, the magnetic material used for saturation recording can only indicate two values, "N" or "S", (like a magnet, see Fig. 4-10). With a

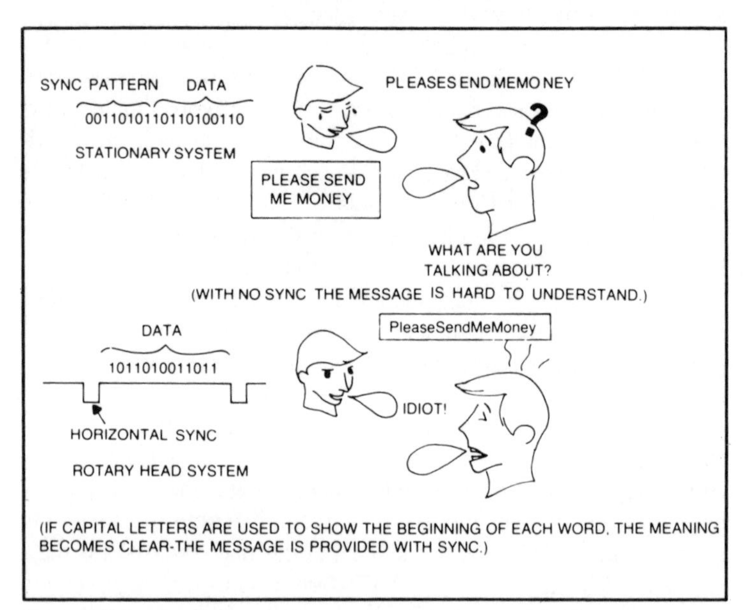

Fig. 4-9. Easily identifiable sync available on rotary head machines because a pseudo video is used.

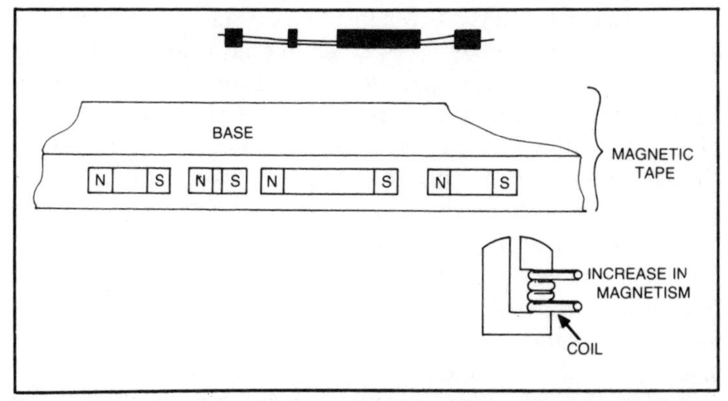

Fig. 4-10. Creation of tiny magnets within the tape by saturation recording.

stationary head recorder, these two values correspond to 1 and 0, but the recording system is such that 1 corresponds to N ⇄ S changes. A rotary head system, however, uses an FM modulator inside the VTR, so that, as shown in Fig. 4-11, changes in frequency allow distinction between 1 and 0. The third frequency shown allows identification of the sync signals.

Figure 4-2 shows the PCM signal as recorded wavelengths corresponding to maximum recorded frequencies. However, because the rotary head machines use an FM signal, the actual recorded wavelengths are shorter than shown (see Fig. 4-11).

A rotary head system carries a very high density, of information, and has a very high head-to-tape speed. Therefore, it is possible to record two channels of data on one track, and the use of FM makes it a simple matter to add sync.

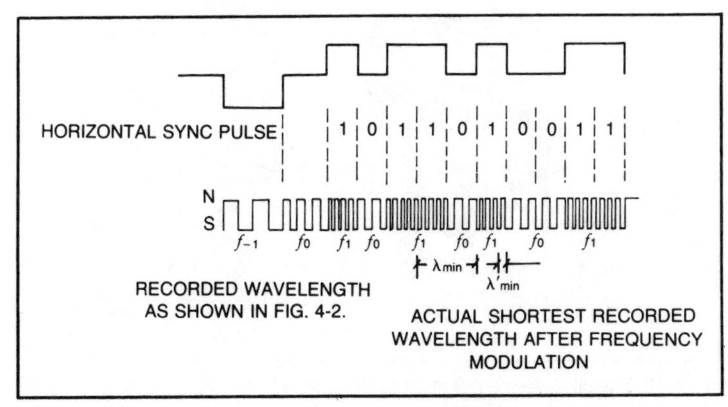

Fig. 4-11. FM recording made witn rotary head system.

Items number seven to ten in Table 4-1 refer to editing. Most rotary head PCM recorders use a cassette, which means that editing by physically cutting and splicing the tape is impossible. To edit a PCM recording made using a rotary head system, some form of electronic editing must be used. Further, because two (or more) channels are recorded on one track, it is not possible to record one channel only while replaying another. Both record and reproduction affect all channels simultaneously. See Fig. 4-12.

Item number 11 in Table 4-1 refers to ease of use. Obviously, it is much more convenient to handle a cassette than an open reel tape for most operations. This can be appreciated very quickly by considering the operational differences between a conventional compact cassette tape recorder and a conventional open reel tape recorder in a domestic situation. From this point of view, a cassette

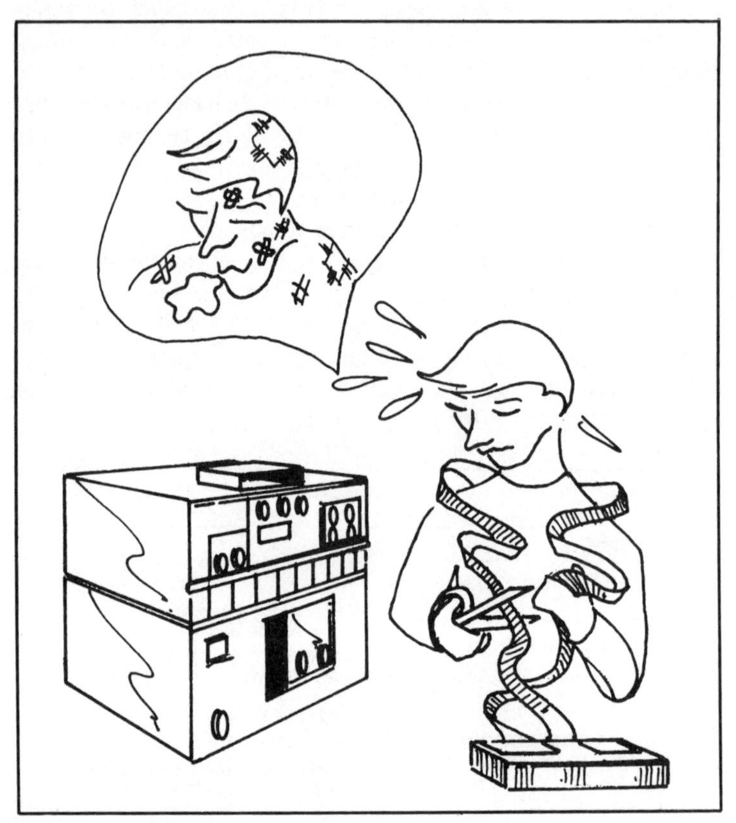

Fig. 4-12. The difficulties of editing a cassette tape by hand—in all other respects cassettes are most convenient.

is much more convenient, because in particular, it protects the tape from finger prints.

The above constitutes a general outline of some of the fundamental principles which underlie the use of rotary head cassette VTRs, using pseudo-video signals for PCM tape recording. However, it is perfectly possible to develop a rotary head PCM tape recorder which carries out "direct recording" in the same way as a stationary head recorder, while having no relationship with a video signal. In this case, as one could imagine from Fig. 4-2, it would be possible to invent a system based on a Betamax VTR, where one cassette could be used to record four times the information at present.

Various Types of Rotary Head PCM Tape Recorders

Table 4-2 gives a general idea of the PCM tape recorders based on rotary head recording systems which have been introduced over the years. No. 1 is the world's first PCM audio recorder, as described in Chapter 1. No. 2 is a development of No. 1, which was actually used for record mastering. No. 2 used a high quality industrial VTR with four heads, while Nos. 3-9 are all based on systems using cassette.

Numbers 1 - 4 use relatively simple code error compensation systems based on double recording. System Number 5 deserves special mention as the first PCM recorder with an error correction system. Standardization of a combination of ORC adjacent codes, CRCC error detection coding, and interleave systems to deal with code error detection, correction and compensation was reached with system No. 9 (see Chapter 8). A detailed explanation of error correction will be found in Chapter 6. See Fig. 4-13.

Numbers 1-6 can be classified as industrial/professional systems, and Nos. 7-9 as domestic systems. The major difference between these two types of systems lies in the VTR used for actual recording. The VTRs used in the industrial/professional systems have a very high degree of reliability, are robust, and the tape can be edited. In addition, the bandwidth is very wide so that the VTR can handle the code error correction data as well as the actual data to be reproduced. The number of quantization bits used is also different; 16 bits for the industrial/professional system, and 13-14 bits for the domestic system. Originally, it was thought that 14 bit linear quantization was sufficient for professional systems (No. 2), but with the passage of time and the growth of experience, this system is now generally classified as for domestic use (No. 9). It is quite possible

Table 4-2. PCM Tape Recorders Using Rotary Head Recording. (As of 1979.)

No.	1	2	3	4	5	6	7	8	9
Manufactures	NHK	Nippon Columbia	TEAC	NHK, Nippon Columbia	NHK, Sony	Sony	Sony	Matsushita, JVC, Mitsubushi, Hitachi, Sharp	EIAJ
Format						A	B		C
VTR used	1" helical scan, tape width 1", head/tape sp. 16m/s, tape sp. 19cm/s	4 head, head/tape speed 40 m/s, tape sp. 38cm/s	Adapter/processor format VTR of U-matic quality and above				VTR of Betamax, VHS quality and above		
No. of channels	2	2/4/8	4	2	2	2	2	2	2
Sampling frequency (kHz)	30	47.25	46.08	47.25	44.1	44.056	44.056	44.056	44.056
Quantization bit no.	12	14	12	14	16	16	13	13	14
Compansion system	5 polygonal lines	—	5 polygonal lines (μ)	—	—	—	3 polygonal lines	3 polygonal lines	—
Modulation system	Pseudo video signal using NRZ-FM (analog)								
Bit transmission rate (Mb/s)	0.96	7.1825		3.07125	3.583125	3.583125	1.764	2.64	2.64
Code error protection	parity (1) interpolation	dual writing parity (2)	dual writing interleave parity (1)	parted dual writing parity (1)	ORC interleave CRCC	crossword code interleave	interleave CRCC interpolation	interleave CRCC parity word erasure	interleave CRCC adjacent crossword code
Notes		used from 1971		used for experiment	used for experiment	marketed April 1978	Marketed Sept. 1977		EIAJ standard June 1979
Model name					PAU-1602	PCM-1600	PCM-1		PCM-10 (and others)

* : Sony, Matsushita, Mitsubishi, JVC, Hitachi, Sharp, Toshiba, Sanyo, Teac, Akai, Pioneer, Nippon Columbia.

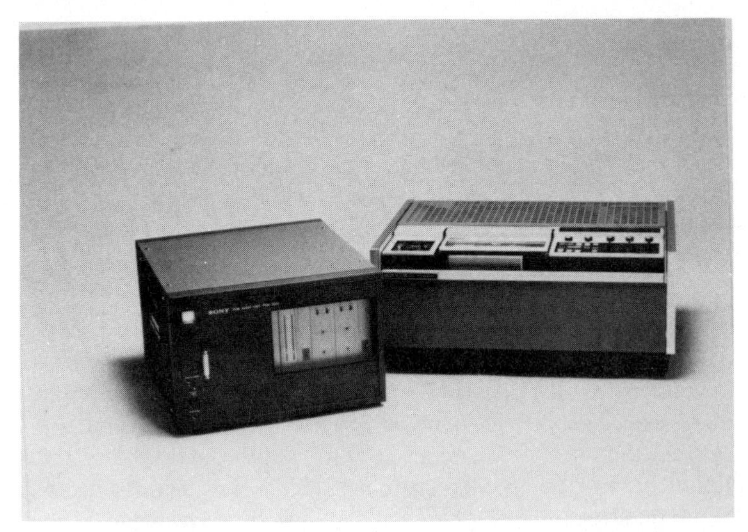

Fig. 4-13. A-Format (professional/industrial PCM processor—PCM1600, with a U-matic standard VTR. Table 4-2 No. 2).

that, in the near future with new systems like the PCM-F1 being developed, the 16 bit system will also be classified as for domestic use.

Number 7 was the first PCM tape recorder that was actually marketed, and played an important part in proving the value of digital recording systems. (Fig. 4-14). Number 6 was the first

Fig. 4-14. B-Format (the first PCM recorder ever marketed. PCM-1 and a betamax VTR. Table 4-2 No. 7).

industrial/professional system to be marketed, and is used in many broadcasting stations and recording studios all over the world.

Number 9 is the format developed under the auspices of the EIAJ (Electronics Industry Association of Japan), and accepted as standard by twelve major companies. In future, the vast majority of domestic PCM systems will be designed to this standard. Chapter 8 contains full details of format No. 9. Formats 6,7 and 9 are known as A-, B- and C-format, respectively.

STATIONARY HEAD PCM TAPE RECORDERS

A stationary head recording system looks very similar to a conventional analog open reel tape recorder (see Fig. 4-3). Many references have been made to stationary head recording systems in the preceding section, in comparisons drawn between rotary and stationary head recording systems. The major features have, therefore, already been covered in detail, so for convenience, only the basic points are noted below:

The Basic Stationary Head Recording System

a) Compared to rotary head systems, the head-to-tape speed is slower.

b) Because the track pitch is greater on stationary head machines, the longitudinal recording density on the tape must be increased to obtain the same recording density as on rotary head systems. As a result, much shorter wavelengths must be recorded.

c) Because of b) above, it is impractical to modulate the PCM system with FM. If FM (and therefore pseudo video) cannot be used, some form of sync must be provided using the two levels, 1 and 0.

d) Because each channel is recorded on independent tracks, record or reproduction may be specified for one channel only.

e) The price is made fairly by the large amount of circuitry necessary for each channel, because one channel may be spread over several tracks.

Certain technical problems are associated with the above mentioned features of stationary head systems. The reduction of recorded wavelengths required (a) and b) above) leads to two further points which must be taken into account:

i) Tape and heads capable of handling these shorter wavelengths must be developed;

ii) More comprehensive error detection and correction systems must be developed to handle the increase in errors due to the use of shorter wavelengths.

Video tape and head technology has made very rapid advances during recent years, and the fundamental technological solutions for (i) above are theoretically available. However, for video purposes, one single head only is necessary, while for stationary head machines, each head stack must be made up of tens of separate heads. To adapt video technology for stationary head PCM recording, much development work is necessary to prevent crosstalk and to achieve increased accuracy.

The problem outlined in (ii) above is somewhat complex. As the tape speed is decreased, the faults on the tape surface (dust, creases, etc.) appear to become longer in time. In addition, the wavelengths used are shorter, and as a result, the code errors appear to increase. For example, as shown in the equation below, during reproduction there will be a fall in output (spacing loss) caused by the separation between tape and head:

$$\text{spacing loss} = 54.6 \times \frac{\text{spacing interval}}{\text{record wavelength}} \text{ [dB]}$$

Equation4-1

Let us assume that some foreign body on the tape surface makes the separation between tape and head 0.2 microns. Then if the record wavelength is 2 microns, the spacing loss will be 5.5 dB; here, the code errors will not be so much increased. However, should the recorded wavelength be 1 micron, the spacing loss will be 11 dB, and an accurate output cannot be expected.

The extent to which the signal is affected depends on the way in which code interference causes errors and how the wavelength is affected. The most effective countermeasure is to use a more comprehensive error detection/correction system and to make the wavelengths recorded as uniform as possible.

Point (c) above, which refers to sync systems, was explained simply in Fig. 4-9, but we shall now examine the possible solutions in more detail.

Synchronization is a method of accurately locating the divisions between separate blocks and bits of data. As shown in Fig. 4-9, if we do not maintain proper "sync" in every day conversations, then misunderstandings arise and we cannot understand one another. Sync for a PCM signal is, however, even more important. The problem is that, should the signal be one bit off, the contents of the recorded signal will be totally different from the original, because there are only two possible values, 1 and 0, which may be used. A stationary head tape recorder uses high density magnetic tape recording, and the signal is expanded and compressed considerably. As a result, the wrong placement of a bit could result in an enormous discrepancy if the digital signal could not be "read" accurately using sync. To prevent any mis-reading, sync is added at fixed intervals (usually referred to as frame intervals), and thus, in addition, helps to nullify the effects of jitter (caused by mechanical inaccuracies in the recording medium).

The sync signal is, of course, made up from ones and zeroes into a pattern which can be immediately recognized as being sync. Let us consider an 8 bit pattern composed entirely of zeros (see Fig. 4-15). Should the first bit of the next signal also be zero (giving a total of nine consecutive zeroes), then the actual position of sync is rendered difficult to identify (Fig. 4-15A). This sync pattern could be improved by making one of the middle bits 1 instead of 0, which would allow accurate identification of the position of the 8 bit sync in relation to the following word (Fig. 4-15B). However, should this bit

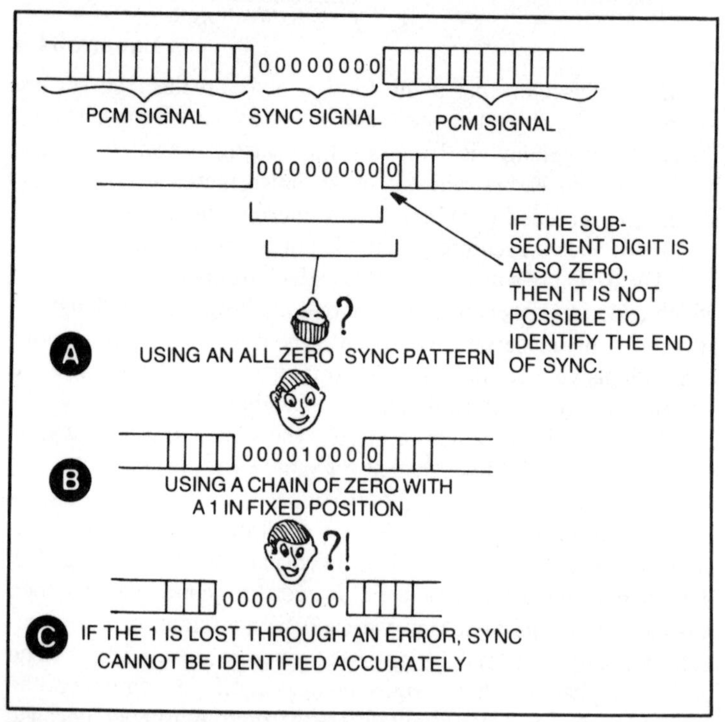

Fig. 4-15. Questionable sync patterns.

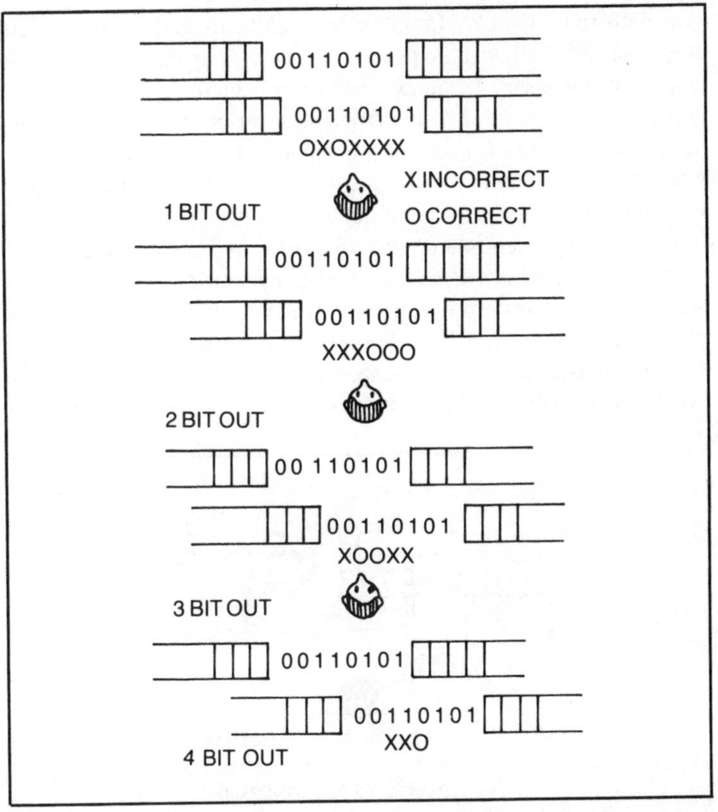

Fig. 4-16. A good sync pattern is one where a discrepancy causes a large number of non-congruous bits.

be lost through a dropout, then the clue for correct sync pattern positioning will be lost (Fig. 4-15C). Thus, the system in Fig. 4-15B is prone to code errors.

Therefore, a sync pattern which is simple to locate, and not affected by code errors has to be developed for stationary head recorders. Actually, this is a very highly specialized field which can only be explained fully through reference to specialized works, but the broad general principles are as follows. Ordinarily, a discrepancy of 1 bit is most likely to occur with that of 2 or 3 bits following. Therefore, as shown in Fig. 4-16, an effective sync pattern has, when the same two sync patterns are matched while shifting 1~3 bits from each other, possesses a large number of incongruous bits.

Even when an effective sync pattern has been decided on, the possibility of the PCM signal generating a pattern the same as the sync pattern is about 2^{-n} (n (bit) of sync pattern), and, once sync has been lost, it is rather difficult to pull it back in again. To re-sync, it would be necessary to know the sync bit number and the frame length, and the type of code error and jitter characteristics would have to be measured accurately. Compared to the rotary head sync system, as shown in Fig. 4-9, sync for stationary head machines is a very difficult process to carry out accurately.

The above mentioned point (d), referring to simultaneous reproduction and record of certain channels, is one of the reasons why stationary head system is required. It is very rare for a pop studio making popular records to gather all the performers together in one studio to make a recording. As shown in Fig. 4-17, each instrument is usually recorded on separate channels, often at different times. In

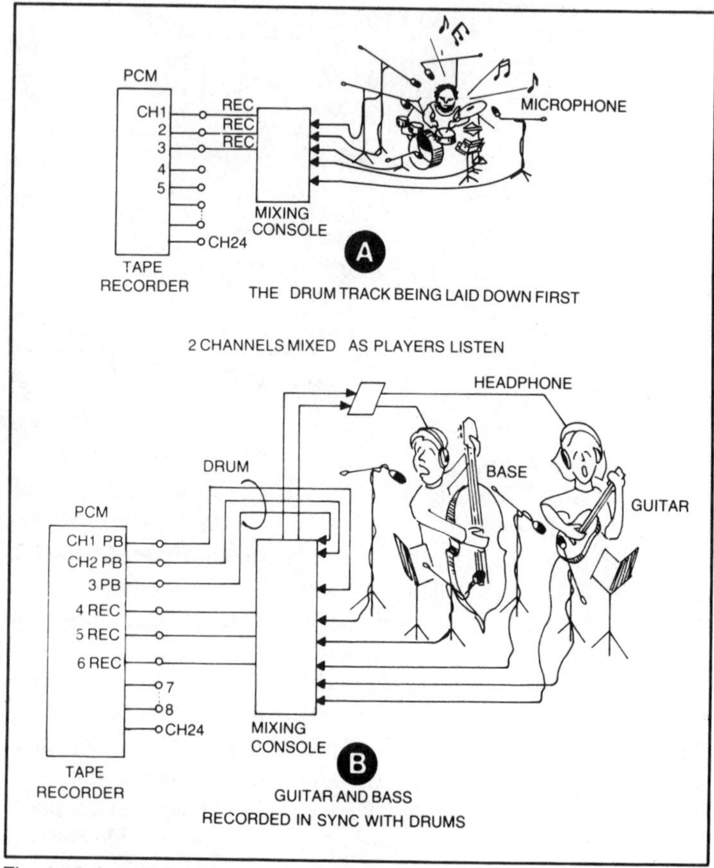

Fig. 4-17. Instruments being recorded separately.

fact, it is not unheard of to record the drums in the U.S.A., and say, the guitar in France. To achieve this successfully, the artists must be able to hear what has already been recorded while they are playing their part of the music, or otherwise, the instruments would not be in time (synchronous recording). Of course, to achieve this, the tape recorder has to have some channels in play mode and some in record at the same time. This type of recording is extremely difficult using rotary head systems, but relatively simple with stationary head.

On some occasions, the artist may not be entirely happy with a short section of his performance, and will want this short section re-recorded. The simplest way of achieving this is to switch the off-ending track from reproduction to record and back to reproduction again as the artist plays the section again. This technique is known as "punch-in-punchout", and of course during the reproduction, record reproduction process all transitions from one mode to another must be smooth and undetectable. Punch-in-punch out is difficult to achieve on rotary head machines because the mode changes might occur in the middle of a field and cause signal disturbance.

On a conventional analog tape recorder there is only one record head and one reproduction head. The record head is used to monitor the signal for synchronous recording and punch in - punch out. Thus, within the record head itself, some channels are in record mode, and some are in the reproduction mode. This however, is not possible with PCM tape recorders owing to problems of signal analysis and delay. A stationary head PCM tape recorder has two reproduction heads so that the normal reproduction head can be used to check the record status and the other can be used as a synchronous monitor. The reproduction signal is digitally delayed on its way to the sync monitor so that the timing of record and reproduction matches exactly. Instead of having two reproduction heads, it is also possible to have two record heads instead.

The last point (e) listed above, refers to the price of stationary head systems; this is in fact one of the major drawbacks. To reduce the price, a system would have to be invented where one channel is laid down on one track, while keeping the tape speed (and therefore, tape consumption) down to manageable levels.

Various Types of Stationary Head PCM Tape Recorders

Table 4-3 shows a list of the various stationary head PCM tape recorders developed by different companies. All of these systems

Table 4-3. Stationary Head PCM Tape Recorders. (As of 1979.)

No.	10	11	12	13	14	15	16	17	18		19			20				21		22		
Manufacturer	Hitachi	BBC	Sony	Toshiba	Matsushita	Sony	Mitsubishi	Sound Stream	3M/BBC		Ampex			Sony				Matsushita		Sony		
Model name			X12-DTC			X22-DTC								PCM-3200						PCM-3300 series		
Channel no.	2	2	2	2	2	2	2	4	2/4	32	2	24	48	2/4	8	24	48	2	24-32	2,4	8/24	24/48
Tape width (inch)	½	½	2	¼	¼	¼	¼	1	¼	1	¼	1	2	¼	½	1	2	¼	1	¼	½	1
Track no./channel	6.5	8	28	16	30	3.5	4	6	1		2			2				4		1,2 switchable		
Tape speed (cm/s)	38	38	76	38	38	38	38	76	114		76			57				38		38/76		
Sampling frequency (kHz)	35.7	32	52	50	49.152	50	44.1	50	50		50			50.4/44.056				50.4		switchable:—32, 44.056, 48, 50, 50.4		
Quantization bit (n./ch.)	12	13	13	14	12	12	15	16	16		16			16				16		16		
Compansion system	—	—	—	7 polygonal line	7 polygonal line	—	—	—	—		—			—				—		—		
Modulation system	MFM	MFM	NRZ	Bi φ	Bi φ	MFM	MFM	MFM	MFM		M^2FM			3PM				MFM		New RLLC		
Rec. density (bit/inch)	5.55k		1.64k	3.3k	3.3k	17.89k	17.64k		28k		25k			30k (20kFRPI)				20.16k		45k (22.5kFRPI)		
Code error protection	parity (2) interpolation		dual writing	parity (1) previous word note	parity (2) dual writing	adjacent code	adjacent code, CRCC		interleave CRCC, parity word, erasure		interleave CRCC, parity word, erasure			interleave, modified crossword code				CRCC, parity word, interleave		cross-interleave CRCC, crossword		
Redundancy (inc. sync)			50%		57.1%	30.5%	35.4%		36%		53.3%			41.7%				33%				
Notes				Thin film head (1T)	Thin film head (1T common bias and MR)		plus two analog tracks							plus two analog tracks, and 1 SMPTE time code track								
	1st generation			2nd generation					3rd generation											4th generation		

are aimed at the professional/industrial market, and so far, no stationary head machines have been developed for domestic use.

Numbers 10-12 are the first generation of stationary head systems, developed as experimental equipment. All have quite low recording densities, and relatively rudimentary error correction systems. Numbers 13-17 are the second generation with high recording densities, and designed for practical use in the field. Numbers 13 and 14 employ thin film heads, and have increased density along the longitudinal axis. Numbers 15 and 16 are provided with improved error correction and detection systems, and their effectiveness against code errors has, therefore, been increased markedly.

Numbers 13-21 constitute the third generation, and were developed as 24-48 channel master recorders, but standardization has been agreed only up to two channels. Thus, no. 18 records one channel on one track, nos. 19 and 20 record one channel on two tracks, and no. 21 records one channel on four tracks. If the tracks used to record one channel are increased, the system becomes more expensive, but the tape speed can be reduced.

Number 22 is a fourth generation machine. New developments in the modulation systems allows much higher densities to be recorded without making the wavelengths recorded on tape unduly short. Any sampling frequency required can be used, and during reproduction, the frequency chosen during record is switched-in automatically.

Sampling Frequencies

Table 4-4 shows all the sampling frequencies which are being put forward for use in PCM recording of audio signals. Frequency no. 1 is for PCM communications for broadcast and telephone rather than for tape recorders. 32 kHz was chosen because around 15 kHz is quite adequate for television broadcasting and FM transmissions.

Numbers 2 and 3 are frequencies suggested for use with PCM equipment utilizing VTRs as the recording medium. These frequencies are more or less totally decided (see Chapter 8). Number 4 has been suggested because it has an integral relationship with No. 1. In the same way, nos. 6 and 7 have an integral relationship with nos. 2 and 3.

Number 5 was put forward because it is very easy to use with sync and all types of VTR, but because subsequent research led to other frequencies mentioned above, which are equally suitable for use with sync, the raison d'etre for this frequency disappeared.

The reason why so much importance is attached to the integrals of sampling frequencies is for conversion. If one wanted to dub a recording between two machines which used different sampling frequencies, the signal would be degraded if any return to analog mode were made (digital signal - D/A - analog - A/D - digital signal). However, if the dubbing process can be carried out in digital mode only, there would be no deterioration of the original. Thus, if the sampling frequencies are simple integrals of each other, or of another frequency, conversion becomes very simple, and can be accomplished at low cost.

Thin Film Head

Thin film heads employ the thin film techniques used to make hybrid ICs. They can be generally classified as heads using hybrid IC winding techniques (generally one turn), and magnetic reluctance elements and hall elements. (This refers to reproduction heads only.) With conventional heads, the price rises steeply as the number of tracks to be handled increases. The number of tracks handled does not, however, affect the price of thin film heads. Futhermore, in comparison with conventional heads, the track pitch and guard band can be made narrower, and the recording density may be increased longitudinally.

References

Materials related to PCM tape recorders in general.

1. Hayashi, Kenji: "PCM recorders", Television Journal, 33, pp. 183-190 (1976-3).

Table 4-4. Sampling Frequencies for Use with Audio PCM.

No.	Sampling frequency	Proposer	Reason for proposal
1	32 kHz	EBU	Related to transmission frequencies used in telephone communication
2	44.056 kHz	SONY	Suitable for use with VTR (NTSC) based PCM recorders (Ch. 8)
3	44.1 kHz	SONY	Suitable for use with VTR (PAL/SECAM) and PCM (Ch. 8)
4	48 kHz	SONY/ POLYGRAM	Integral relationship (2:3) with No. 1
5	50 kHz	3M	Could be dropped
6	50.35 kHz	SONY	Integral relationship (7:8) with No. 2
7	50.4 kHz	SONY	Integral relationship (7:8) with No. 3

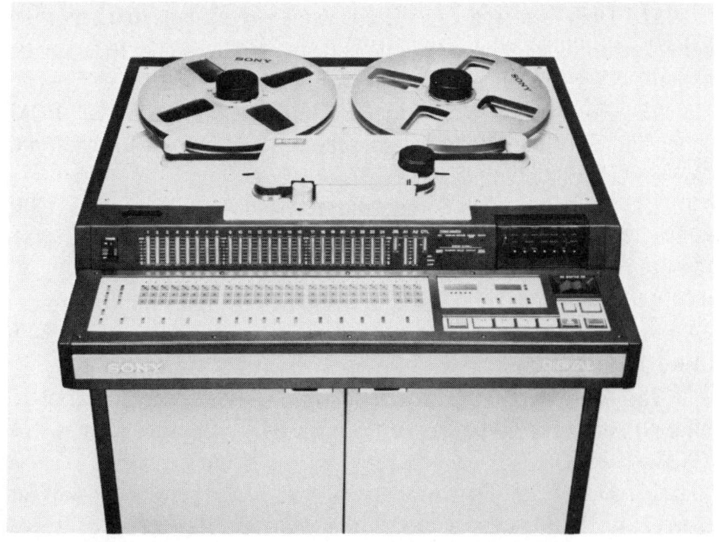

Fig. 4-18. Sony 24 channel digital audio recorder (PCM-3324).

2. Matsumura, Sango: "PCM record and playback" Shingaku Jiho, EA 78 -24 (1978-7).

3. Doi, Toshitada: "PCM recorders", Television Journal, 33, 1 pp. 17-25 (1979-1).

Materials related to rotary head PCM recorders.

4. Hayashi, Kenji: "Stereo recorders", NHK Giho, 12, 11, pp., 12-17 (1969).

5. Anazawa, Mori: "PCM recording equipment using 4 head VTRs", Television Society Recording Materials, 11-4 (1975-3).

6. Anazawa, T., K. Yamamoto, S. Todoroki, and A. Takasu: "Improved PCM (Pulse Code Modulation Recording Systems", AES 56th Convention, No. 1206 (F8) (1973).

7. Oba, Kubo, Doi, Tsuchiya, Miyashita, Todoroki: "Prototype PCM record playback adaptor for broadcast use", Television Society Recording Materials, VR 25-2 (1977-9).

8. Iga, Odaka, Ogawa, Hashimoto, Masaoka, Yasuda, Yokota, Doi: "Domestic PCM audio unit for connection to a home VTR", Nihon Ongakkai koron 3-2-9 (1977-10).

9. Tsuchiya, Doi, Otsuki, Kazami: "A PCM audio unit using error correction codes", Nihon Ongakkai koron 3-3-10 (1977-10).

10. Yamaguchi, Utsumi, Nakamura, Mawatari: "Industrial Standard 4 channel PCM record playback processor", Shin Gakkai Jiki Kenshi MR 77-23 (1977-11).

11. Doi, Tsuchiya, Iga: "Concerning a record playback method using techniques for conversion of an audio PCM signal to an image signal", Shih Gakkai Jiki Kenshi, MR 77-24 (1977-11).
12. Kosaka, M., K. Odaki, M. Tsuchiya, and R. Wada: "PCM recording with error correction scheme", AES 60th Convention. No. E8 (1978-5).
13. Ishida, Y., N. Tomikawa, S. Nishi, and S Kunii: "VTR based PCM recorder with error correction scheme", AES 60th Convention E10 (1978-5).
14. Nakajima, H., T. Doi, Y. Tsuchiya, A. Iga., and I. Ajimine: "A new PCM audio system as an adaptor for VTRs", AES 60th Conv., E11 (1978-5).
15. Ajimine, Otsuki, Kazami, Anju, Okude: "An industrial 16 bit two channel PCM recorder using a VTR", Shingaku, Jiho EA 78-35 (1978-7).
16. Doi, T., Y. Tsuchiya, and A. Iga: "On several standards for converting PCM signals into Video signals," J. AES, 26, 9, pp. 641-649 (1978-9).
17. Matsuoka, Yamada, Yamasaki; "Listening tests for a PCM record playback system using a VTR", Shingaku Jiho, EA 78-32 (1978-7).
18. Mori, Sasada, Juso: "Dropout analysis in PCM recorders using domestic VTRs", Shingaku Jiho, EA 78-33 (1978-7).
19. Fukuda, Odaka, Doi, Hamada: "Simple error correction on a PCM adaptor connected to a domestic VTR", Nihon Gakkai Koronshu, 3-9-15 (1978-10).
20. Naito, Yokota, Yasuda, Odaka, Doi, Sekiguchi: "Rotary head PCM decks", Shingaku Jiho, 3-9-18 (1978-10).
21. Doi, T. K. Odaka, and G. Fukuda: "A new format for digital audio processors for home use VTRs" AES 64th Conv. (1979-11).
22. Doi. Fukuda, Odaka: "A comparison of decoding systems in various PCM recorders with VTR", Nihon Ongakkai Koron (1979-6).
23. Doi: "Crossword coding of interleaved adjacent code", Nihon Ongakkai Koron (1979-6).
Materials related to stationary head PCM tape recorders.
24. Sato, Ishimatsu: "Multi-channel PCM recorders", Television Society Recording materials, VR 11-15 (1975-3).
25. Takayama, Umeda: "Stationary head PCM magnetic record playback equipment," Television Society recording materials, VR 11-6 (1975-3).
26. Kanazawa, Umemoto, Tosai, Nakamura: "A prototype

stationary head PCM recorder", Nihon Ongakushi, 31, 10, pp., 585-592 (1975).

27. Kunii: "Stationary head PCM recorders", Television Society recording materials, VR 25-1 (1977-9).

28. McCracken, J.A.: "A high performance digital audio recorder", AES 58th Conv., No. 1268 (N3) (1977-11).

29. Sonoda, Tsuchiya, Ishida, Nakai: "Stationary head PCM recorder using error correction", Shingaku Zendai, No. 72 (1978).

30. Onishi, Osaki, Kawabata, Tanaka, Sato: "Head editing on a stationary head PCM recorder," Shingaku Jiho, EA 78-29 (1978-7).

31. Matsushima, Shimemoto, Kihara, Mima, Kanai: "Stationary head PCM recorders," Shingaku Hiho, EA 78-30 (1978-7).

32. Tsuchiya, Y., T. Sonoda, J. Nakai, M. Ishida, and T. Doi: "A 24 channel stationary head digital audio recorder", AES 61st Convention, No. 1412 (F-2) (1978-11).

33. Engberg, E.W.: "A digital audio recorder format for professional applications," AES 61st Convention (F-1) (1978-11).

34. Tanaka, K., T. Furukawa, K. Onishi, T. Inoue, S. Kunii, and T. Sato: "A 2 channel PCM tape recorder for professional use," AES 61st Convention (F-3).

35. Doi, T., Y. Tsuchiya, and G. Fukuda: "Statistical analysis of error correcting schemes of stationary head digital audio recorders" Joint meeting of ASA and ASJ, No. MM 13 (1978-11).

36. Matsushima, H., K. Kanai, T. Mura and T. Kogure: "A new digital audio recorder for professional applications", AES 62nd Convention. No. 1447 (G-7) (1979-3).

37. Doi, T., Y. Tsuchiya, T. Sonoda, and M. Tanaka: "A stationary head digital audio tape recorder with multiple sampling rates" AES 63rd Convention (1979-5).

38. Doi, Tsuchiya, Kanai, Tanaka: "A PCM stationary head recorder with multiple sampling frequencies," Nihon Ongakkai Koron, (1979-6).

39. Kaminaka, Kanai, Nochi, Nomura, Hirota: "PCM recording using thin film heads," Shingaku Kenshi MR77-41 (1977).

Chapter 5

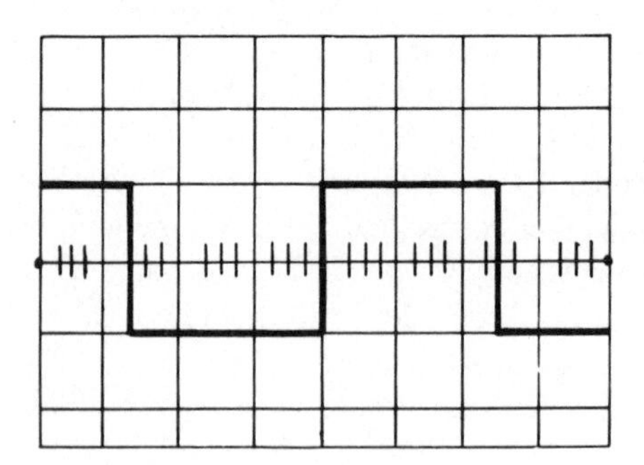

Digital Audio Disc Systems

In the previous chapter we discussed the necessity of recording much higher frequencies for PCM than are commonly used for analog recording. In other words, a much wider bandwidth is required to make a pulse code modulated recording. To put it another way, recording a wide frequency spectrum is the price which has to be paid for the numerous benefits of PCM and for the excellence of its electronic characteristics. For this reason, PCM tape recorders were developed using wide bandwidth VTRs.

Exactly the same conditions apply to the Digital Audio Disc. However, rather than develop the conventional analog record to cope with the necessary high frequencies, it was more appropriate to utilize the technology developed for the video disc. Thus, this chapter contains a brief explanation of the basic principles underlying the design of the Digital Audio Disc, including methods of signal detection, as well as actual construction.

A GENERAL OUTLINE OF VIDEO DISC SYSTEMS

Table 5-1 gives an overview of the various types of video disc which have been introduced. Methods of signal detection and the merits and demerits of each system are also included. Specifications such as the modulation systems used for the video signals and actual disc dimensions have been omitted as they are not directly related to our purpose, which is to consider how the video disc system was adapted for audio use.

Table 5-1. Video Disc System.

	A Philips/MCA	B Thomson	C JVC	D RCA	E Matsushita	F TED
Signal detection system	Laser system		Capacitance system		Mechanical	
Pick-up	No physical contact		Physical contact			
Signal recording surface	one side (buried)	Both sides (surface)				
Cutting	Laser (photo-resistance)			Mechanical (Cu)		
Grooves	None (dynamic tracking)			Grooved		
Disc material	Transparent		Added carbon		Optional	
Disc coating	yes (A1)	no	no	no (but surface has foil coating)		
Advantages	Long disc and stylus life				cost	
	High resistant to dirt and scratches Accessibility	Long playing time	Accessibility		Disc material	
			Duct exclusion	Low dropout (mechanical cut)		?
Disadvantages	cost		stylus and disc life			
		low speed revolution difficult	stylus surface?		Stylus construction?	
	one side only	subject to scratching				

Systems A to E are exactly the same in terms of recording density, while that of system F is several grades lower. Although recording density is the same, the content is different; systems A to C use a laser for cutting, and so have a long record wavelength and a narrow track pitch. Systems D and E, on the other hand, use mechanical cutting (piezoelectric element), and have the opposite characteristics.

There are three types of signal detection system: laser, capacitance and mechanical (piezoelectric). The laser system detects the signal on the disc with no actual physical contact. The other two systems both require some type of pick-up element to make physical contact with the disc before the imprinted signal can be read.

With reference to where the signal is recorded on the disc, system A uses a disc which has a recording on one face only, while all the other systems use both faces of the disc. In system A, the laser beam shines through a transparent upper layer onto the signal surface on the opposite side of the disc, and is then reflected back. Thus, only one side of the disc may contain recorded material. As a result, this system is highly resistant to dust, dirt and scratches on the record surface. In system B, the signal is recorded on the upper face on both sides of the disc. The disc material must be transparent because the laser beam must be able to shine through it.

Systems A to C use a glass plate coated with a photo resistive substance and a high output laser beam for cutting. As a result, the signal is recorded as unevenness (pits) in the surface corresponding to the operation of the high output laser. Systems D and E, on the other hand, use a copper plated base which is etched by the cutter.

The existence of grooves on a video disc is related to the cutting system used (see Fig. 5-1). Mechanical cutting for video discs is fundamentally the same as for conventional analog audio discs, as the signal is cut at the same time as the grooves. With mechanical systems, it is extremely difficult, if not impossible, to cut the correct signal depressions without cutting the groove as well. However, with a laser system, it is a simple matter as the laser output is only either ON or OFF. Grooved video discs must of course, use some kind of pick-up assembly to trace the grooves, just as a conventional analog audio record does. The ungrooved discs, on the other hand, do not use physical contact, but uses some form of dynamic tracking to trace the signal correctly.

Dynamic tracking is a system for accurately tracing the tracks laid down on the record: a servo is used to detect and correct any misalignment between the signal track and the pick-up device (in

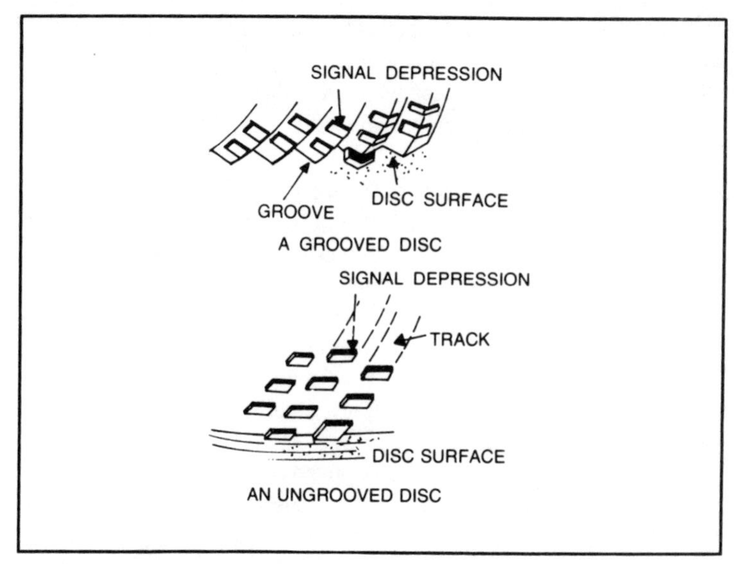

Fig. 5-1. Ungrooved discs are also used for video discs.

this case, a laser beam). The following types of dynamic tracking have been developed:

A. Multi-pick up system: Apart from the main signal detection beam, other beams are located to the right and left. Their outputs are compared, and any misalignment is detected and corrected. In principle, this system is also applicable on capacitance players, provided that three electrodes are used.

B. Wobbling system: The pick up element (laser beam) is caused to oscillate slightly from left to right in a sinusoidal fashion. Basically, the servo senses that the output is uniform in 90° and 270° phase. Instead of causing the pick-up to oscillate, the tracks may be laid down in an undulating pattern.

C. Tracking signal system: Side signals of fixed, slightly different frequencies, f1 and f2 are laid down on either side of the main signal track. The servo then detects movement to the right as an increase in f1, and movement to the left as an increase in f2.

As for the materials used for making video discs, some carbon is added for systems C and D to improve conductivity. The materials used for discs for systems A and B must be transparent, while that for E and F are optional. The next item on Table 5-1, is related to what happens to the disc in subsequent processing. For system A, a reflective metal coating is added (usually aluminum) to improve the reflective capabilities of the disc as well as a protective coating on both sides of the disc. A lubricant is added for discs to be used on systems D and E so that stylus and disc life is increased.

The advantages and disadvantages of the various systems are briefly compared: The greatest advantage of the systems which do not employ a direct contact pick-up device is the enormous increase in disc life. On the systems using direct contact between stylus and record, the life of both stylus and record is limited, although, as a corollary, the price of the system is lower; the price decreases progressively for systems C, D, E, and F. Because the signal surface is buried for system A, its major advantage is high resistance to dirt, dust and scratches. However, it is only possible to make a recording on one side of the disc. Of course, it would be possible to join two single sided discs together, but the price would be more than doubled.

System B relies on transmission, so that by adjusting the focus of the objective lens, the signal on the upper or lower surface can be read. As a result, an extremely long consecutive playing time can be achieved, as both the upper and lower surfaces can be read without turning the disc over. With systems A and B, the disc is made stable by the high speed of revolution, and as a result, it is difficult to make these discs play at a slower speed to increase playing time.

SIGNAL DETECTION SYSTEMS

Figure 5-2 shows the basic principles involved in signal detection with an optical video disc. After the direction of the beam from the laser source has been altered by the mirror, it is focused onto the signal surface of the disc through an objective lens. Because the signal surface is coated with a reflective layer, the reflected beams then travel back along the same path. They are then deflected to the light sensor by the beam splitter. The strength of the reflected beam is altered by the presence or absence of signal depressions (pits) in the signal surface. The output of the light sensor then corresponds to the signal cut into the record surface.

Optical Systems

Because the beam is not focused at the point where it hits the lower surface of the record, any dust or dirt on the underside of the record has very little effect on the signal output. This will become self-evident if actual figures are given: after the beam has been focused on the signal surface of the record, the beam spot has a diameter of 1 micron. However, at the lower surface of the disc, the diameter is 1 mm. Thus, there is a difference of 1,000,000 to 1 in the

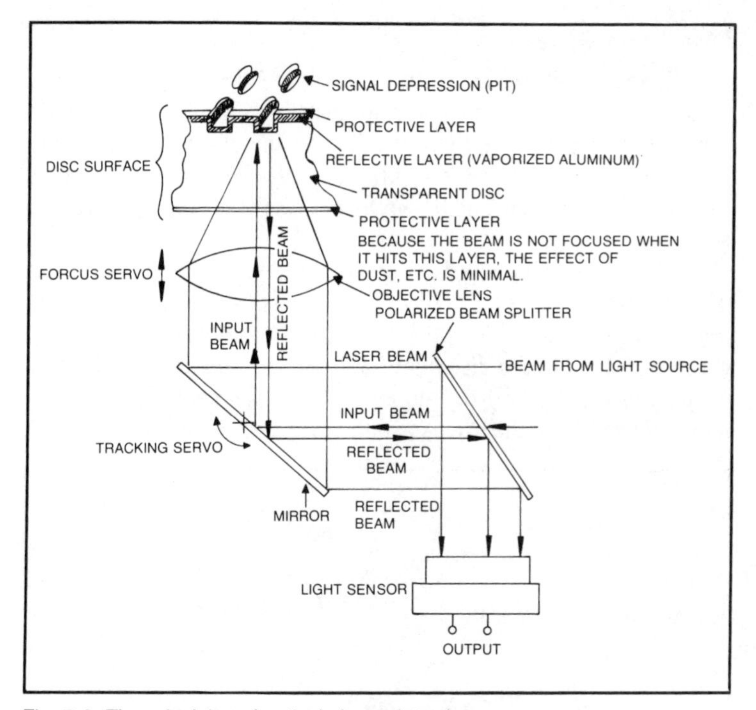

Fig. 5-2. The principles of optical signal detection.

two areas covered. The effect of any dust, dirt or scratch marks on the lower surface will therefore be reduced to a millionth.

Both the upper and lower surfaces of the disc are coated with a protective layer. It is not necessary for the protective layer on the upper surface to be transparent, so this layer can be made very strong and thick. As a result, the reflective layer can be effectively protected against damage through mishandling.

The objective lens is so constructed that it can be moved up and down, so that, should any deviation be detected in the disc, the lens can be maintained at a fixed distance from the surface. The movement of the lens is controlled by the focus servo.

The mirror can be rotated from right to left, so that the positioning of the beam, which might be affected by any eccentricity in the disc, can be controlled accurately. This constitutes dynamic tracking, and is also known as a tracking servo.

A He-Ne laser, or a semiconductor laser can be used as a laser source. The explanation above specifically refers to reflective optical systems, but the basic principles of the transmission optical system are the same.

Capacitance Systems

Figure 5-3 shows the basic principles of signal detection using a capacitance system. The stylus itself is made from sapphire, and a thin metal electrode is then attached to the surface. The record surface is provided with signal pits, in exactly the same way as discs for the optical system, and the stylus tracks these across the record surface.

The special characteristic of the capacitance system is its use of the conductivity of the disc. In the past, a metal coating was applied, as for system A (see Table 5-1), but a new method in which carbon is added to the disc material itself as a way of increasing its conductivity has been developed. The capacitance between the metal electrode on the stylus and the disc changes according to the

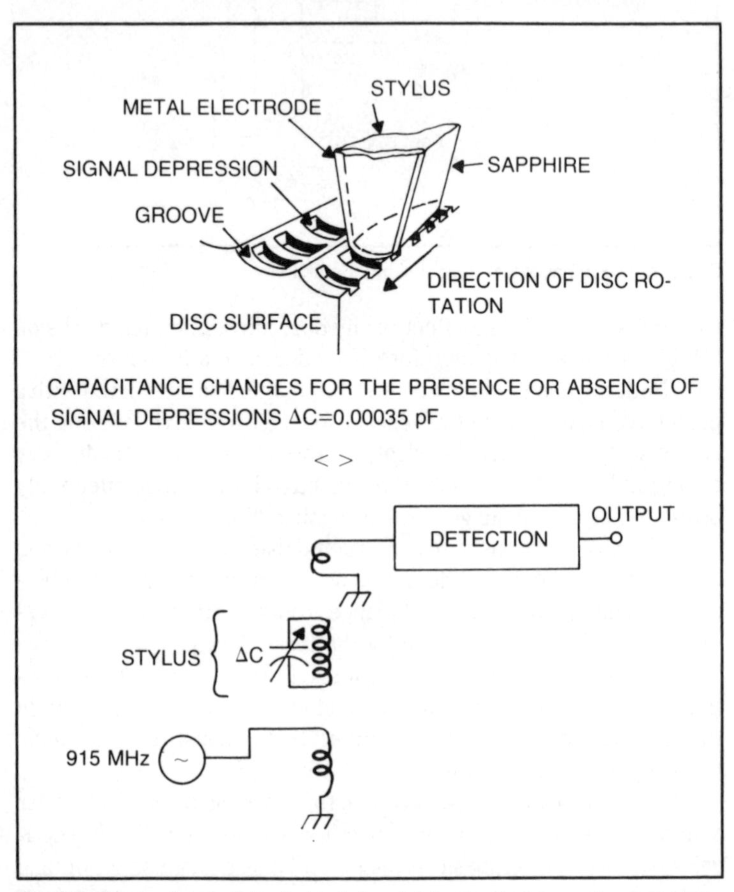

Fig. 5-3. The basic principles behind signal detection in a capacitance system.

presence or absence of signal pits. These changes are read as changes in resonant frequency. Since capacitance changes are minute, it is necessary to make the oscillator frequency used for detection sufficiently high.

Figure 5-3 shows the grooves in a disc for system D, but the principles involved also apply to systems which use dynamic tracking, such as C. The price increases for units where dynamic tracking is used. As there are no grooves, the stylus can move freely from side to side, if required, which means that a desired track can be found very quickly. Also, a large stylus spanning several tracks may be used, so that any rubbish on the surface is simply swept aside. The major drawback of this is the extremely sophisticated level of technology required to produce it.

Mechanical Systems (Pressure, Piezoelectric)

Figure 5-4 shows the basic principles involved in signal detection for mechanical video disc systems. This is based upon very simple principles: the stylus oscillates as it traverses the signal pits on the surface of the record, and these movements are then translated into an electrical signal by the piezoelectric element. If one looks at the various systems in detail, it is clear that they are substantially identical in terms of signal propagation, whether by pressure waves from the stylus, or the use of the piezoelectric elements.

Because the basic principles are so simple, the range of materials which can be used to make the disc is fairly wide, although it

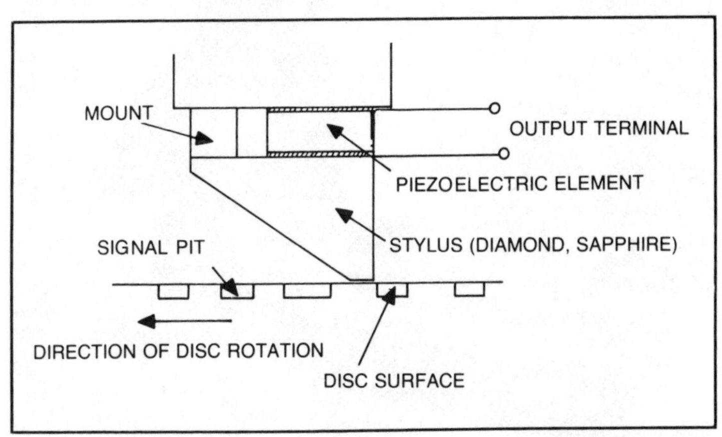

Fig. 5-4. The basic principles for signal detection in a mechanical (pressure) system.

must be a substance whereby a fair degree of accuracy can be maintained for the signal pits. Mostly discs with grooves are used, and in order to increase the life of both stylus and disc, the disc surface is coated with a lubricant.

Table 5-2 shows the various digital audio disc formats which have been announced by manufacturers in the past few years. All three types of system, optical, capacitance and mechanical, have been employed by different manufacturers.

Numbers 1, 3, 5, 7, 11, and 12 use a video signal as the modulation system. This is exactly the same as explained in the previous chapter for rotary head VTRs used for PCM recording. These discs use a pseudo-video format to modulate the signal.

Not a great deal of information about error detection and correction systems for video discs has been introduced. System number 9 (Fig. 5-5) uses cross-interleave error detection and correction to attain a very high level of accuracy.[5] System number 9 is unique in having an extremely long playing time (two and a half hours); actually, since the disc used is 30 cm across, the theoretical maximum playing time could be ten hours. Developments along this line will no doubt continue in the future. System number 4 employs CLV (Constant Linear Velocity), a system involving changes in rotational speed, so that a very small disc may be used (Fig. 5-6).

Fig. 5-5. A digital audio disc player (system No. 9)

Table 5-2. Types of Digital Audio Disc Systems.

	1	2	3	4	5	6	7	8	9	10	11	12
Manufacturer	Mitsubishi, Teac, etc.	Sony	Hitachi, Nippon Columbia	Philips	Matsushita	JVC.	Pioneer	Hitachi, Nippon Columbia	Sony	Mitsubishi	Sanyo	Toshiba
Signal detection system	Optical				Mechanical	Capacitance	Optical					Capacitance
Disc diameter (mmϕ)	–300			115	–300							
No. of modulation (rpm)	1800	900	1800	CLV1.5 m/s	450	900	1800	600	450	450	1800	450
Playing time (hours)	0.5	1.0	0.5	1.0	0.5×2	1.0×2	0.5	1.5	2.5	2.0	0.5	1.0×2
Modulation	Video	MFM	Video	MFM	Video	NRZ-FM	Video	MFM	3PM	MFM	Video	Video
Quantization (bit)	12N	13N	14	14	13N	14	14	16	16	?	13N	14
Channel no.	2											
Sampling frequency (kHz)	46.08	44.056	47.25	44.33	44.056			47.25	44.056			
	'77 Aug., Sept.			'78 Mar.	'78 Sept.							

Fig. 5-6. A prototype Compact Disc player.

MOVEMENT TOWARDS STANDARDIZATION

Using a magnetic tape recorder, it is possible to make recordings, as well as reproduction. So even if full standardization is not carried out and a number of different systems are originally introduced, the success of the idea will not be materially affected. A disc system, on the other hand, is passive. Thus, unless a full standardization is carried out before manufacturers start production, the market will become chaotic. To try and avoid the introduction of incompatible digital audio discs, an organization known as the DAD (digital audio disc) Convention was organized by 35 Japanese manufacturers in September 1978.

Table 5-3 shows all the various prospective formats which have been collated by the DAD Convention after thorough investigation. In future, all developments by the manufacturers concerned will be carried out along these lines.

THE COMPACT DISC DIGITAL AUDIO SYSTEM (7), (8), (9)

In June of 1980, Sony and Philips jointly proposed a system named "The Compact Disc Digital Audio" which was a combination of the product concept of (4) (Table 5-2), and the signal processing technology from Sony (9).

The quantization was increased into 16-bit linear, with a slight increase in the diameter of discs (120 mm). The reliability and the yield of discs are greatly improved from all of the experimental systems shown in Table 5-2, by developing the new error correcting code named CIRC (Cross Interleave Reed-Solomon Code) which is a combination of Cross Interleave Code (Chapter 6) and the Reed-Solomon Code.

Packing density was increased greatly from all of the systems shown in Table 5-2, by developing the new channel coding named EFM (Eight to Fourteen Modulation). Already, more than 40 companies of hardware as well as software joined this system. Figure 5-7 shows the final version of Compact Disc and its player. Figure 5-8 shows an example of optical cutting systems. The specifications of the systems are described in the following.

GENERAL INFORMATION ON COMPACT DISC DIGITAL AUDIO (SONY CORPORATION/N.V. PHILIPS)[(8)]

1 Disk (Fig. 5-9)

Playing time, single side, 2 channels	Maximum 74 min
Scanning velocity (2 channels)	1.2-1.4 m/s
Sense of rotation seen from reading side	Counterclockwise
Track pitch	1.6 μm
Disc diameter	120 mm
Disc thickness	1.2 mm[1]
Diameter of center hole	15 mm
Starting diameter of program area	50 mm

2 Signal Format

Number of channels	2 and/or 4
Quantization, per channel	16 bits linear
Encoding	2's complement
Sampling frequency	44.1 kHz
Error-correction code	CIRC[2]
Channel modulation code	EFM[3]
Channel bit rate	4.3218 Mb/s

[1]Double-sided disc optional
[2]*C*ross *I*nterleave *R*eed *S*olomon *C*ode
[3]*E*ight-to-*F*ourteen *M*odulation

3 Frame Format

	Data Bits	Channel Bits
Synchronization		24
Control and display	8	14
24 data symbols	192	336
8 error-correction symbols	64	112

Table 5-3. Targets for DD (Digital Audio Disc) (DAD Study Group, Joint Sessions W62,3: 17 November 1978.)

Contents of paragraphs		Requirements for WG1[*3]	Brief for WG2,3[*4]	
Paragraph	Contents		Aims	Notes
1. Specifications	(1) freq. characteristics (2) dynamic range (3) distortion (4) separation (5) wow and flutter (6) noise uncorrected uncompensated	average digital master with max. effort none	(Dc)-20 kHz better than 90 dB less than 0.03% better than 80dB less than 0.003% less than 10times/min. less than 0.5 times/hr.	high freq. distortion for 4 separate channels
2. Facilities	(1) access	minimum, separate tracks. (repeat play if possible)	ability to address once/sec, with above detail	
	(2) copy protection	protection against digital copying	anti-copy (change standard freq. for domestic units)	analog anti-copy: not possible. Copy from RF : success in protect by addition of complex addresses
3. Disc spec.	(1) disc diameter (2) channel (3) playing time (4) recording system	30 cm max. (20 better) 4 (4Xmono,2X stereo) continuous 40 min. 1 pc. 80 min.	Recording density : min 160 ch./side for 30 cm record	WG2, 3 brief stops at indicating volume for disc. WG1 is to re-investigate actual use

4. Signal format	(1) signal bit (2) sampling freq. (3) correction, compensation (4) control signal (5) modulation system		16 slots to watch master with correction address ident. code (par 2-i) pending	"compensation only" not considered WG1 to consider anything else necessary MFM, 3PM,FM,Video FM, etc.
5. Reliability	(1) disc life (2) anti-dust (3) handling (4) accuracy	more than 100 rev. same as analog under consideration	more than 100 rev. the same as, or better than analog not decided	various methods hardware accuracy : JIS S 8502
6. Compatibility with video		common video/audio player to make DAD a world-wide standard	min. limit detection & rotation system common	switching possible

*1 This table was prepared from DAD WG1 materials (WG1-53-3,4,5), after discussions with WG2,3.
*2 The specifications show how the disc itself can be made to perform accurately.
*3 Requirements from software providers.
*4 Mutual agreements among hardware manufacturers.

Fig. 5-7. Sony digital audio Compact Disc reference player (CDP-701ES).

Merging and low-frequency suppression	102
Total frame	588

4 Optical Stylus

The wavelength λ and the numerical aperture NA have to fulfill the requirement

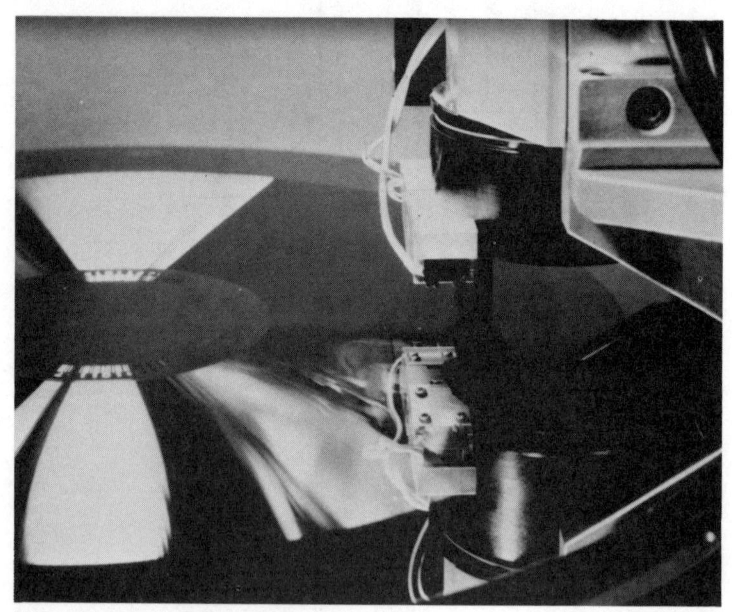

Fig. 5-8. An example of optical cutting system.

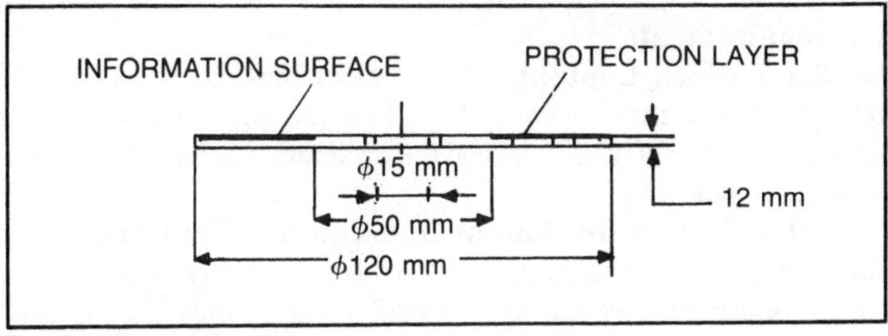

Fig. 5-9. Dimensions of disc.

$$\frac{\lambda}{NA} \leqslant 1.75\ \mu m.$$

The stylus should be diffraction limited, and the information is viewed through a transparent plane parallel plate of 1.2-mm thickness (refractive index $\approx$ 1.5).

The system is optimized for a wavelength of 0.78 μm (e.g., laser wavelength of A1GaAs). The depth of focus of the optical stylus is $\pm 2\ \mu$m.

The method of radial tracking is differential, and the method of high-frequency detection is integral.

5 Modulation System

The NRZ signals from the A/D converter and the error-correction parity generator may have a high dc content and are not self-clocking (the run length is not limited).[4] Therefore they cannot be used on the disk. The signals have to be converted into another code which should meet some special requirements.

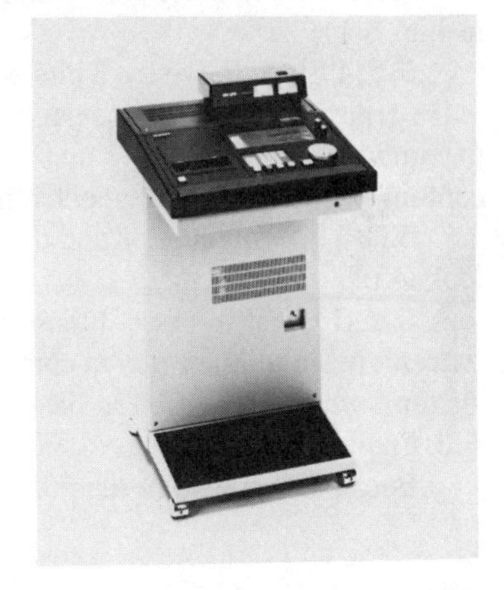
Fig. 5-10. Sony professional Compact Disc player (CDP-5000).

5.1 Requirements

5.1.1 Clock Content. The bit clock must be regenerated from the signal after readout. Therefore the signal must have a sufficient number of transients and the maximum run length must be as small as possible.

5.1.2 Correct Readout at High Information Densities. The light spot with which the disk is read out has finite dimensions. These dimensions give rise to intersymbol interference. This effect can be minimized by making the minimum run length as large as possible. However, too large a value has a negative influence on the clock content of the signal.

5.1.3 Servo. The modulation code must be dc free, because the low frequencies of the spectrum give rise to noise in the servo systems.

5.1.4 Error Propagation. The error propagation of the modulation system must be as small as possible.

5.2 Eight-to-Fourteen Modulation Code (EFM)

5.2.1 Each block of 8 data bits is mapped onto 14 channel bits. To each block of 14 channel bits 3 extra bits are added, 2 bits for merging the blocks and 1 redundancy bit for LF suppression.

5.2.2 The information is contained in the positions of the transients. For mapping 8 data bits 256 combinations of channel bits are needed.

5.2.3 The code is generated in such a way that the minimum distance between 2 transients is 3 channel bits ($\approx$ 1.5 data bits), and the sampling window or eye pattern is 1 channel bit ($\approx$ 0.5 data bit). This yields a good compromise between intersymbol interference and clock accuracy (phase jitter). The maximum run length within the blocks is 11 channel bits ($\approx$ 5.5 data bits). An example is shown in Fig. 5-11.

5.2.4 Since the extra 3 bits do not contain any information, an extra transient may be inserted in these bits. In this way the maximum run length T_{max} in two successive blocks and the dc content of the frequency spectrum can be controlled.

5.2.5 The modulator and demodulator can be realized with a lookup table in a ROM.

5.2.6 Because of the block structure this modulation code is extremely suitable for use in conjunction with the error-correction system, whose operation is based on 8-bit blocks.

5.3 Frame Format

Because the system must be self-clocking, synchronization is

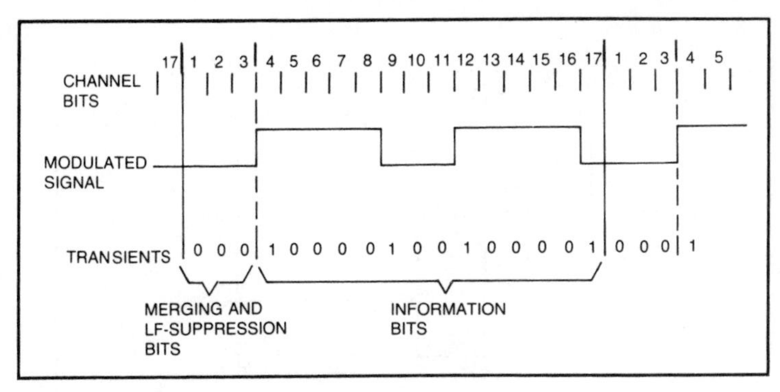

Fig. 5-11. Modulation code.

necessary. Therefore the data stream is split up into frames. Each frame contains:

1) A synchronization pattern of 24 bits
2) 12 data words of 16 bits each
3) 4 error-correction parity words of 16 bits each
4) A control and display symbol of 8 bits.

The data and error-correction words are each split up into two 8-bit blocks, which are fed into the modulator circuit. After modulation each block is converted into 3 + 14 channel bits.

The total number of channel bits per frame is:

Sync pattern	24	channel bits
Control and display	1 × 14	channel bits
Data	12 × 2 × 14	channel bits
Error correction	4 × 2 × 14	channel bits
Merging and LF suppression	34 × 3	channel bits
Total	588	channel bits

6 Error-Correction System (Figs. 5-12 and 5-13)

An efficient error-correcting system, named CIRC, has been developed with different decoder strategy possibilities. A simple 4-frame correction to a more complex 16-frame correction is possible, keeping full compatibility.

6.1 Requirements

1) High random error correctability
2) Long burst error correctability
3) In case burst correction is exceeded, a graceful degradation

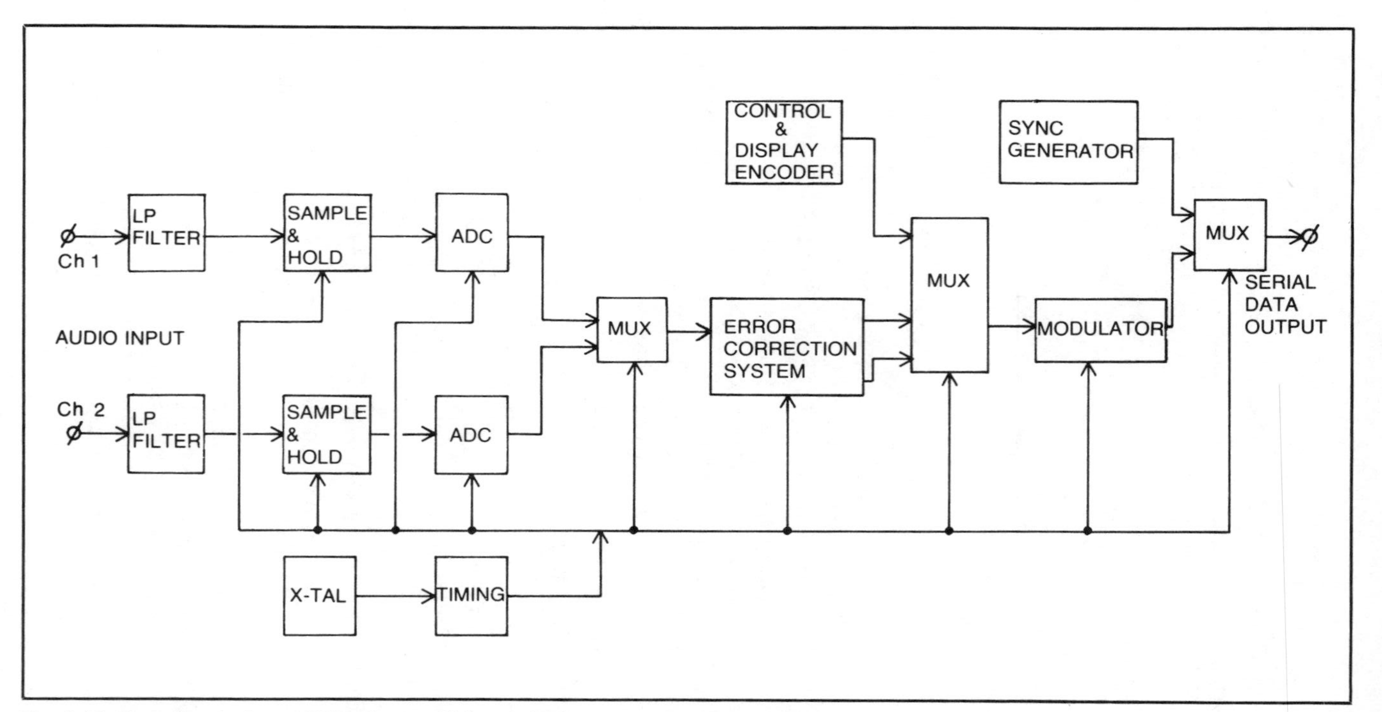

Fig. 5-12. Encoding system. MUX—time multiplexer; ADC—analog-to-digital converter.

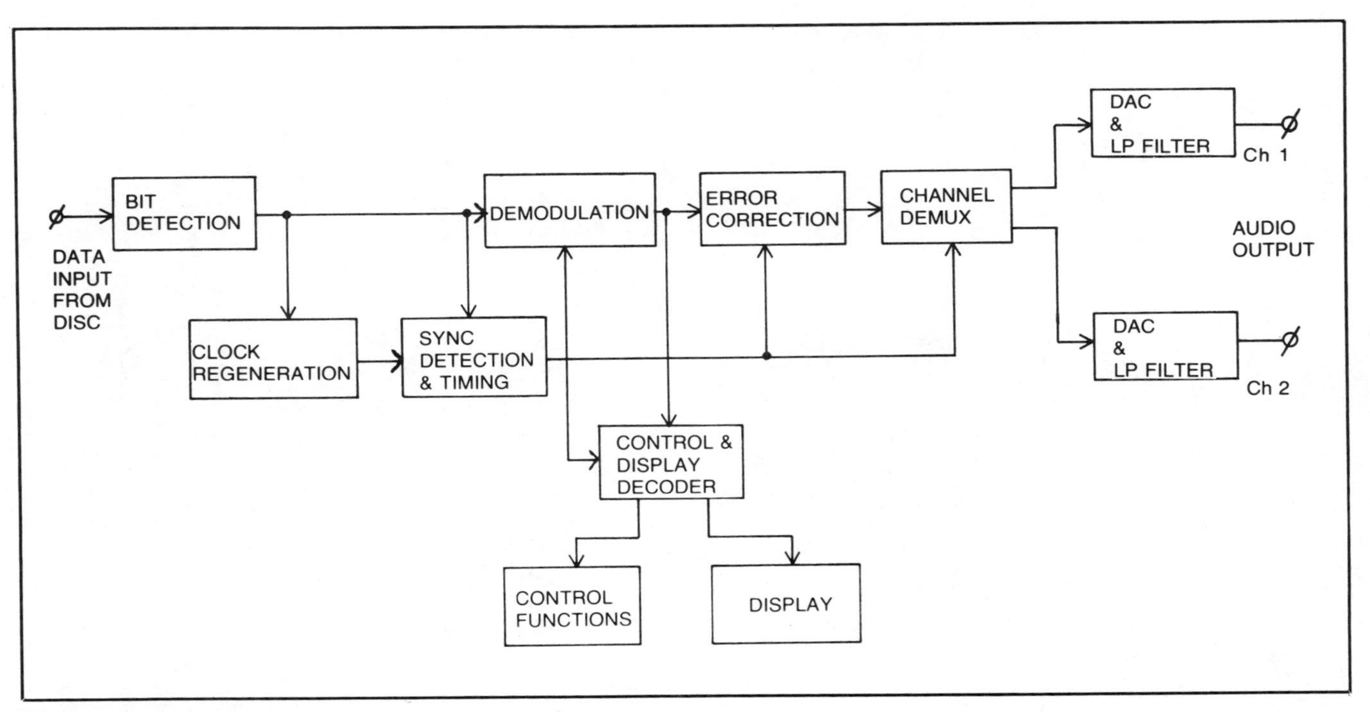

Fig. 5-13. Decoding system.

4) Simple decoder strategy possibility with reasonably sized external random access memory

5) Redundancy as low as possible (not much parity should have to be added)

6) Possibility for future introduction of four audio channels, without changes in the decoder chip.

6.2 Cross Interleave Reed Solomon-Code (CIRC)

6.2.1 The code corrects most errors that occur on the disk. However, some error patterns are not correctable. In this situation the error is detected and the decoder reconstructs the sample value by means of interpolation.

6.2.2 The performance of the CIRC is such that 1000 samples per minute (out of 2.6 million samples per minute) will have to be interpolated at 10^{-3}.BER. If the BER is 10^{-4}, only 1 sample per 10 hours will have to be interpolated. However, an average BER of 10^{-5} is typical.

6.2.3 Since the probability that an uncorrectable error is not detected is nonzero, which may lead to a click, the detection capability of the code was designed to ensure less than 1 click per month at 10^{-3} BER.

6.2.4 A disk that is handled very roughly might have scratches. Because of that the code should be capable of dealing with long burst errors. CIRC can fully correct burst errors up to 4000 bits (2.5 mm).

6.2.5 The decoder complexity of the CIRC has been reduced considerably by splitting up the decoder into two main parts:

1) A special-purpose decoder LSI
2) Standard 2k words of 8 bits.

6.2.6 CIRC has an efficiency of ¾, which means that 3 data bits will result in 4 bits after encoding.

6.2.7 The signal format has been designed in such a way that 4 channels are possible in the future, without changes in the decoder chip.

6.3 CIRC Encoder and Decoder (Figs. 5-14 and 5-15)

The CIRC consists of a C1 and a C2 Reed Solomon code as follows: C1 is a (32, 28) Reed Solomon code over GF[(5)] (2^8), and C2 is a (28, 24) Reed Solomon code over GF (2^8). The horizontal blocks between C1 and C2 represent 8-bit-wide delay lines of unequal lengths (interleaving). Before the C2 encoder a delay of one symbol is inserted in the even words to facilitate concealments in simplified decoder versions. After the C1 encoder a delay of one symbol (8 bits) is inserted in the even symbols (scrambling).

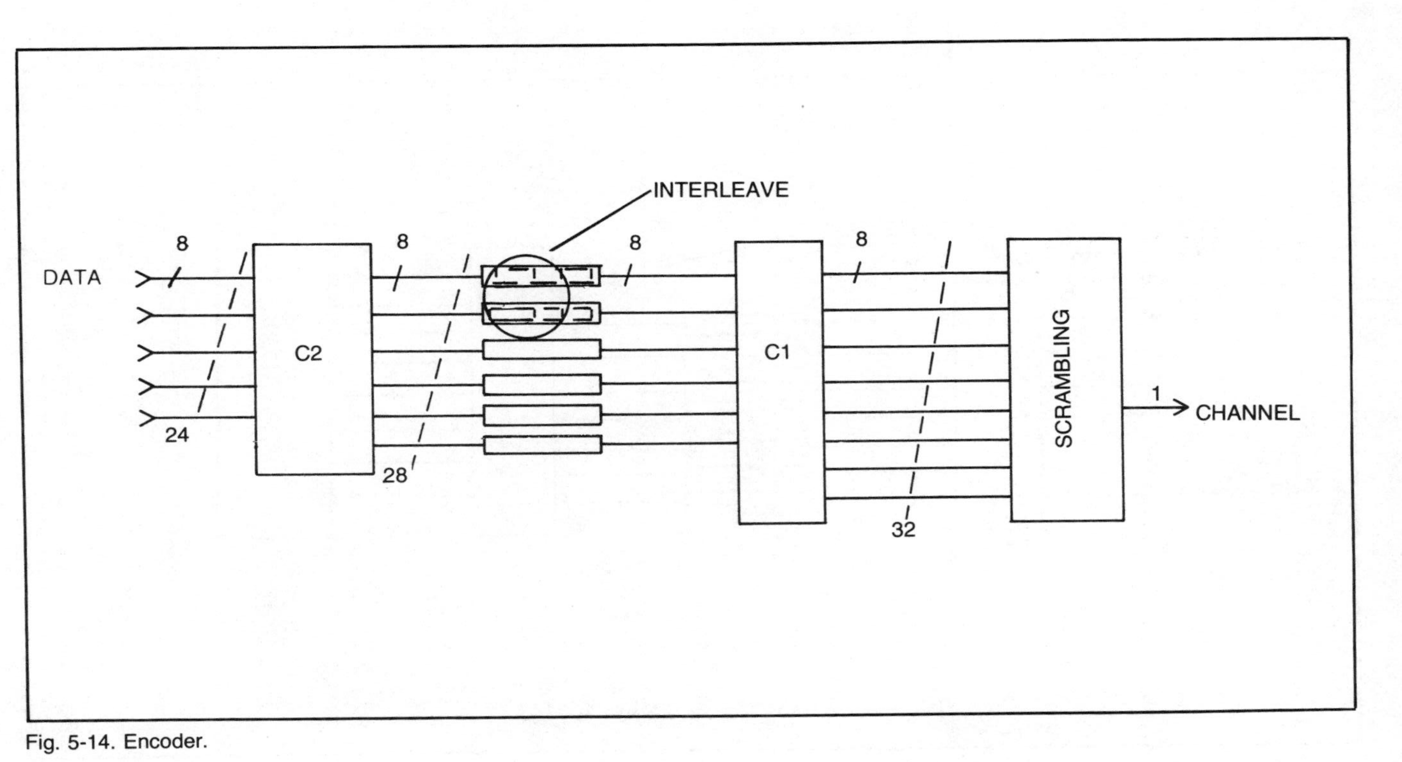

Fig. 5-14. Encoder.

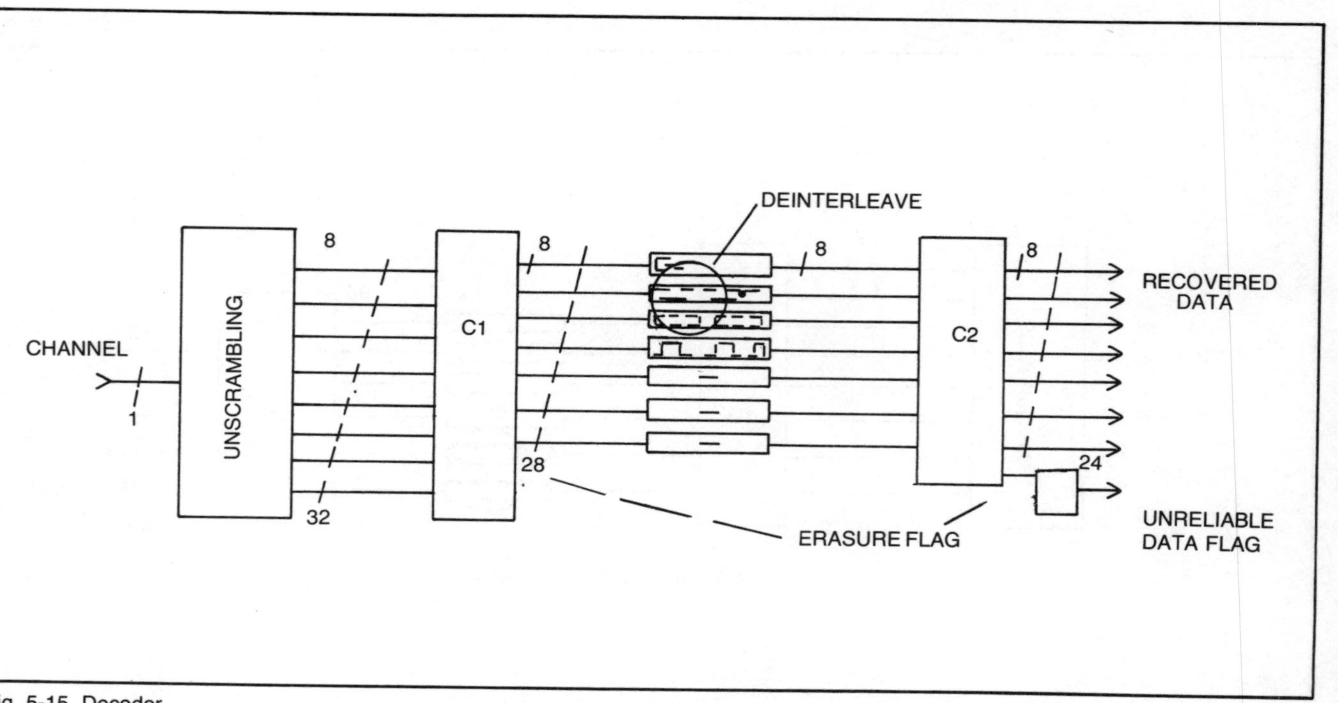

Fig. 5-15. Decoder.

The decoder operates as follows: The C1 decoder accepts 32 symbols of 8 bits each from which 4 parity symbols are used for C1 decoding. The parity is generated according to the rules of Reed Solomon coding, and because of that the C1 decoder is able to correct a symbol error in every word of 32 symbols. If there is more than one erroneous symbol, then regardless of the number of errors, the C1 decoder detects that it has received an uncorrectable word. If this is the case, it will let all 28 symbols pass through uncorrected, but an erasure flag is set for each symbol to mark that all symbols from C1 are unreliable at that moment.

Because the delay lines between the C1 and the C2 decoders are of unequal lengths, the symbols marked with an erasure flag at one instant arrive at different moments at the C2 decoder input. Thus the C2 decoder has for every symbol an indication whether it is in error or not. If a symbol does not carry an erasure flag it is error free. If no more than 4 symbols carry an erasure flag, then the C2 decoder can correct a maximum of 16 frames.

In cases that even the C2 decoder, cannot correct, it will let the 24 data symbols pass through uncorrected, but marked with the erasure flags that had originally been given out by the C1 decoder.

7 Audio Performance

Frequency response	20-20 000 Hz ±0.5 dB
Quantization, per channel	16 bits linear
Signal-to-noise ratio	> 90 dB
Dynamic range	> 90 dB
Channel separation	> 90 dB
Harmonic distortion	< 0.01%
Wow and flutter	equal to crystal oscillator accuracy

References

1. Kagami, Fukugawa, Nakata, Sato: "PCM audio disc players", Television Journal, 33, pp 26-31 (1979-1).

2. Itoya, A., M. Nakaoka and T. Kubo: "Development of the PCM Laser Sound Disc Player", IEEE Trans CE, CE-24, 3, pp 443-452 (1978-8).

3. Ogawa, Ito, Doi: "The construction of the optical digital audio disc", Shingaku Kenshi, EA 78-27 (1978-7).

4. Ito, Ogawa, Doi, Sekiguchi: "Revised construction of the digital audio disc system". Nihon Ongakkai Koron, 3-P-13 (1978-10).

5. Doi, T., T. Ito and H. Ogawa: "A long play digital audio disc system", AES 62nd Conv., No. 1442 (G-4) (1979-3).
Other materials relating to the video disc:
6. "Special edition: Video discs", Television Journal, 32, 1(1978-1).
7. L. B. Vries, K.A. Immink, J.G. Nijboer, H. Hoeve, T.T. Doi, K. Odaka, H. Ogawa; "The Compact Disc Digital Audio System: Modulation and Error Correction", AES 67th Conv., No 1674 (H-8), Nov. 1980.
8. T.T. Doi; "General Information on a Compact Disc Digital Audio", JAES, *Vol. 29,* No. 1/2, pp 60-66, Jan./Feb. 1981.
9. T.T. Doi; "The Signal Format of Compact Disc Digital Audio", IECE of Japan, Tech. Report, No. EA81-23, July, 1981.

Chapter 6

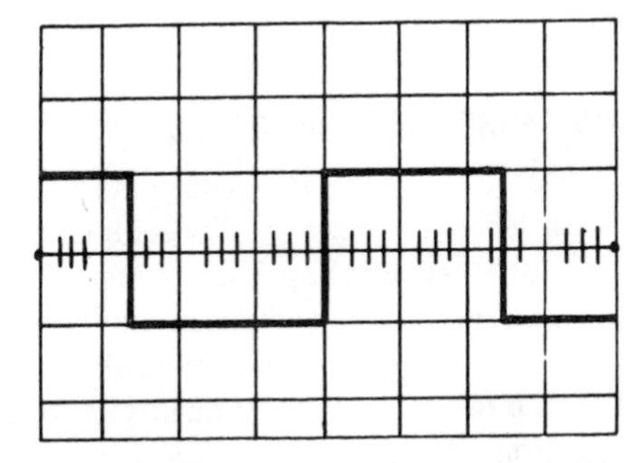

Code Errors and Their Countermeasures

The success of digital audio systems, particularly PCM tape recorders and CD (Compact Disc) depends to a large extent on the success of the measures taken to prevent code errors. As mentioned earlier, these code errors occur during the recording process, when the digital signal is being transferred to the recording medium, whether it be magnetic tape of some other format. In 1948, Shannon published his communications theories[1], which were a theoretical investigation of code errors. The second theory in particular, which is also known as the communication path coding theory, is highly pertinent to the field of digital audio. This theory demonstrates the possibilities of a code (error detection, error correction) which can convert an imperfect communication path containing errors into a high quality communication path with few errors. Since the publication of Shannon's theories, a great deal of research has been carried out on code theory[2], and has achieved great success in practical fields: for instance, the phenomenal advances made in electronic calculators over the past few years and also in data communications.

However, generally speaking, the generation of errors in digital audio equipment is astronomically high when compared to the above fields of development. Unfortunately, very stringent conditions are demanded by Hi-Fi equipment, because certain types of noise can be extremely objectionable. Thus, code errors must be kept to an absolute minimum, and the advances outlined in the preceding paragraph cannot be directly used for digital audio. This chapter outlines the special techniques and development work necessary to adapt code error theories for practical use in digital audio equipment.

THE NATURE AND CAUSES OF CODE ERRORS

Table 6-1 indicates the basic causes of sound quality reduction in PCM tape recorders and digital audio equipment in general. Quantizing noise is an unavoidable primary cause of objectionable noise, but can be pushed beyond audible limits by increasing the number of bits used. Devices such as A/D and D/A converters used in the circuitry may, if not of the highest quality, also contribute to the level of distortion. However, it is vitally necessary to take comprehensive countermeasures against code errors when designing digital audio equipments. This is because these errors may cause distortion in the form of pronounced clicks after reconversion of the signal into an analog form. Thus, unless effective countermeasures are taken, this type of distortion would amount to a substantial diminution of quality.

However, the form that these code errors take, and the reason why they are caused depends on the recording medium, and on the way the equipment is constructed. Before effective countermeasures can be taken and compensation carried out, it is necessary to consider afresh the nature and causes of error code generation.

The Causes of Code Errors

Dropout.[4] The most significant cause of code errors when using either a PCM tape recorder or a CD (Compact Disc) is dropout. When using any tape recorder, microscopic particles of dust, measured in tens of microns, may adhere to the surface of the magnetic material with which the tape is coated.

Equally small scratches may also be caused in this coating, and as a result, a "gap" will occur between the magnetic head and the surface. This phenomenon may be catastrophic for a PCM tape recorder (where the rotary heads are running the tape at between 7 and 10 meters per second), unless some code error protection is available. On a PCM tape recorder, the "gap" caused by the dust or dirt is in effect magnified many times, and on playback, the signal level will momentarily fall dramatically. (This is known as "spacing loss"). Thus, dropout is a temporary loss of information caused by spacing loss. In general, when dealing with VTRs, the appearance of a bright line running like a scratch across the TV picture in a

Table 6-1. Primary Factors Effecting Sound Quality.

Type	Cause	Effect	Degree	Counter measure
Fundamental	Quantization noise	Limits dynamics range, S/N ratio	Small effect	Increase number of quantization bits
Quasi-fundamental	Aperture effect Aliassing	Frequency response S/N affected		Frequency response compensation LPF with near ideal characteristics
Caused by circuit devices	Sample and hold A/D converter D/A converter LPF	Worsening of distortion Freq. response, S/N		Ensure that circuits are high speed, high accuracy Use ideal characteristics
Caused by recording chain	Jitter etc.	Code errors		Correction on temporal axis
Caused by recording medium	Dropouts etc.	Code errors	Larger effect	Error compensation (de-tect, correct, interpolate)

horizontal direction is known as dropout. The dropouts we are referring to with reference to PCM are essentially the same kind of phenomenon. The dirt, dust or scratches on the surface of the tape are caused during the production process as well as when the tape is actually being used (Figs. 6-1 and 6-2).

During tape manufacture, the coating of the tape base with magnetic material is, of course, carried out in isolation in a "clean room". However, during this process, it is possible for microscopic dust motes, measuring several microns across, to become embedded in the magnetic coating. (Minute shavings from the tape base material might, for example, be floating around). It is an extremely time consuming process to check and correct every part of the reel of the tape. (A standard Betamax cassette, the L-500, measures ½ inch by 500 feet).

After the tape base material has been coated with magnetic material, the tape is slit to the required width before being wound on the cassette hub or tape reel. At this stage, slight differences in level between the hub or reel and the leader tape may occur at the beginning or end of the tape winding. This will result in a slight imperfection which will, in turn, cause a dropout. Generally the extreme ends of a magnetic tape will have a higher level of dropouts than the middle section.

While the tape is actually being used, a number of factors may combine to cause small scratches on the surface. For instance, the

Fig. 6-1. Microscopic particles of dust can become attached to the surface of the tape during production (5μm/cm).

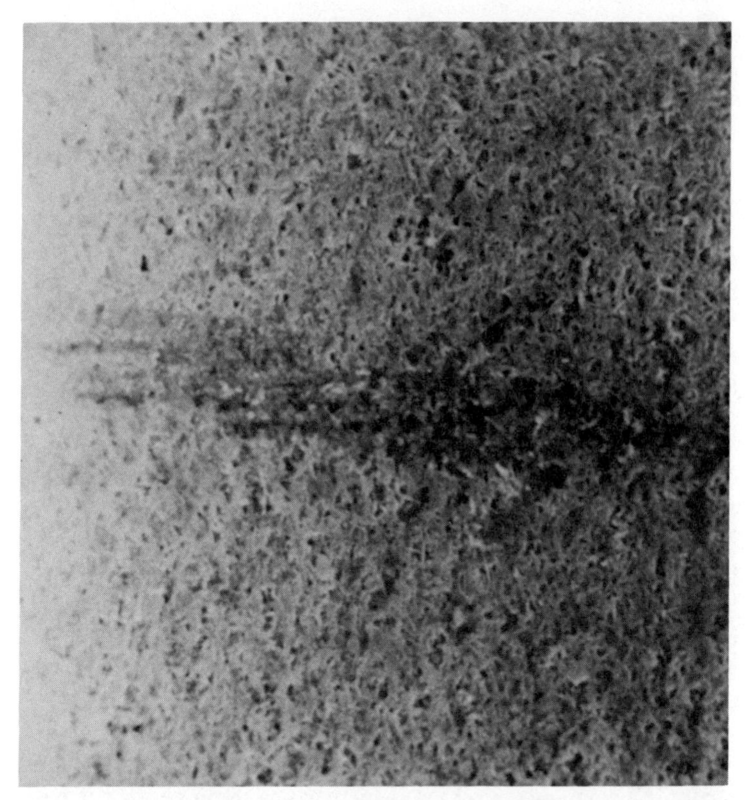

Fig. 6-2. Scratches in the magnetic coating can occur while the tape is running.

small fluctuations caused by the cassette and tape guides, dust from the operating environment, or tiny pieces of waste embedded in the magnetic coating (as described above) may cause scratches on the tape as the head passes over them, and thus generate a dropout.

However, dropouts of the type so far described, which are due to spacing loss, do not occur with digital audio disc systems, particularly the optical system. Dropouts are caused on CD records by incorrect "pits" on the record surface; these occur either during the stamping process, or during pressing.

Thus, the primary factors which cause the creation of dropouts on digital equipment are very varied, ranging from deficiencies in the recording medium to problems in the manufacturing process. "Burst errors" are the most common type of code error found when using magnetic tape where a continuous code error lasting for several tens of bits on average occurs. A burst error happens once in about 10^3 to 10^6 bits. An optical digital audio disc, on the other hand, is prone to "random errors", where sporadic code errors of only one bit on average occur once in about 10^2 to 10^4 bits. Errors on mechanical and capacitance DAD systems are not, however, limited to such low figures.

Jitter. The phenomenon of jitter consists of variations of the time axis during the reproduction of the digital signal (i.e. it is similar to wow and flutter). A PCM tape recorder, just like any other tape recorder, depends upon a mechanical system to run the head along the tape. Thus, mechanical imperfections cannot be avoided, and if time variations exceeding operational limits occur during reproduction of the digital signal, code errors will occur.

For convenience, an ideal reproduction waveform using the NRZ (Non Return to Zero) system is shown in Fig. 6-3. One bit cell represents one data bit on the NRZ system, and signal A represents the correct situation where no fluctuation has occurred. If the signal level is measured at the center of the bit cell, the data bit can be reproduced correctly. However, should time fluctuations occur, as shown in signals B and C, the data bit can still be reproduced accurately, as with signal A, provided that the fluctuation does not exceed the limits:

$$\Delta t = \pm \frac{\tau}{2}$$

"Δt" is the jitter margin, and if jitter exceeding this value occurs, random code errors will increase to the level that it becomes impossible for the machine to read the signal correctly. Of course, the reproduction waveform will, as a rule, deviate from the ideal and this in itself will reduce the jitter margin and lead to an increase in code errors.

Jitter may be divided into types according to the cause. Generally speaking, there are two major divisions: low frequency fluctuations (less than 1Hz, often called drift), and fluctuations of high frequency components.

The former is ultimately due to changes in the mains frequency because any irregularities in tape reel revolution or in the recorder are controlled by the servo, which in turn is synchronized to the mains frequency. The latter often occurs in rotary head PCM tape recorders, and is caused by irregularities in the operation of the rotary heads or the capstan. On stationary head PCM tape recorders, it may be caused by oscillation in a horizontal direction along the tape as a result of stretching. Jitter on a Digital Audio Disc is caused by a lack of uniformity in the rotational system.

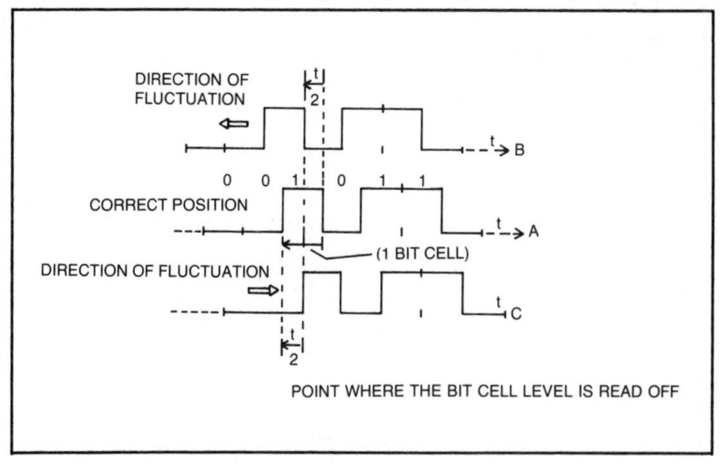

Fig. 6-3. Jitter margin (with an ideal NRZ reproduction waveform).

Intersymbol Interference. When using a PCM tape recorder to record high density digital signals, it is possible that the quality of the reproduction signal may be degraded if the codes which make up the recorded bit pattern overlap and cause mutual interference. This phenomenon is usually caused by deficiencies in the record chain as a whole, as well as in the tape, the heads and their relative spacing. This type of interference makes it difficult to retrieve the digital signal accurately by the generation of code errors. Mutual interference between codes is known as intersymbol interference.

Figure 6-4A shows an ordinary record-reproduction system using differential response magnetic heads (similar to standard ring heads.) The reproduction output waveform corresponding to the unit function record current waveform becomes a gauss distribution isolated waveform as shown in Fig. 6-4B. Should the bandwidth of the record chain be insufficient, a more blunted reproduction waveform is produced.

When the record waveform is sufficiently long when compared to the bandwidth of the record chain, as shown in Fig. 6-4C, the reproduction waveform peak separation T_1 corresponds almost exactly to the length of the record current waveform To. However, when the waveform is too short, as shown in Fig. 6-4D, the two adjacent isolated waves cause mutual interference. The peak separation T_1 is clearly markedly different from To. It would be accurate to think of this as a situation where the reproduction waveform is more or less identical to a composite waveform constructed from

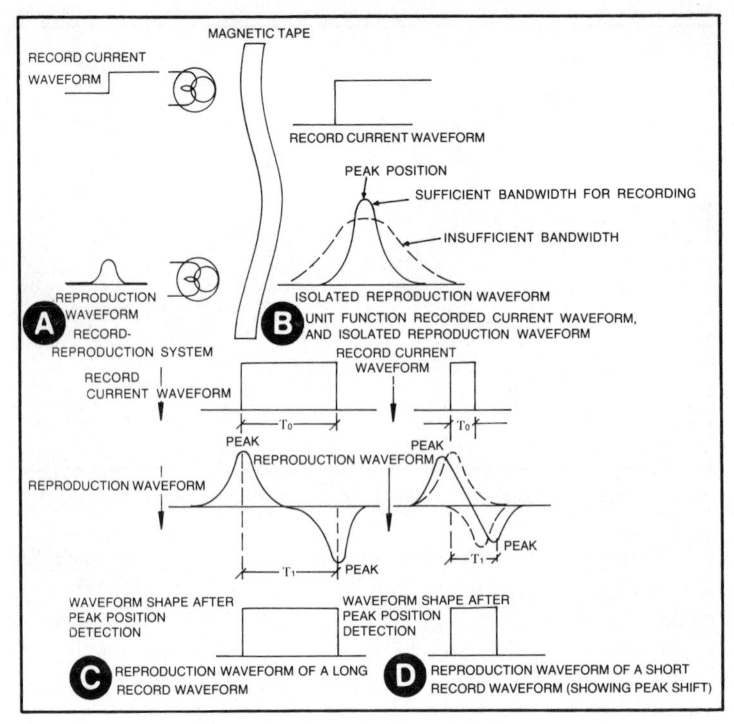

Fig. 6-4. Peak shift caused by intersymbol interference.

the superpositioning of two isolated waveforms. See Figs. 6-5A and 6-5B.

This phenomenon caused by intersymbol interference is known as peak shift. The process can be demonstrated as in the following equation where peak shift ratio is T:

$$T = \frac{T_1 - To}{To} \times 100\ (\%)$$

Peak shift caused by intersymbol interference is thus an important factor in determining the limits of record density. In addition, as the peak shift ratio increases, the reproduction signal deteriorates, and it becomes progressively more difficult to extract the digital signal accurately. And of course, code errors are generated.

Noise. There are an extremely large number of factors which may cause noise; some of the major ones are as follows:

1. Effects generated by the recording medium itself, such as hiss noise on tape.

2. Crosstalk between channels generated within the multichannel magnetic head, or crosstalk caused within the wiring system.

3. Other factors not directly linked to the signals path, e.g., power source.

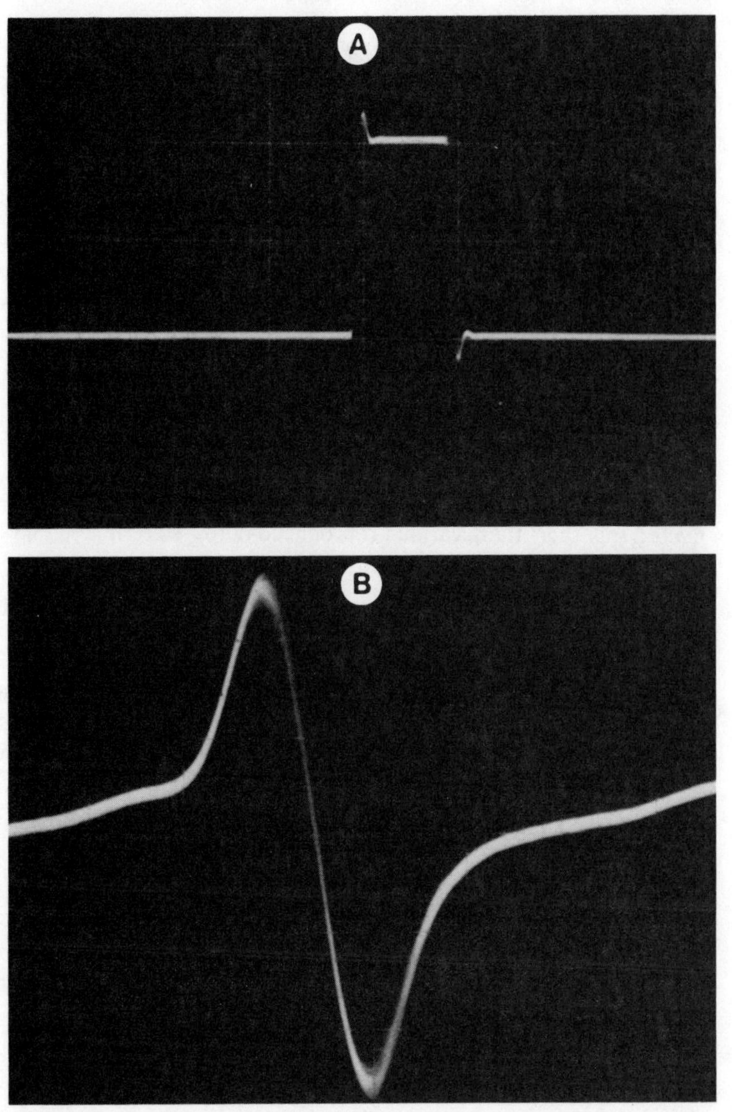

Fig. 6-5. (A) Record current waveform, (B) Reproduction waveform showing peak shift.

All these sources of noise combine together with the other factors already mentioned (dropout, jitter, peak shift and so on) to multiply the number of code errors caused. Figure 6-6A shows a diagrammatic representation of this. For example, if there is a dropout and, therefore, an insufficient S/N ratio during reproduction when the digital signal is being retrieved and reconverted, as shown in Fig. 6-6B, then there will be a loss of signal. This is because the noise referred to above is, in effect, added to the threshold level used for level detection, and thus the actual signal level is too low to be identified. As a result, drop in signal output caused by a relatively small spacing loss is intimately linked to dropout, and once again, this will cause code errors. These errors can then be compounded by a concurrent decrease in the jitter margin or an increase in peak shift caused by intersymbol interference.

Code Errors Caused During Editing. Some form of editing to achieve top-quality software is as necessary for PCM tape recorders as it is for analog tape recorders.

Editing can, for practical purposes, be divided into two types:

1. Manual editing, where a recorded tape is physically cut and attached to another piece of recorded tape.

2. Electronic editing, where several tape recorders are connected together; their various reproduced outputs are linked electronically, and editing is carried out by recording a new tape.

The ability to momentarily switch sources, so that one tiny section is added on top of an existing recording, is called "punch in, punch out" in electronic editing.

Thus, when designing a PCM tape recorder which may have a maximum record density of 30 kb/inch, it is a real headache to

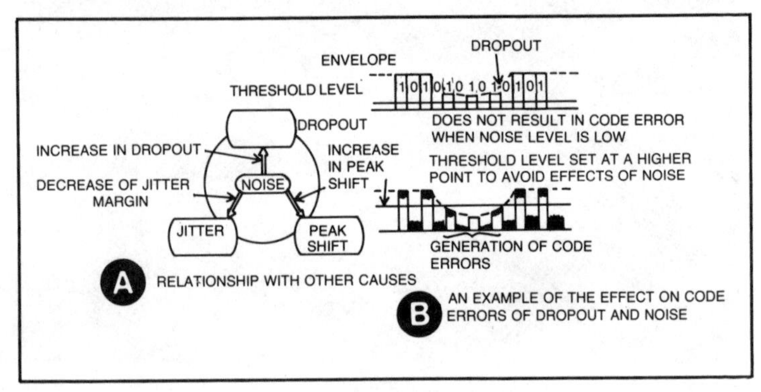

Fig. 6-6. The effect of code errors caused by noise.

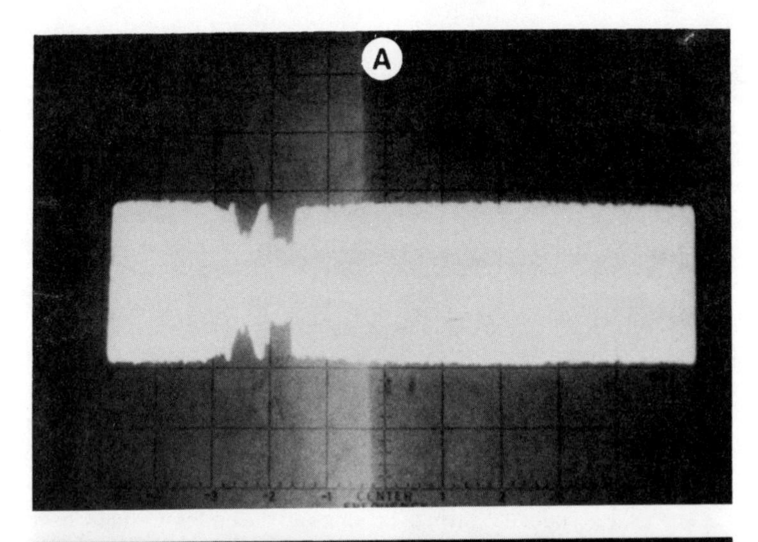

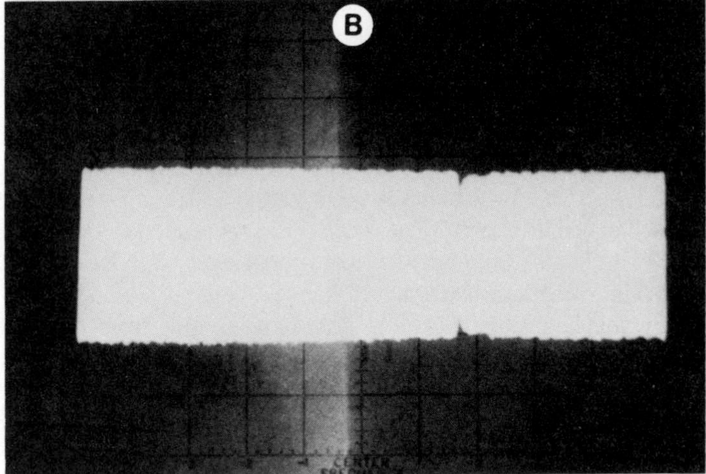

Fig. 6-7. An example of information loss caused by manual editing (stationary head, record density 30 kb/inch, carrier envelope visible).

produce an effective editing system. A PCM editing system must, on the one hand, provide all the facilities necessary for the successful production of software, and yet on the other, it must not generate code errors at the actual edit point.

Manual editing may result in a loss of recorded information at the point where the tape is actually cut, as shown in Fig. 6-7. In addition, it is necessary to touch the magnetic coating when making the edit, so the surface may become dirty. This dirt will, of course, cause code errors. In fact, finger-printing is the most tenacious kind of "dirt" most likely to cause dropouts.

It is clear from the foregoing that most of the causes of code errors generated while editing are, in fact, necessary evils. Thus, countermeasures have to be designed into the editing system, while maintaining all the other required facilities.

The Nature and Measurement of Code Errors

In order to design effective countermeasures against code errors into digital audio equipment, all types of code errors likely to be generated within the system have to be considered, measured and analyzed. All the factors mentioned so far which contribute to code errors must be examined, and the different types of code error caused by each must be statistically investigated.

The Use of "Eye Diagrams". When a pseudo-random code (M series, PN code, etc.) which contains all the codes normally used for recording is generated, as shown in Fig. 6-8, the corresponding reproduction waveform may be viewed on an oscilloscope. An eye shaped pattern may then be observed, known as an "eye diagram" or "eye pattern".

An eye diagram is not suitable for examining informatic losses of relatively long duration, such as dropouts. It may, however, be effectively used for examining the combined qualitative effects of jitter, peak shift and noise on the record reproduction chain as a whole. It is also an extremely convenient and effective tool for observing waveform equivalence and the general performance of the system being investigated. If jitter or peak shift are affecting the record reproduction chain, then the "eye" opening deteriorates. The general effects of this are shown in Fig. 6-9. Ideally, the existence of jitter or peak shift will cause a deterioration in the opening; amplitude fluctuation is indicated by Δa, temporal variations by Δb.

The amount of deterioration caused may be calculated as shown below where p represents the "eye opening ratio":

$$p = \frac{a_2}{a_1} = \frac{(a_1 - 2\Delta a)}{a_1}$$

Measurements relating to the digital signal must be taken from the center of the opening in the eye diagram. Accompanying deterioration of the opening, one will find a decrease in the margin of the threshold level which is used to set the timing and level, and of

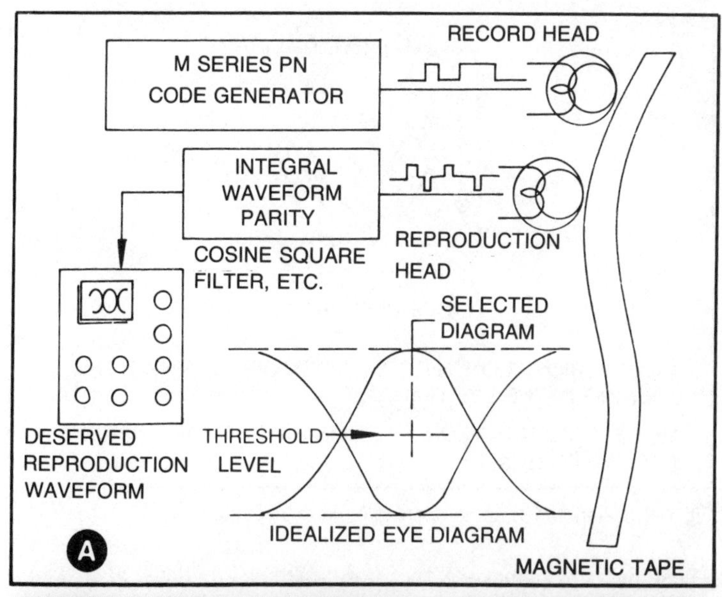

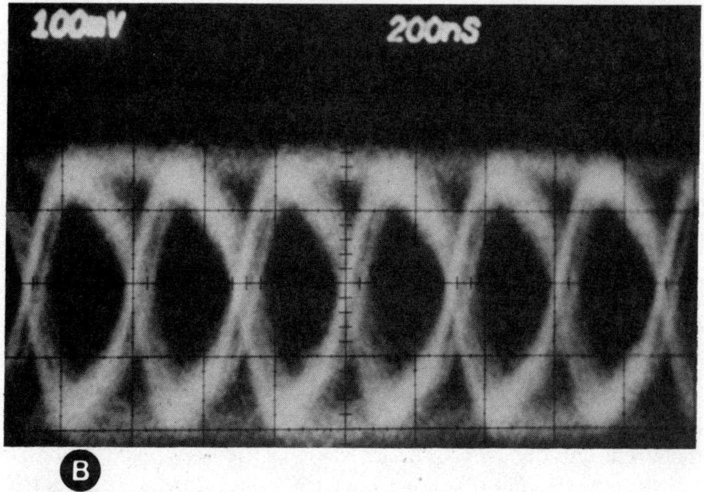

OBSERVED EXAMPLE (DAD, DIGITAL AUDIO DISC)

Fig. 6-8. Eye diagram.

course, an increase in the probability that code errors will be generated.

Measurement of Code Errors Using Special Equipment. Quantitative measurement of dropout, the primary cause of code errors, cannot be carried out by observation of an eye diagram.

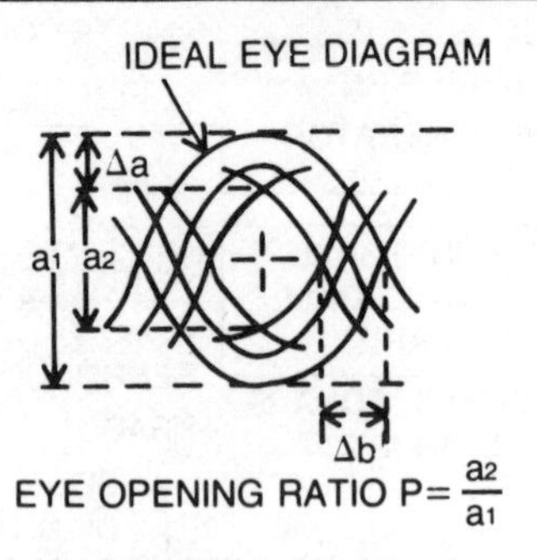

Δa: DETERIORATION IN OSCILLATION DIRECTION OF THE OPENING IN THE EYE DIAGRAM

Δb: DETERIORATION ON THE TIME AXIS OF THE OPENING IN THE EYE DIAGRAM

Fig. 6-9. Deterioration of the opening in an eye diagram.

These measurements are, therefore, taken using the special equipment introduced below.

System 1 [5]

This system was devised for the measurement of dropouts caused by spacing loss between the tape and heads, and for dropouts caused by the recording medium, magnetic tape and so on. It cannot, however, be used to measure code errors generated by jitter or peak shift, so it is desirable to use this system in conjunction with an eye diagram.

The construction of this system is shown in Fig. 6-10A. First of all, a fixed level carrier signal is recorded on tape, and the envelope of the carrier signal (marked (i) in the diagram) is then reduced to a fixed known value during reproduction. Any dropouts caused may then be detected. The dropout signal thus obtained (marked (ii) in the diagram) is then fed through a waveform shaper in such a way that areas without dropout (marked *) that are less than one-sixth of a line (a horizontal TV line about 63.5 μsec long in EIA/NTSC systems) are considered as being part of a continuous dropout. Dropouts less than one-sixth of a line in duration (marked **) are, however, regarded as what they are: dropouts. The signal is divided up into units one-sixth of a line long. This measurement is used because the memory and response speed of the mini-computer is sufficient to handle it.

After waveform shaping, the dropout signal is fed to a counter run by a highly accurate clock, and the length of the dropouts is

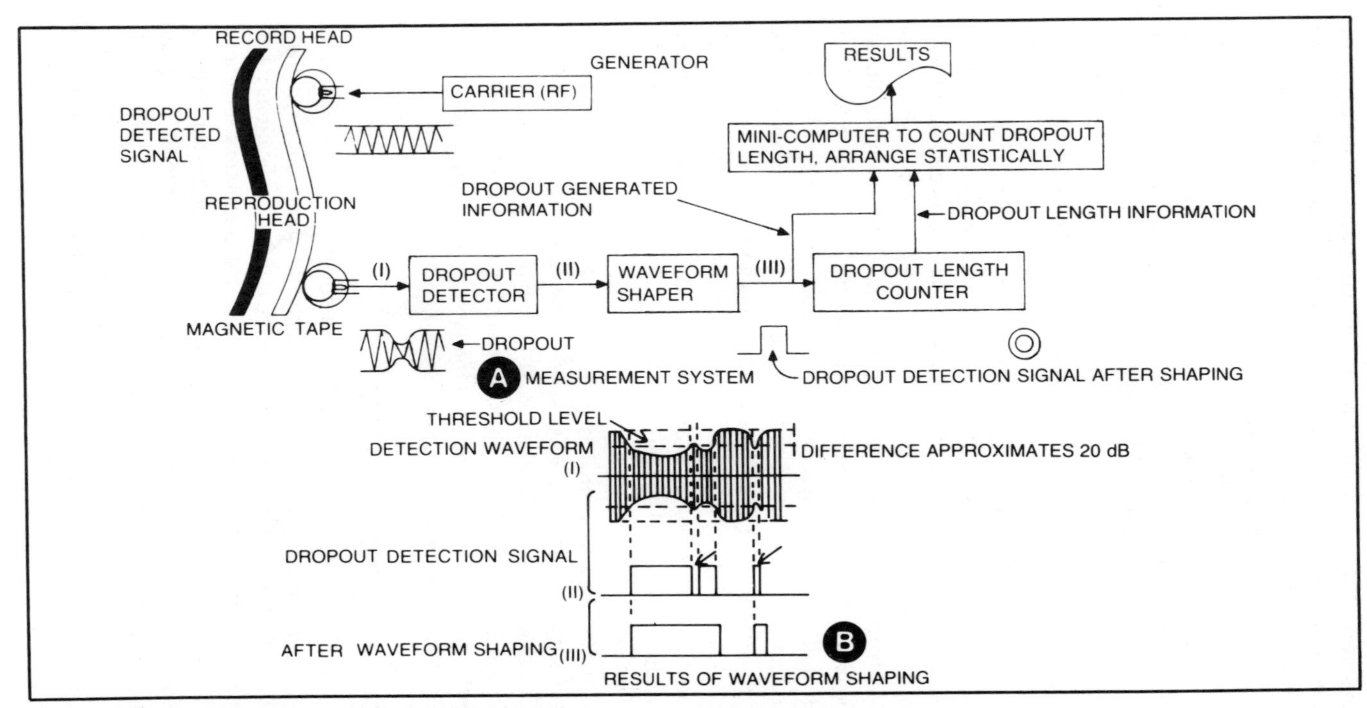

Fig. 6-10. Measurement system for detection of a reduction in the carrier envelope (No. 1).

measured. Then it is fed into a mini computer in order to be processed statistically. Figure 6-11A shows the dropout measurements taken for six video cassette tapes, each one hour in length, used with a rotary head PCM recording system. The differences in dropout occurrence on each of the six cassettes measured is shown in Fig. 6-11B. From the data provided, it is evident that by far the greatest number of dropouts have a length of less than one horizontal TV line (63.5 μsec EIA/NTSC). Dropouts longer than 9 lines (about 6 msec) are rather rare. However, it is also clear that the length and number of dropouts vary widely from tape to tape.

System 2

Figure 6-12 shows a diagram of the second measurement system used. This system differs from the one outlined above in so far as detection bits for error detection are recorded along with information bits with a suitable content. The presence or absence of code errors in each block unit may then be identified by the detection bits at the time of playback, but after the digital signal has been selected. Using this system, it is possible to make accurate measurements of all types of code errors, whether caused by dropout, jitter, peak shift or noise occurring in each one block unit.

A record is made up of the measurement results for each block by using a mini computer to assign sequential numbers to each block containing errors. Thus, timing data of the code errors may be obtained for examination.

Using this system it is possible to measure the mutual relationship of code errors between channels on a stationary head PCM tape recorder. In addition, it is also possible to examine the correspondence between physical position and measurement data for the dropouts on the surface of the tape. An example of data taken using this system is shown in Fig. 6-13.

COMPENSATION OF CODE ERRORS

Code error compensation has traditionally been a matter of great concern and interest in the field of code communication links. It is carried out in a large number of wide ranging applications for all types of data communication: memories for computers, magnetic discs and drums, and of course, magnetic tape. Code error compensation systems may be broadly divided into two types: ARQ and FEC. ARQ (Automatic Repeat Request) systems detect the occurrence of code errors in the data transmitted and then send a request for the re-transmission of the section of data containing information

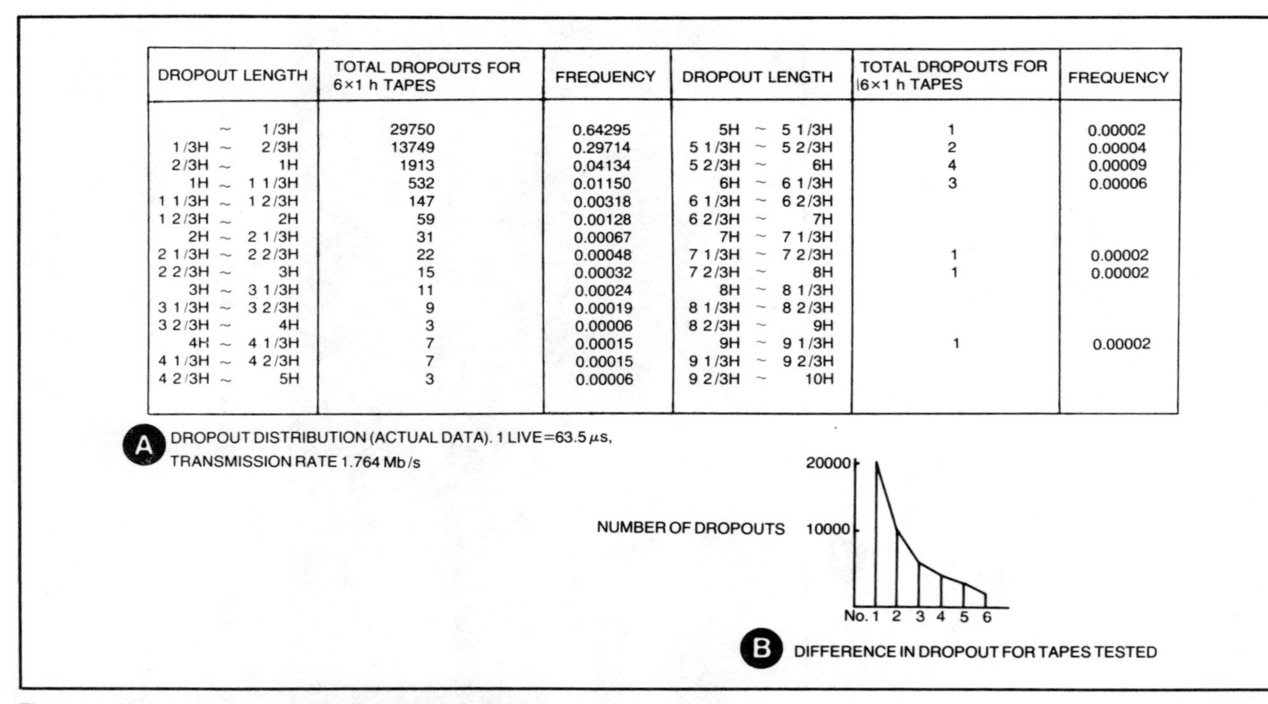

DROPOUT LENGTH	TOTAL DROPOUTS FOR 6×1 h TAPES	FREQUENCY	DROPOUT LENGTH	TOTAL DROPOUTS FOR 6×1 h TAPES	FREQUENCY
~ 1/3H	29750	0.64295	5H ~ 5 1/3H	1	0.00002
1/3H ~ 2/3H	13749	0.29714	5 1/3H ~ 5 2/3H	2	0.00004
2/3H ~ 1H	1913	0.04134	5 2/3H ~ 6H	4	0.00009
1H ~ 1 1/3H	532	0.01150	6H ~ 6 1/3H	3	0.00006
1 1/3H ~ 1 2/3H	147	0.00318	6 1/3H ~ 6 2/3H		
1 2/3H ~ 2H	59	0.00128	6 2/3H ~ 7H		
2H ~ 2 1/3H	31	0.00067	7H ~ 7 1/3H		
2 1/3H ~ 2 2/3H	22	0.00048	7 1/3H ~ 7 2/3H	1	0.00002
2 2/3H ~ 3H	15	0.00032	7 2/3H ~ 8H	1	0.00002
3H ~ 3 1/3H	11	0.00024	8H ~ 8 1/3H		
3 1/3H ~ 3 2/3H	9	0.00019	8 1/3H ~ 8 2/3H		
3 2/3H ~ 4H	3	0.00006	8 2/3H ~ 9H		
4H ~ 4 1/3H	7	0.00015	9H ~ 9 1/3H	1	0.00002
4 1/3H ~ 4 2/3H	7	0.00015	9 1/3H ~ 9 2/3H		
4 2/3H ~ 5H	3	0.00006	9 2/3H ~ 10H		

Fig. 6-11. An example of dropout measurement results.

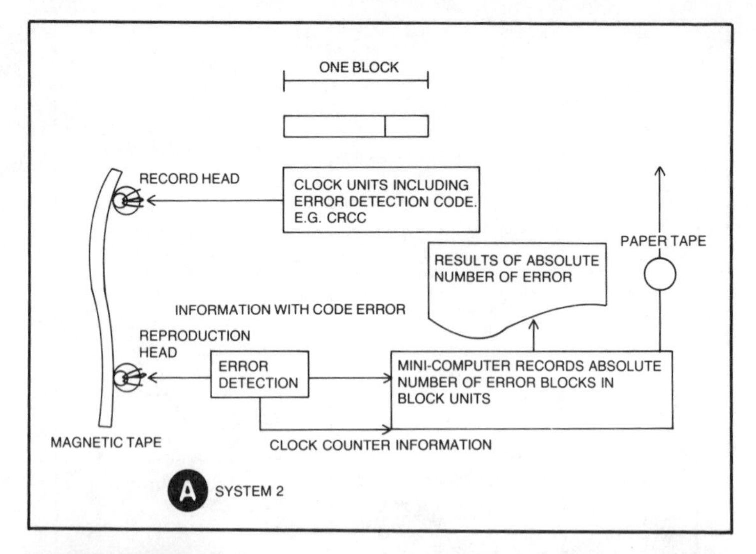

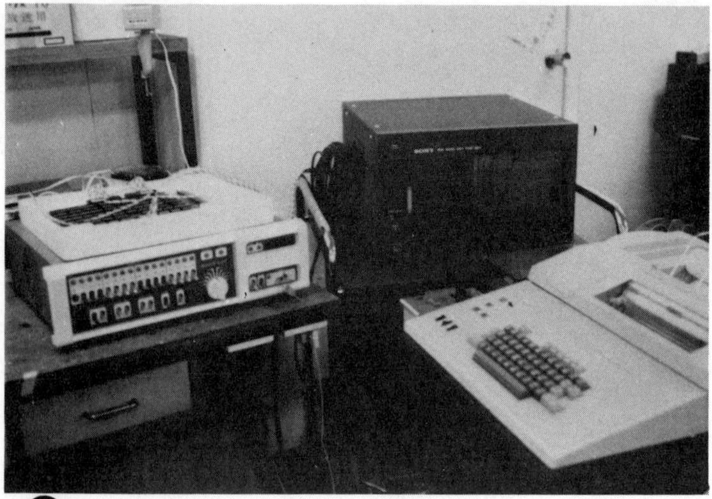

B EQUIPMENT USED

Fig. 6-12. A system for detecting and measuring accurately the code errors in block units.

losses. FEC (Forward Error Correction) systems detect code errors and correct them in real time.

Signal processing is carried out in real time when using digital equipment such as PCM tape recorders. Therefore, ARQ systems are not suitable for error compensation, and as a rule, FEC systems are the basis for error compensation.

```
****************************
* ERROR MEASUREMENT SYSTEM *
****************************

          RUN DISTRIBUTION

L(FR)       GOOD        BAD
    1          0        109
    2          0         21
    3          0         23
    4          0         14
    5          0          1
    6          1          0
    7          0          0
    8          0          0
    9          0          0
   10          0          2
   20          0          0
   30          0          0
   40          0          0
   50          0          0
   60          1          0
   70          2          0
   80          1          0
   90          0          0
  100          0          0
  200          2          0
  300          2          0
  400          5          0
  500          2          0
  600          1          0
  700          1          0
  800          1          0
  900          2          0
 1000        149          0

 TOTAL       170        170

 AVE(G)   0.5639666E+04

 AVE(B)   0.1770586E+01
```

Fig. 6-13. An example of measurements taken for code errors in block units.

Code Error Compensation For Digital Audio Equipment

Figure 6-14A shows the general construction of a code error compensation system using error correction codes. Redundancy must be added to the information to be recorded in order to correct errors which may occur in the recording during real-time processing. Therefore, detection bits (n-k (bit)) are added according to a fixed pattern depending on the content of the information bits (k (bit)). Then a block of n (bit) is recorded containing the new information. If code errors occur within the block, deviations from the fixed pattern mentioned above will appear. The occurrence of code errors is detected in the decoder from the appearance of these deviations, and the information is then corrected.

The particular pattern used is different for every type of error correction code. In a way, one could regard this pattern itself as being the error correction code. Error correction codes vary greatly in the block codes and convolutional codes used to create the information pattern. As shown in Fig. 6-14B, although encode and decode for correction of the block code is carried out for the block units, error correction is not concluded at the end of each block. This is because the detection bits or detection word for the convolutional code is formed from a uniform period of past information. In general, therefore, the degree of redundancy required for convolutional code is smaller than that needed for a block code. On the other hand, attention must be paid to the fact that if an uncorrectable error were to appear it would then be transmitted to the following block.

Table 6-2 shows examples of the correction efficiency of several error correction systems. These are the basic error correction codes used in computers and data communications.

There are major differences between the error compensation methods used for digital audio equipment and those widely used today in computers and data communications. For example, the transmission path used (e.g. magnetic tape) is greatly inferior in quality, and therefore, the number of code errors likely to occur is much greater, while it is not possible to allow any errors to remain after compensation has been carried out. An error compensation system for audio use must, in fact, satisfy two preconditions: firstly, no errors must remain in the signal, and secondly, the compensation system used must not cause any noise generation during reproduction of the audio signal. It is dangerous to rely solely on the traditional error correction codes outlined in Table 6-2 for code error compensation. The error correction code shown in Fig. 6-15A, for example, is unsuitable for use in digital audio equipment

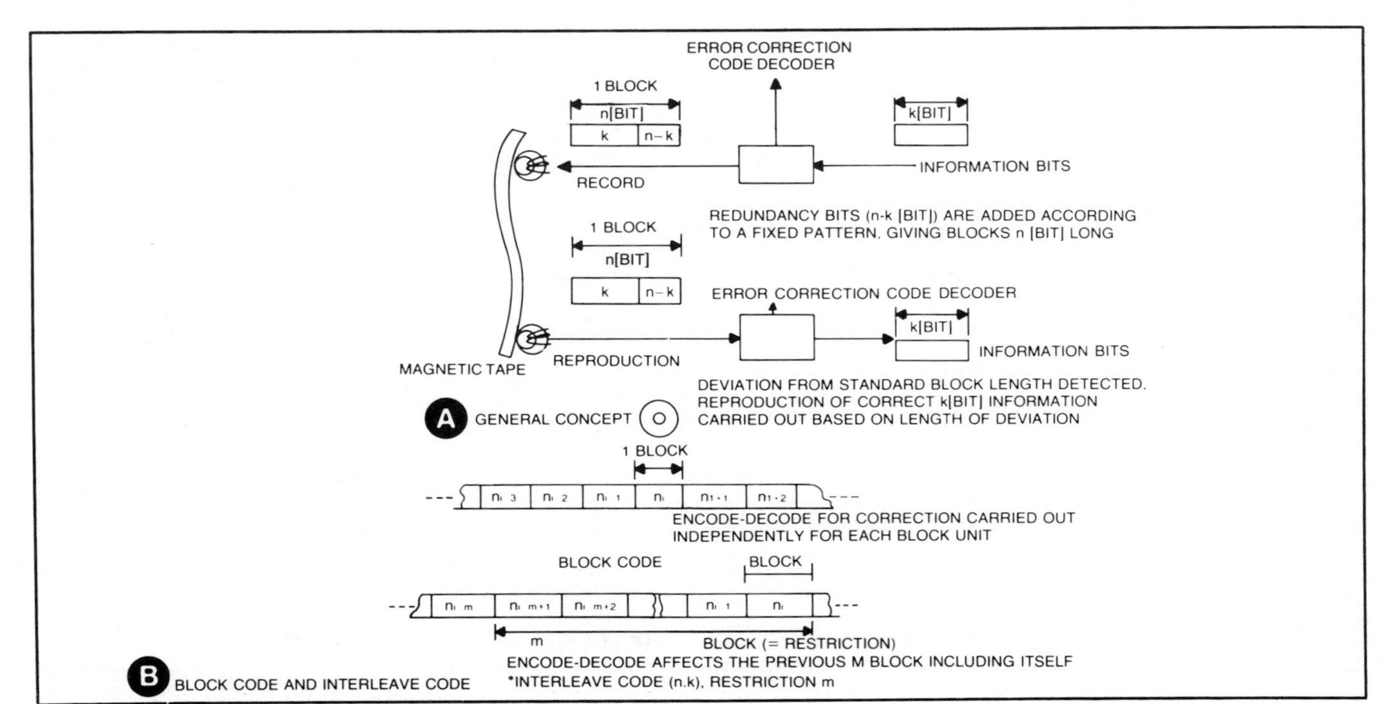

Fig. 6-14. General error correction code.

Table 6-2. Examples of Different Types of Error Correction Codes.

Type of error	System	Examples where the largest number of bits which can be corrected is 3.
Effective against random errors	Block code	BCH code (n,k) ex n=15, k=3 correction of 3 bit random errors anywhere in a 15 bit block.
	Convolutional code	Self correcting code (n,k), restriction = m ex n=3, k=2, m=120 where the error bits are less than 3, and when other errors do not occur within the restriction period, full correction.
Effective against burst errors	Block code	Abridged cyclic redundancy check code (n,k) ex n=15, k=9 correction of a 3 bit burst error anywhere in a 15 bit block.
	Convolutional code	Iwadare code (n,k), guard space = g ex n=3, k=2, m=27, g=26 correction of 3 bit burst errors separated by 26 bit guard spaces.

n : 1 block length = information bits (k) + redundancy bits (n-k)
k : information bit
m : restriction (bit)
g : guard space (bit)

which requires much more complex codes. For although this system can detect and correct 100% of the code errors which occur within the scope of its correction capabilities, should errors which exceed the system's correction performance occur, then correction and, of course, detection become very difficult. The simplest type of error correction system which could be used in digital audio equipment is shown in Fig. 6-15B.

In this example, no error correction is carried out, but if we compare error detection codes with error correction codes from the point of view of the amount of redundancy required, the former is far more effective. This is because a relatively small amount of redundancy will allow very effective error detection. If error compensation is carried out using interpolation (described below) then a degree of compensation may be achieved which renders errors inaudible.

The most desirable type of error compensation system for use in digital audio equipment is shown in Fig. 6-15C. Using this system, any code errors which occur may be detected and corrected provided that they fall within the limits of the capabilities of the error correction code. Code errors which exceed the performance of the error correction system are, however, still detected by the error detection code, and are then compensated. Not all the required redundancy is used for error correction alone. A small amount of the error correction performance is sacrificed, because part of the redundancy is diverted to the detection of errors which exceed the performance of the error correction code. The highly complex nature of the code errors which occur in PCM tape recorders necessitates this kind of system, so that compensation may be carried out in high quality equipment.

Code Error Detection[6]

There are, broadly speaking, two types of error detection system used in digital audio equipment.

Code Error Detection Using Parity Check. The most basic one bit parity check code error detection system is shown in Fig. 6-16A.

The parity bit P is constructed adding together all the information bits, K (bit), using modulo 2 addition. Thus, blocks of a length K + 1 (bit are created by adding the parity bit to the information bits. Code error detection is carried out during reproduction, when each block of information bits is again totaled using modulo 2 addition. The result is then checked for accuracy with the existing value of

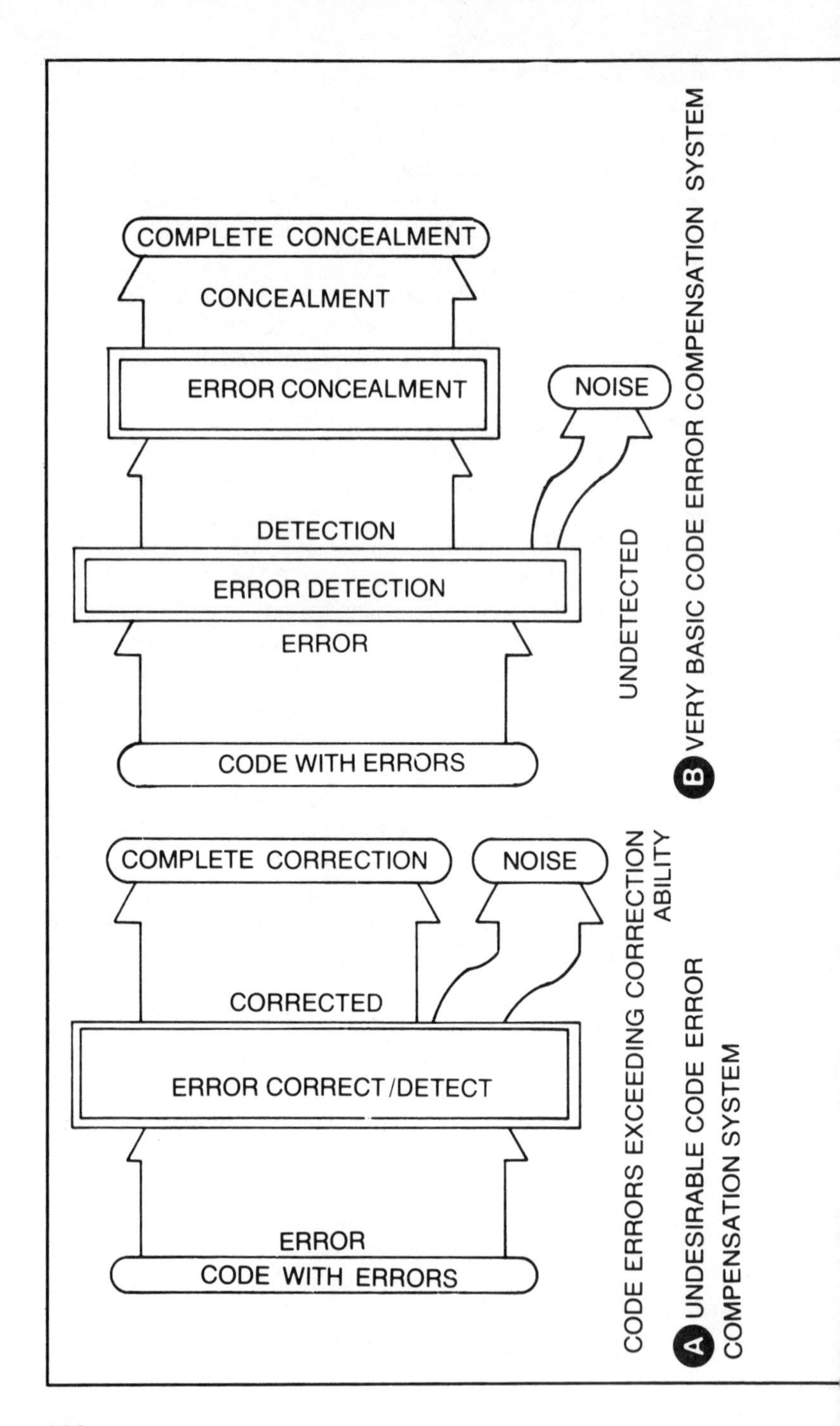

A UNDESIRABLE CODE ERROR COMPENSATION SYSTEM

B VERY BASIC CODE ERROR COMPENSATION SYSTEM

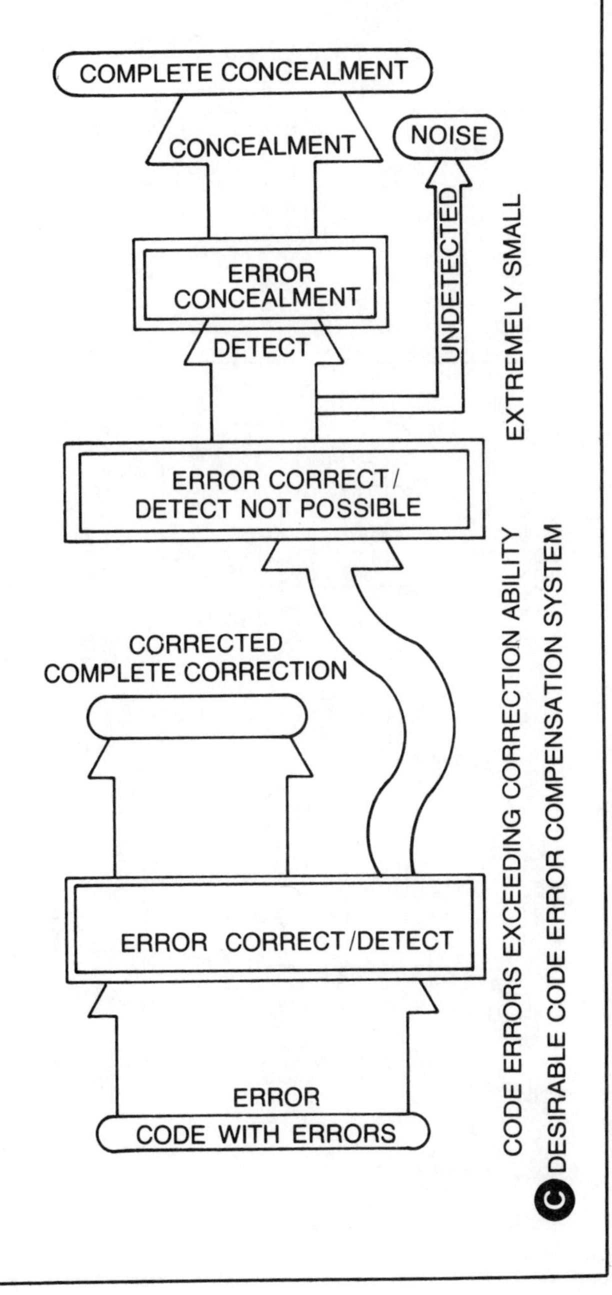

Fig. 6-15. Error Compensation for Digital Audio Equipment.

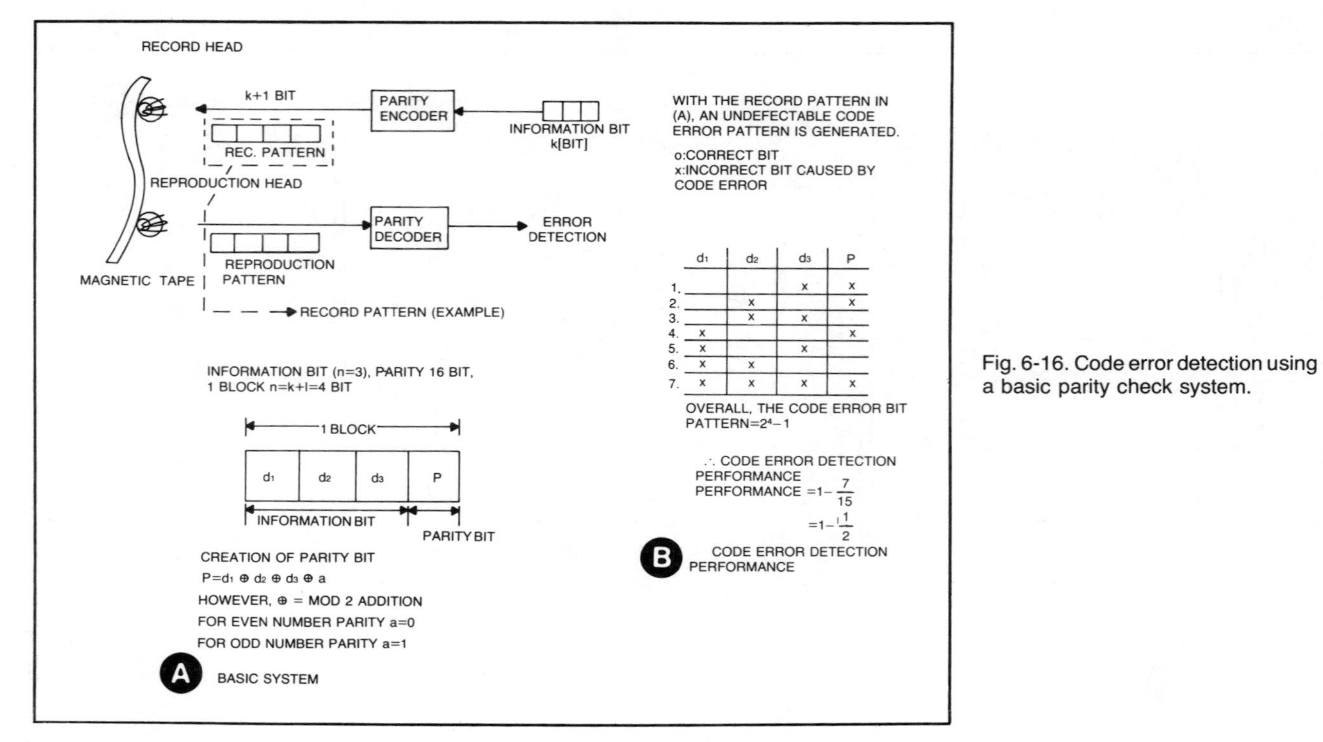

	d_1	d_2	d_3	P
1.			x	x
2.		x		x
3.		x	x	
4.	x			x
5.	x		x	
6.	x	x		
7.	x	x	x	x

Fig. 6-16. Code error detection using a basic parity check system.

the parity bit P. Even if only this simple detection system were used, the rate of detection would be 100% for a one bit code error in each block. Even on a 9 track computer tape which has 1600 bits/inch, the likelihood of a code error occurring at the same instant in time on several tracks is extremely small. For this reason, it is quite practicable to use one eight bit block of information bits plus one parity bit for the vertical axis of the tape.

However, as shown in Fig. 6-16B, the detection code performance is halved if one considers all the possible patterns for code errors which could occur in one block. It is, therefore, not suitable for use with PCM tape recorders which are extremely susceptible to the occurrence of multiple code errors within one block.

An extended parity check system is shown in Fig. 6-17A. One block is made up of m-1 information words and one parity word (where one word = n (bits)). The parity word is constructed using modulo 2 addition of the bits composing the m-1 information words in sequential order. It is then recorded along with the information words. Error detection is achieved by determining whether or not the block of information words and parity words conform to the relationship outlined above. In these circumstances, correction performance for all code error patterns in one block is 1-1/2n. As shown in the simple example in Fig. 6-17B this value is not dependent on the number of words m, but on the number of bits n from which one word is constructed. If the number of bits n, which make up one word, is sufficiently large (for example, n = 16), it is then possible to use this type of system in a PCM tape recorder.

However, if a continuous code error occurring at the same bit position in each word is generated during the record-reproduction process, the code error detection performance decreases dramatically. It is, therefore, necessary either to alter the bit order for each word, or to use bit sequential even and odd parity.

Code Error Detection Using CRCC (Cyclic Redundancy Check Code). Cyclic redundancy check codes are very widely used in various fields for code error detection, but they were first used for magnetic computer tape with a packing density of 800 bits/inch. Much research into the mathematical nature of CRCC has been carried out leading to extremely precise theoretical analysis. In the following pages, we shall discuss actual code error detection systems which employ CRCC theory.

In the following explanation, we will assume that we are using the polynomial expression shown in Table 6-3 for the bit chain containing the information to be transmitted. The basic process

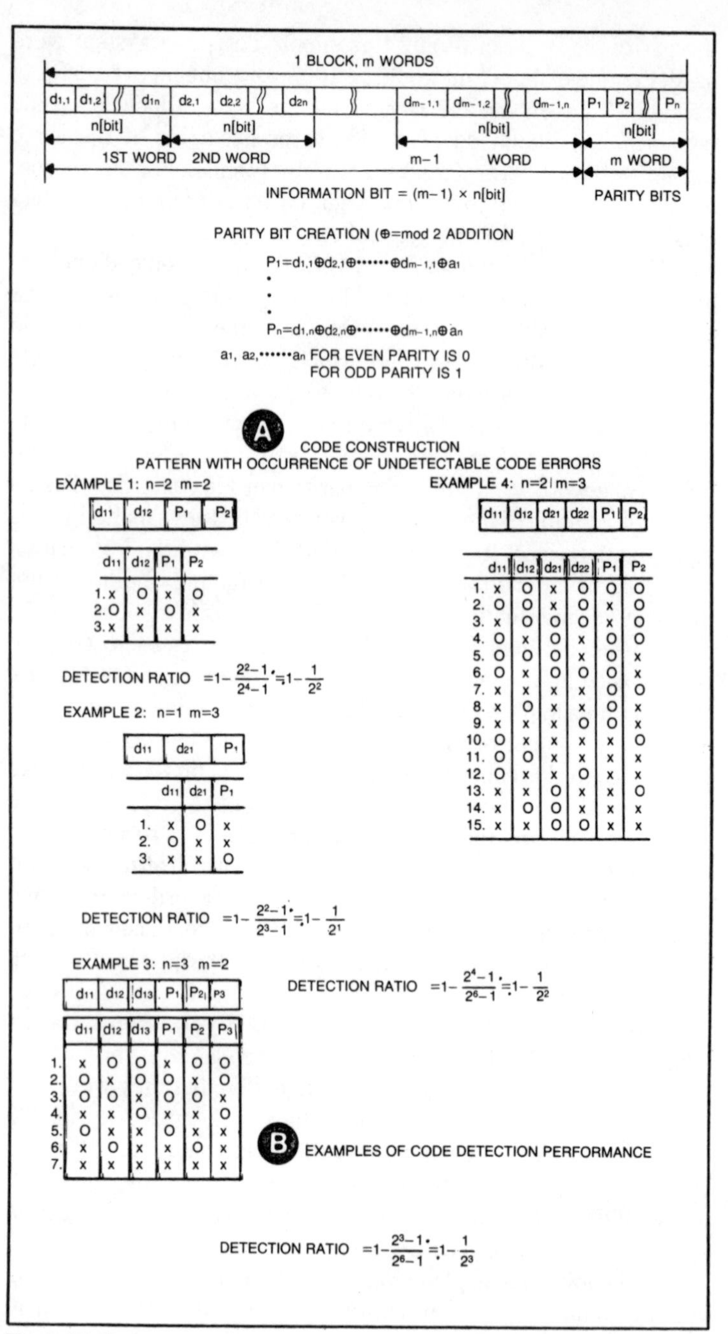

EXAMPLE 1: n=2 m=2

	d_{11}	d_{12}	P_1	P_2
1.	x	O	x	O
2.	O	x	O	x
3.	x	x	x	x

EXAMPLE 2: n=1 m=3

	d_{11}	d_{21}	P_1
1.	x	O	x
2.	O	x	x
3.	x	x	O

EXAMPLE 3: n=3 m=2

	d_{11}	d_{12}	d_{13}	P_1	P_2	P_3
1.	x	O	O	x	O	O
2.	O	x	O	O	x	O
3.	O	O	x	O	O	x
4.	x	x	O	x	x	O
5.	O	x	x	O	x	x
6.	x	O	x	x	O	x
7.	x	x	x	x	x	x

EXAMPLE 4: n=2 m=3

	d_{11}	d_{12}	d_{21}	d_{22}	P_1	P_2
1.	x	O	x	O	O	O
2.	O	O	x	O	x	O
3.	x	O	O	O	x	O
4.	O	x	O	x	O	O
5.	O	O	O	x	O	x
6.	O	x	O	O	O	x
7.	x	x	x	x	O	O
8.	x	O	x	x	O	x
9.	x	x	x	O	O	x
10.	O	x	x	x	x	O
11.	O	O	x	x	x	x
12.	O	x	x	O	x	x
13.	x	x	O	x	x	O
14.	x	O	O	x	x	x
15.	x	x	O	O	x	x

Fig. 6-17. Code error detection using extended parity check.

Table 6-3. Polynomial Expression of the Bit Chain.

Shows a bit chain of n [bits] as the n-1 polynomial below with x as variable

A Expression

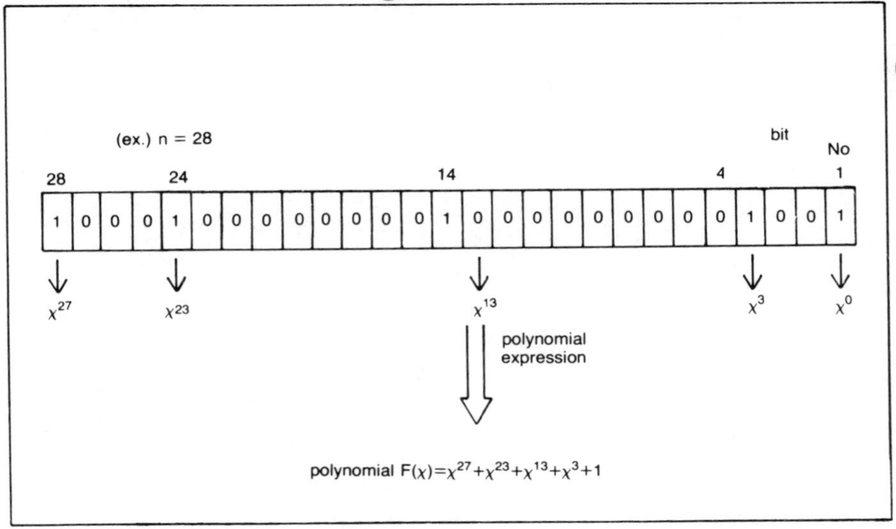

(ex.) n = 28

bit No: 28, 24, 14, 4, 1

1	0	0	0	1	0	0	0	0	0	0	0	0	1	0	0	0	0	0	0	0	0	0	1	0	0	1

χ^{27} χ^{23} χ^{13} χ^{3} χ^{0}

polynomial expression

polynomial $F(\chi)=\chi^{27}+\chi^{23}+\chi^{13}+\chi^{3}+1$

•All addition signs indicate module 2 addition

B Basic operation of a polynomial expression

	contents
1.	$1\chi^a+1\chi^a=0\chi^a$
2.	$1\chi^a+0\chi^a=0\chi^a+1\chi^a=1\chi^a$
3.	$0\chi^a+0\chi^a=0\chi^a$
4.	$-1\chi^a=\chi^a$

whereby code errors are detected using CRCC is shown in Fig. 6-18. The transmitted data is made up of the information bits k and the detection bits n - k; this is shown as the transmission polynomial $U(\chi)$, which is generated in the following manner.

First of all, the data or message polynomial $M(\chi)$, which contains the information is multiplied by χ^{n-k}, then it is divided by the specific polynomial $G(\chi)$ which includes the degrees of n - k order. Assuming that the quotient and remainder thus obtained are $Q(\chi)$ and $R(\chi)$, respectively, their relationship may be shown thus:

$$\chi^{n-k} M(\chi) = Q(\chi) G(\chi) + R(\chi)$$

The transmission polynomial $U(\chi)$ may be calculated by adding the remainder $R(\chi)$ to the above equation, as shown below:

$$\begin{aligned} U(\chi) &= \chi^{n-k} M(\chi) + R(\chi) \\ &= Q(\chi) G(\chi) + \underbrace{R(\chi) + R(\chi)}_{0 \text{ (as shown in Table 6-3(B))}} \\ &= Q(\chi) G(\chi) \end{aligned}$$

Thus, the data polynomial $M(\chi)$ is divisible by the abovementioned specific polynomial $G(\chi)$, and can therefore be converted into the transmission polynomial $U(\chi)$ which is subject to fixed conditions. It can then be recorded. The series of numerical values used to actually generate the transmission polynomial $U(\chi)$ is shown in Fig. 6-18B.

The transmitted data which has been recorded in this way is reproduced as the received data polynomial $V(\chi)$. This is produced by the addition of the error polynomial $E(\chi)$, which indicates the code error patterns produced during record.

Code error detection is then carried out by dividing the received data polynomial $V(\chi)$ by the polynomial $G(\chi)$, which is the same as that used at the time of transmission. Remainders of zero or not zero are then detected, and hence errors can be identified.

$G(\chi)$ is known as the generation polynomial, and it is selected so that the error polynomial $E(\chi)$, which consists of the pattern of errors created, cannot be divided by $G(\chi)$.

In many instances, digital audio equipment uses polynomial expressions based on the recommendations of the C.C.I.T.T. (Comité Consultatif International Télégraphique et Téléphonique). Therefore, a one chip IC which uses the recommended polynomials is marketed; details are given in Fig. 6-19.

Occasionally, there may be a failure to detect a code error, where for example, the error polynomial $E(\chi)$ can be divided by

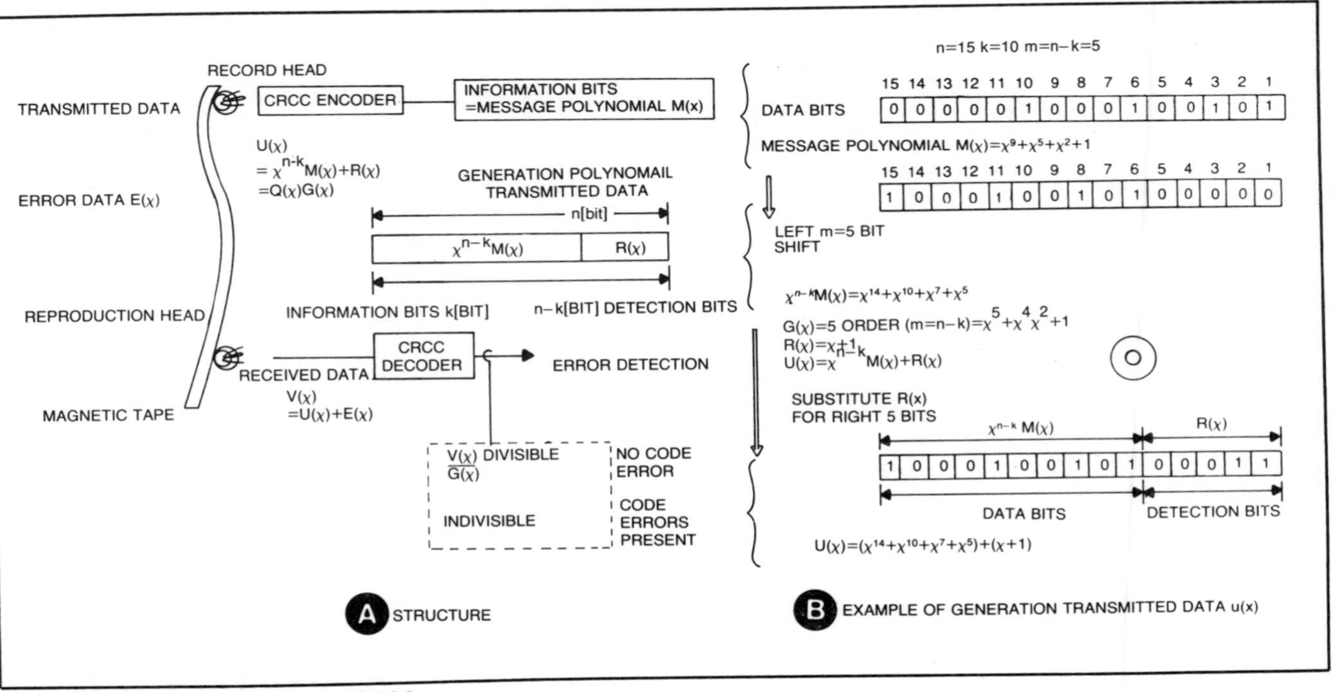

Fig. 6-18. Code error detection using CRCC.

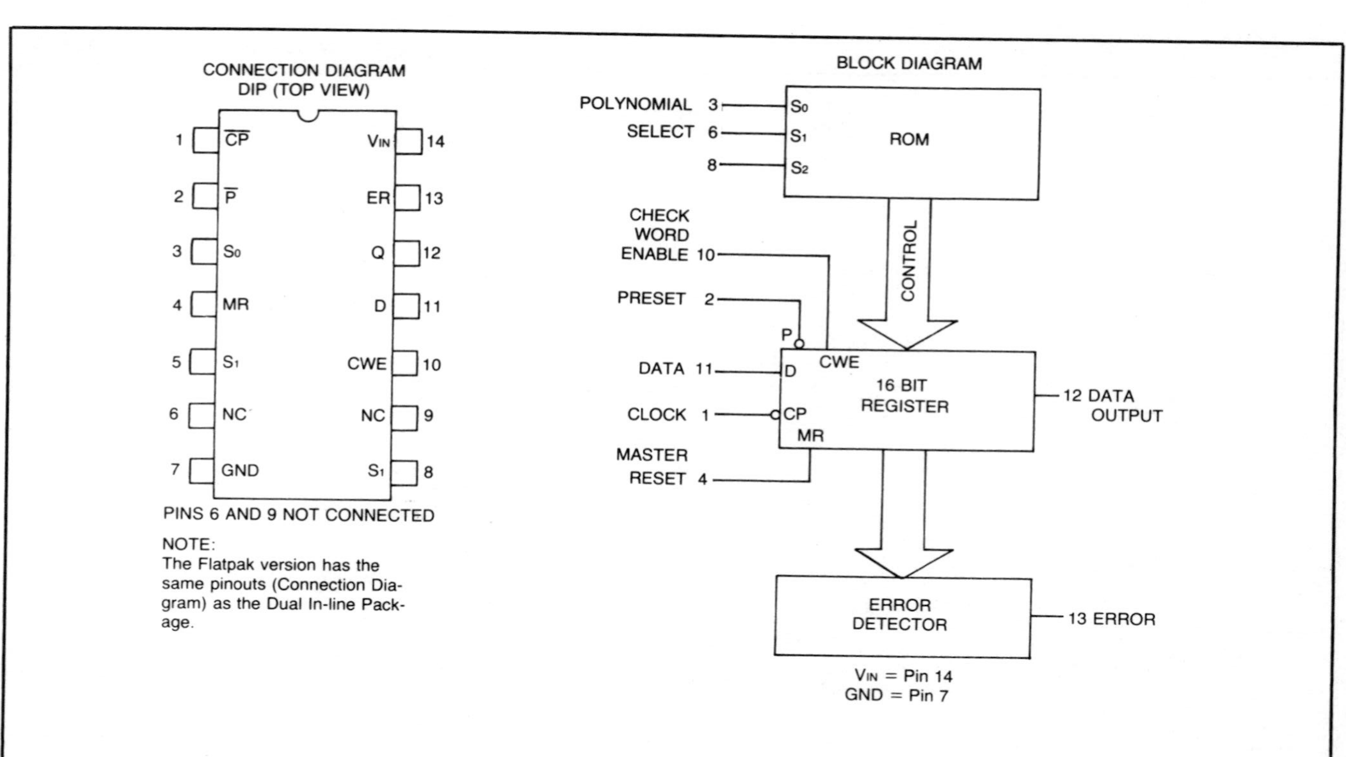

CONNECTION DIAGRAM
DIP (TOP VIEW)
1 $\overline{CP}$
2 $\overline{P}$
3 S_0
4 MR
5 S_1
6 NC
7 GND
V_{IN} 14
ER 13
Q 12
D 11
CWE 10
NC 9
S_1 8
PINS 6 AND 9 NOT CONNECTED
NOTE:
The Flatpak version has the same pinouts (Connection Diagram) as the Dual In-line Package.
BLOCK DIAGRAM
POLYNOMIAL 3
SELECT 6
8
S_0
S_1
S_2
ROM
CONTROL
CHECK WORD ENABLE 10
PRESET 2
DATA 11
CLOCK 1
MASTER RESET 4
P
D
CWE
CP
MR
16 BIT REGISTER
12 DATA OUTPUT
ERROR DETECTOR
13 ERROR
V_{IN} = Pin 14
GND = Pin 7

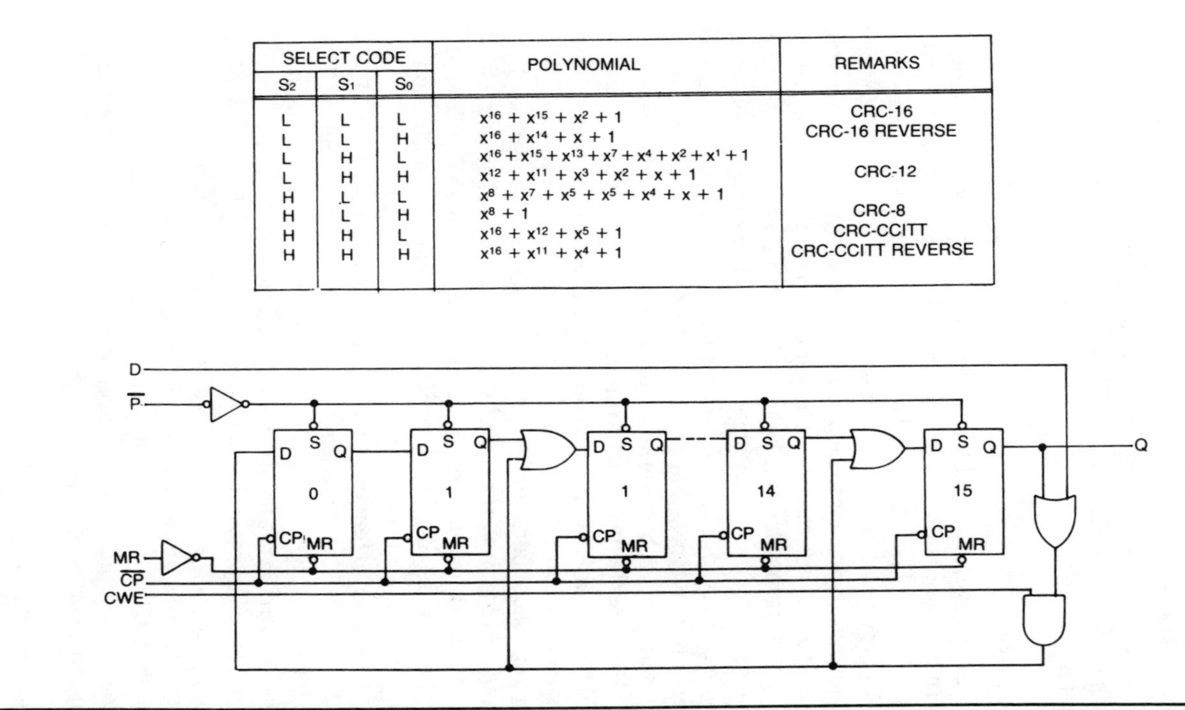

SELECT CODE			POLYNOMIAL	REMARKS
S_2	S_1	S_0		
L	L	L	$x^{16} + x^{15} + x^2 + 1$	CRC-16
L	L	H	$x^{16} + x^{14} + x + 1$	CRC-16 REVERSE
L	H	L	$x^{16} + x^{15} + x^{13} + x^7 + x^4 + x^2 + x^1 + 1$	
L	H	H	$x^{12} + x^{11} + x^3 + x^2 + x + 1$	CRC-12
H	L	L	$x^8 + x^7 + x^5 + x^5 + x^4 + x + 1$	
H	L	H	$x^8 + 1$	CRC-8
H	H	L	$x^{16} + x^{12} + x^5 + 1$	CRC-CCITT
H	H	H	$x^{16} + x^{11} + x^4 + 1$	CRC-CCITT REVERSE

Fig. 6-19. An example of using an IC CRCC generator checker (Fairchild 9401).

$G(\chi)$ despite the fact that there is an error. This may happen in a situation where a received data polynomial $V(\chi)$ is generated, which can be divided by $G(\chi)$ despite the fact that it includes an error polynomial $E(\chi)$ which is not zero. The likelihood of this situation is $\frac{1}{2}^{n-k}$. However, the efficacy of code error detection is not decreased by the occurrence of error patterns which appear to have regularity when examined by parity check. Thus, this system is still a viable one for error detection.

Interleave

Interleave, or scramble as it is sometimes called, is a method of altering the order of the data composing the PCM code chain according to a fixed pattern. The data is then recorded in this new interleaved sequence, and is subsequently de-interleaved (or unscrambled) during reproduction so that it returns to its original state. Using this technique, the nature of code errors generated by the recording medium may be altered (i.e. burst errors will become random errors). This technique is, therefore, often used as a supplementary aid to achieving effective error compensation. There are two types of interleave described below, which may be used in this way: bit interleave and block (or word) interleave.

Bit Interleave. The incorporation of bit interleaving into a PCM tape recorder allows the conversion of burst errors into errors of a random nature. This then enables random error correction codes or short burst error correction codes to be used for the correction of what were originally long burst errors. The basic block diagram for a bit interleave circuit is shown in Fig. 6-20A. Its effectiveness is illustrated in Fig. 6-20B and (C) which show a simulation on a computer.

Block Interleave. In bit interleave the order of the bits is altered, and in block interleave the same principle is applied. The position of the blocks or words, which are of uniform length, is altered according to a fixed pattern. Using block interleave the burst errors caused by recording onto magnetic tape may be sequenced in such a way that, on playback, correct words alternate with error words at a fixed distance. A typical pattern is shown in Fig. 6-21. As described in the next section, this procedure is invaluable in aiding code error concealment. However, it is rarely used in conjunction with error correction codes.

Code Error Concealment [7]-[10]

Code error concealment, like code error detection, is a funda-

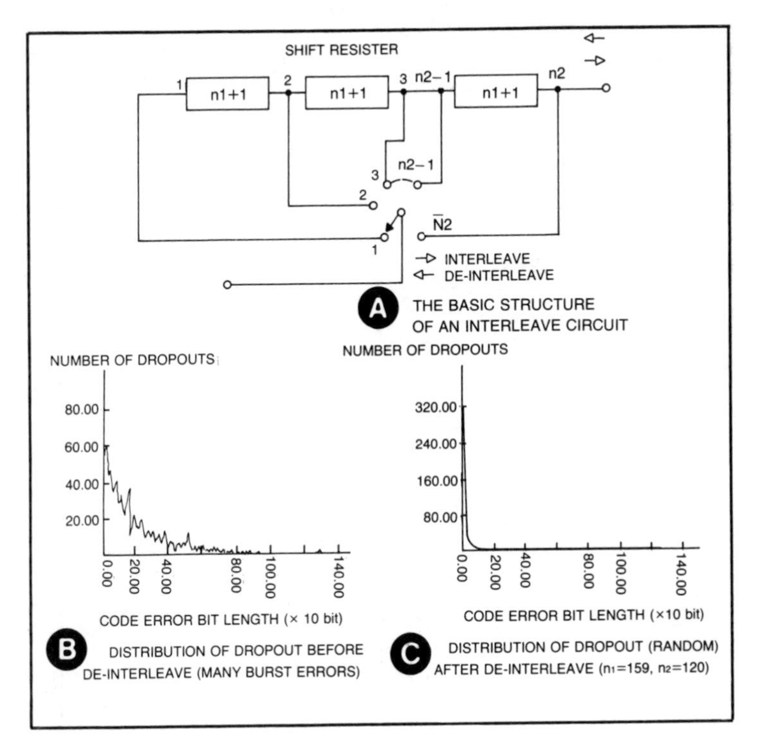

Fig. 6-20. Bit interleave.

mental technique for code error compensation. It is shown in Figs. 6-15B and C.

Methods of Code Error Concealment. The reproduction sound quality may be protected from deteriorating by using some method of concealment for words containing code errors. It is, in fact, possible to design an effective system which uses no error correction, but instead employs concealment for error words. The basic principles for an error concealment system are shown in Fig. 6-22.

Using a concealment system, a continuous code error during reproduction caused by a dropout, for example, would be dealt with in the following way. The correct words and the error words would be re-positioned so that the burst error is split up, and then concealment would be carried out by taking the sampled value of the preceding correct word.

If we then assume that, for the sine wave g(t) in Fig. 6-22, gi^{θ} is the original sampled value for each word i with a phase of θ, we can calculate gi^{θ} using the following equation:

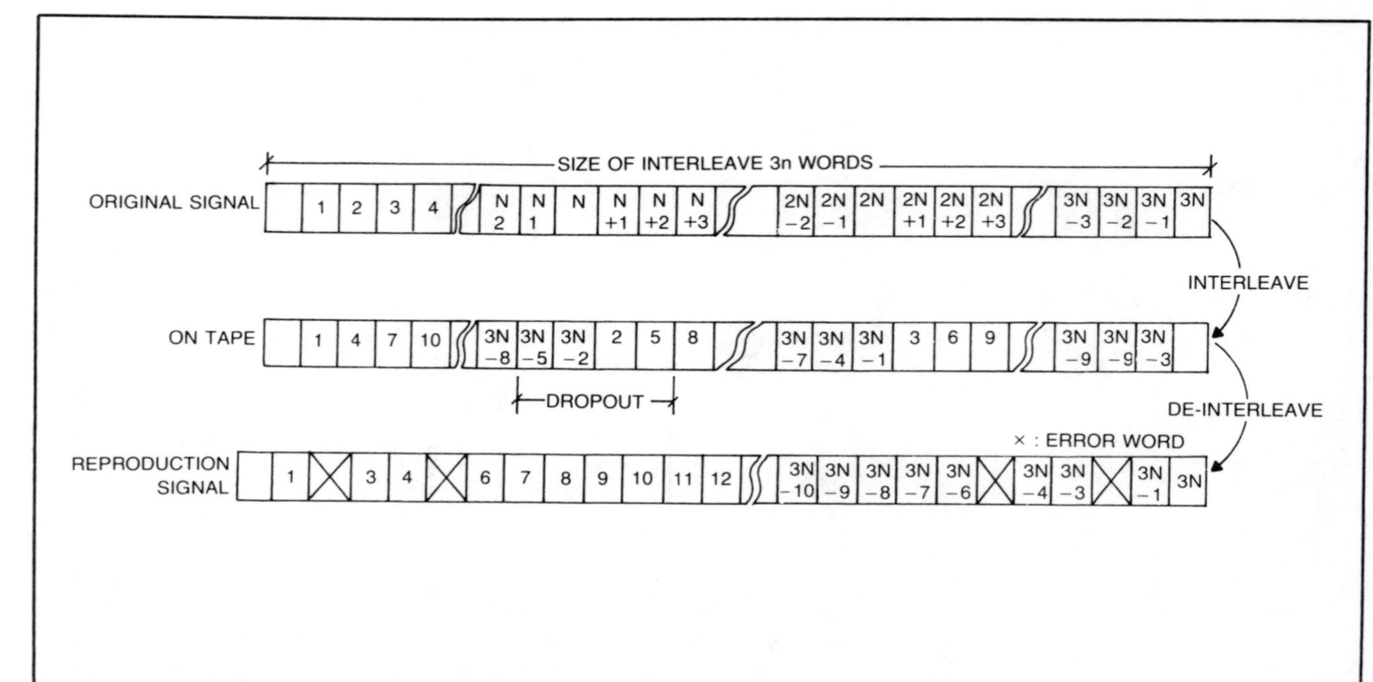

Fig. 6-21. An example of the structure of block interleave.

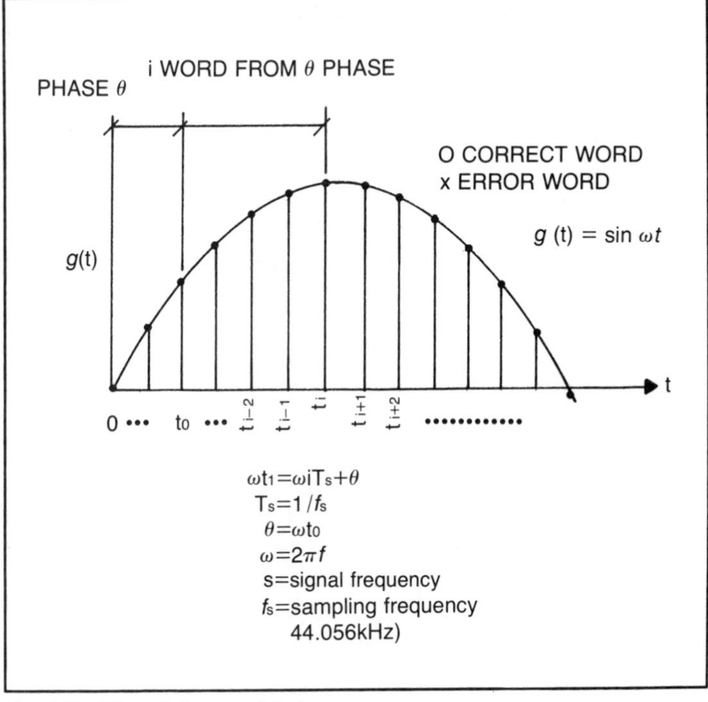

Fig. 6-22. Interpolation and interleave.

$$gi^{\theta} = g(ti) = \sin(\omega ti) = \sin(\omega iTs + \theta)$$

Then, by using whichever concealment method seems most appropriate, we can calculate $\hat{g}i^{\theta}$, which is the concealment value for the error word i, in the following manner:

1. Muting
   Substitution of a value of 0 for the error word. $\hat{g}i^{\theta} = 0$
2. Previous word hold (O order polynomial interpolation)
   Substitution of the previous sampled value.

$$\hat{g}i^{\theta} = gi-1^{\theta}$$

3. Linear interpolation (1st order polynomial interpolation).

$$\hat{g}i^{\theta} = a_{-1}\, gi-1^{\theta} + a_{1}\, gi + 1^{\theta}$$

   where $a_{-1} = a_{1} = 1/2$
4. Third order polynomial interpolation
   $\hat{g}i^{\theta} = a_{-3}\, gi-3^{\theta} + a_{-1}\, gi-1^{\theta} + a_{1}\, gi + 1^{\theta} + a_{3}\, gi + 3^{\theta}$

   where $a_{-3} = a_{3} = 1/16$, $a_{-} = a_{1} = 9/16$

5. Fifth order polynomial interpolation

$$\hat{g}i^{\theta} = a_{-5}\, gi-5^{\theta} + a_{-3}\, gi-3^{\theta} + a_{-1} gi-1^{\theta} + a_{1}\, gi+1^{\theta} + a_{3} gi+3^{\theta} + a_{5}\, gi+5^{\theta}$$

where $a_{-5} = a_5 = 3/256$, $a_{-3} = a_3 = -25/256$, $a_{-1} = a_1 = 150/256$

The effects caused in the reproduction signal when using concealment may be demonstrated by the calculations detailed below.

For the purposes of appraisal, we define the power difference between a correct sine wave and its compensated reproduction signal, and the power ratio R of the correct sine wave as the error noise power ratio.

Pi is the error noise power when code errors are generated at the code point in one sample which is the ith word from phase θ, as in Fig. 6-22. Pi may be calculated approximately as shown in the following equation, from the response characteristics of an ideal low pass filter. This is shown in Fig. 6-23, and the impulse contains correct sampled values as well as the wave height of interpolated sampled values.

$$Pi^{\theta} = \left\{gi^{\theta} - \hat{g}i^{\theta}\right\}^2 \int_{-\infty}^{\infty} \left\{\frac{\sin(\pi f_{st})}{\pi f_{st}}\right\}^2 at = \left\{gi^{\theta} - \hat{g}i^{\theta}\right\}^2 T_s$$

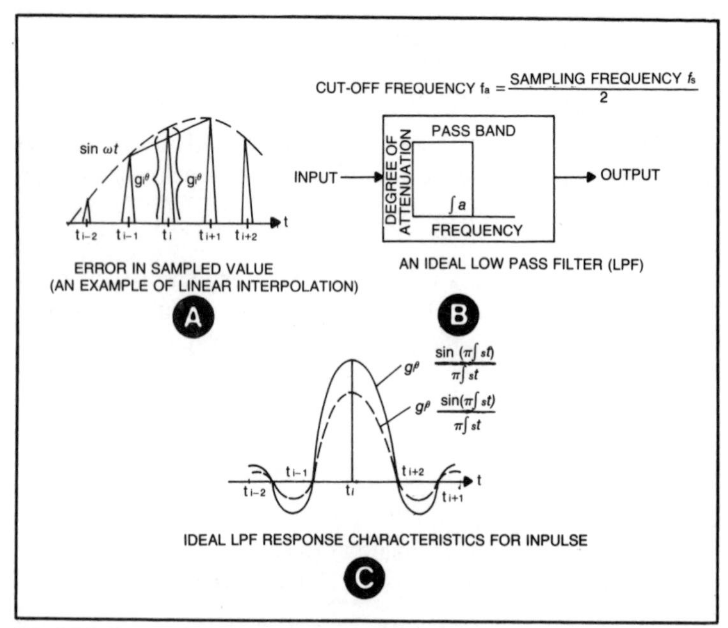

Fig. 6-23. Error noise power at one sampling point.

$*Pj^\theta$, which is the error noise power for continuous code errors for j words from phase θ, can be calculated as follows:

$$*Pj^\theta = \sum_{i=0}^{J\text{-}i} Pi^\theta$$

Phase θ where code errors are generated has the same probability of occurrence within the scope of $0 \leqq \theta \leqq 2\pi$.

Therefore, we may demonstrate $\overline{Pj}$, which is the anticipated value of $*Pj^\theta$, in the following manner:

$$\overline{Pj} = \frac{1}{2\pi} \int_0^{2\pi} *Pj^\theta \, d\theta$$

If we assume that, for a burst code error of N bits where code errors of m_1 through m_2 words have been generated, the probability of code error generation for j words or $m_1 \leqq j \leqq m_2$ is qj, then the error noise power $\overline{P}$ may be calculated thus:

$$\overline{P} = \sum_{j=m_1}^{m_2} \overline{Pj}\, qj$$

Then, the required error noise power ratio R may be expressed as follows:

$$R = 10 \log_{10}\left\{\overline{P}/\int_o^T \sin^2 \omega t \, dt\right\} = 10 \log_{10} (2f\overline{P})$$

Using the above equation, Fig. 6-24 indicates the results which may be expected when a 100 bit burst error (N=100) occurs, and shows the efficacy of the various concealment techniques. These types of concealment are very effective for error compensation of audio signals because the interphase values of the sampling point periods are very robust.

The Effect of Concealment Length. When errors are corrected using concealment words distributed along the time axis, the nature of this dispersion or the time points of dispersal has an effect on the reproduction audio signal. It is, therefore, necessary to understand the characteristics of the changes which are caused in the reproduction signal by concealment.

Let us consider an example where the error words are concealed using linear interpolative techniques, and where these words are dispersed over time periods ranging from 100 Hs to 2.5 mins after the operation of various types of interleave. In this case, the

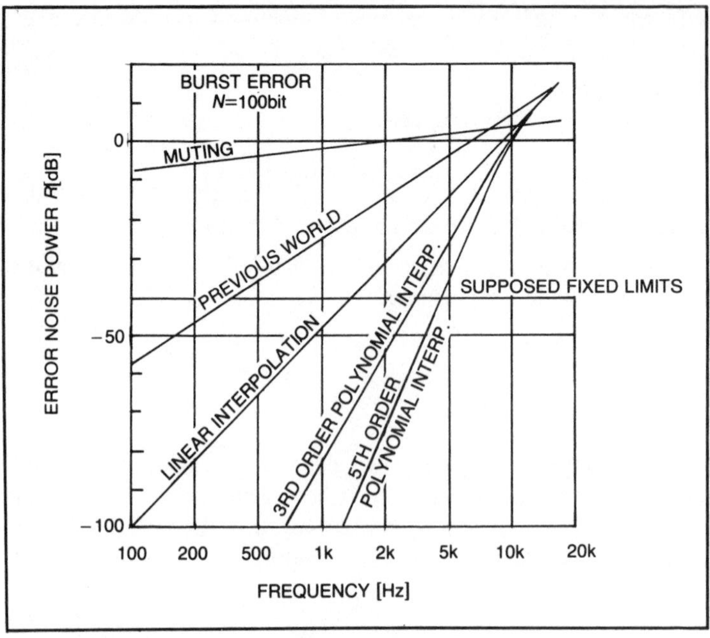

Fig. 6-24. Error noise power ratios for all types of interpolation.

fixed frequency limits obtained from auditory experiments are given in Fig. 6-25 along with a summary of the experiment.

First, we must examine the changes apparent in the fixed frequency limit in Fig. 6-24 caused by the differences between concealment techniques. If we then compare this to the differences in the dispersal nature or dispersal time of the concealed words, we can see that the changes are rather small. We may, therefore, conclude that when investigating the concealment techniques which may be used for code errors, sufficient attention must be paid to their selection in order to make the investigation effective.

Error Concealment Codes Used In Digital Audio Equipment

At present, there are a number of effective code error concealment techniques used in digital audio equipment. Below, the following concealment systems are described: pointer erasure, crossword codes, cross-interleave codes and adjacent codes.

The Pointer Erasure Method. Hardware design for this technique is comparatively simple, so it is a practical code error concealment system which has excellent error concealment performance. A typical example of code structure is shown in Fig. 6-26.

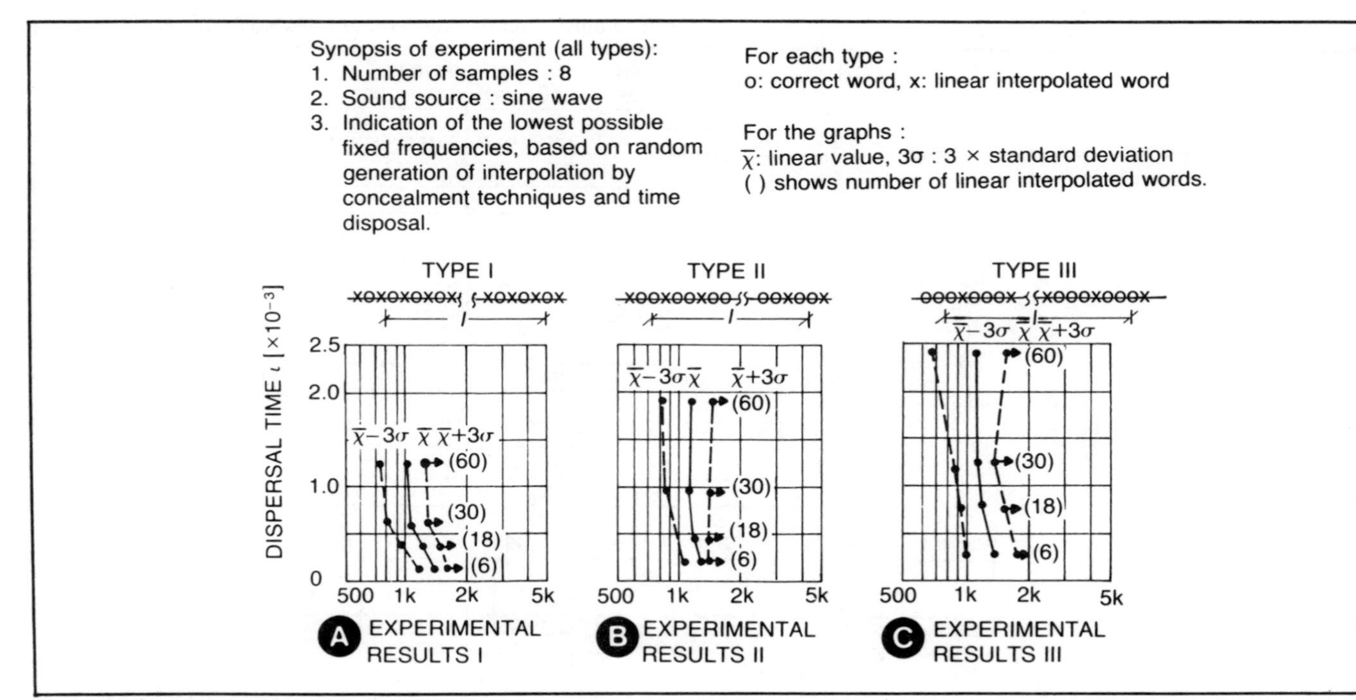

Fig. 6-25. Auditory experiment results.

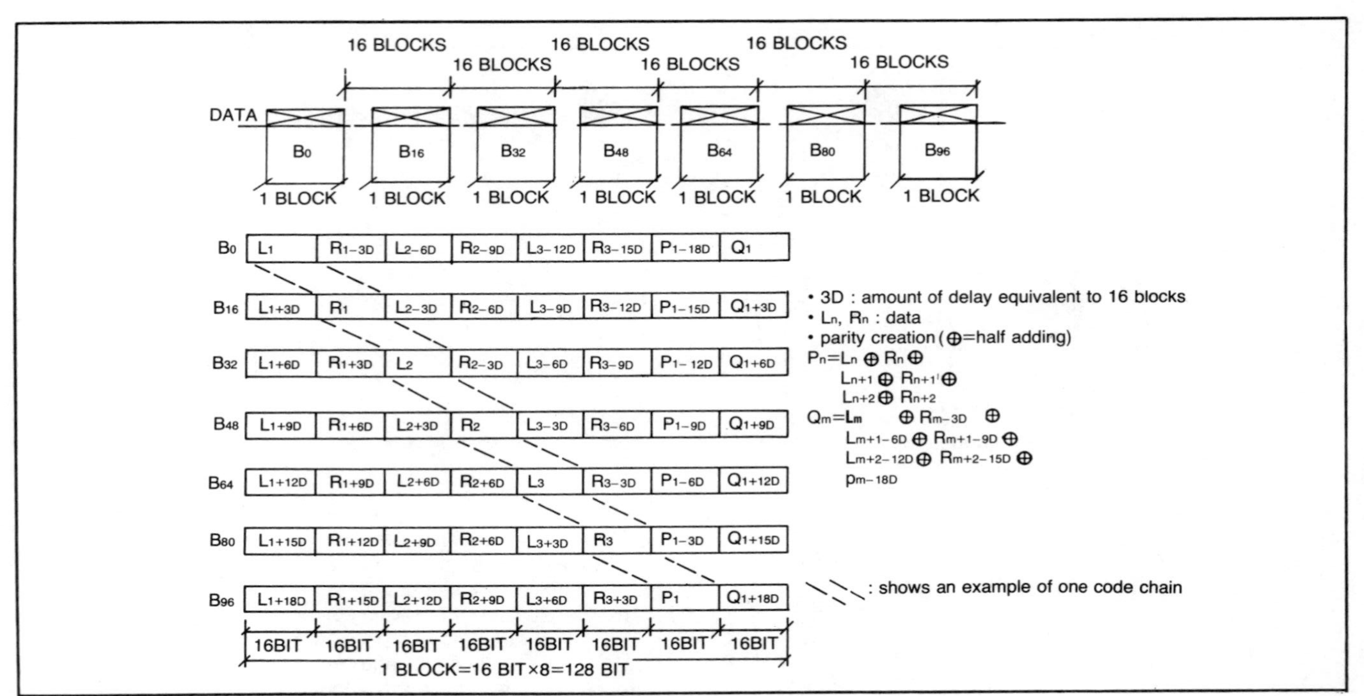

Fig. 6-26. Code error concealment using the pointer erasure method.

The PCM code chain (in the diagram: the left and right channels are each represented by 3 words: L_n, R_n, L_{n+1}, R_{n+1}, L_{n+2}, R_{n+2}), and the parity words (Pn), which are created from the PCM code chain are re-positioned with a 16 block delay, as shown in the diagram. Then new parity words (Qn) are created from the information within the blocks. These form one block, and are recorded on magnetic recording tape, along with the other data.

The following operations are carried out when a code error is detected in the blocks (B0 to B96) where the encoded data (L_1, L_2, ... P_1) and the parity words are recorded. (1) Concealment is possible when there is a code error in one of the seven blocks shown in Fig. 6-26. Should a code error occur in block B_{16}, this is detected by parity word $Q_1 + 3D$. Furthermore, the fact that R_1, for example, is the wrong value may also be ascertained. The concealment process is then carried out using the information from parity word P_1.

If we assume that the code pattern for R_1 when the data is incorrect is R_1, then the correct value for R_1 may be calculated in the following manner.

$$R_1 = R_1' \oplus (R_1' \oplus L_1 \oplus R_1 \oplus L_2 \oplus R_2 \oplus L_3 \oplus R_3 \oplus P_1)$$

Here the sign $\oplus$ indicates half-adding.

It is evident from the above equation that burst errors of less than 16 blocks may be concealed using this method. (2) In some cases a code error may occur in two adjacent blocks, for example, when we assume that there is a code error in both B_{32} and B_{48} in Fig. 6-26. The existence of incorrect data at locations L_2 and P_2 can be detected by the use of $Q_1 + 6D$ and $Q_1 + 9D$, but concealment of these two errors is not possible. In this situation, the only way to reduce the effect of the errors would be to employ linear interpolation. One word concealment using interpolation would be carried out by using the preceding and following correct words. For example, L_1 and L_3 would be used to approximate L_2; R_1 and R_3 would be used to approximate R_2.

$$L_2 = (L_1 \oplus L_3)/2$$
$$R_2 = (R_1 \oplus R_3)/2$$

Should errors occur in three or even four adjacent blocks, the code errors themselves would be detected in the manner outlined above. It would then be possible to carry out one or two word concealments again, using the preceding and following correct words and one of the various concealment techniques. This would, however, lead to a certain degradation of quality.

The pointer erasure method of code error correction is, therefore, a system with an extremely simple construction. It is particularly well-suited for dealing with burst errors because of its capacity to carry out efficient error correction and concealment. It is, in fact, a code error correction code suitable for use in PCM tape recorders.

Crossword Code[(4),(11)~(14)] When designing a digital audio equipment it is of the utmost importance to select an error correction code which is capable of detecting code errors which exceed the correction performance of the error correction code. This principle is outlined in Fig. 6-15C.

To achieve an error correction code which may detect errors which exceed its correction performance, it is possible to use an error correction code combined with an error detection code such as a Cyclic Redundancy Check Code (CRCC). However, a more sophisticated solution to this problem may be found in crossword codes, which achieve the requisite levels of detection and correction using only one code.

Crossword codes were developed specifically for use in digital audio equipment: their particular advantage lies in their ability to accurately detect code errors which exceed their intrinsic correction performance. In addition, crossword codes are constructed in such a way that they are restricted to minimum limits for words which need concealment.

This type of code is considered to be particularly suited to the correction of short random burst code errors where the average length of error is less then one word. In other words, they are suitable for the correction of code errors generated in long bursts. The basic principles behind crossword codes are outlined in Fig. 6-27.

The method of coding itself is relatively simple, as can be seen from section (b). Parity checking is carried out horizontally and vertically, in the same manner as shown previously in Fig. 6-17. The detection bits thus created are appended to the information bits, and recorded as one block.

The particular advantage of the crossword code lies in the decoding process. As shown in Fig. 6-27C, a syndrome word is created from the blocks of reproduction data, which is possible to carry out all detection and correction processes using only the information from this syndrome word.

If there is no code error in the reproduction block, then all the bits making up all the syndrome words will be zero. However, as shown in the diagram, the existence of a code error will be indicated by the fact that the bits composing the relevant syndrome word will not all be zero. As a result, the correction process is then carried out as shown in the following example.

The example in Fig. 6-27 shows syndrome words with the following relationships:

$$S_1 = S_3 \neq 0 \text{ and } S_2 = S_4 = 0$$

From this we can deduce that a code error exists in U_1. In these circumstances, from the operations

$$U_1 \oplus S_1 = M_1 \oplus E_1 \oplus E_1 = M_1$$

or:

$$U_1 \oplus S_3 = M_1 \oplus E_1 \oplus E_1 = M_1$$

We can correct these code errors. This process is shown in Fig. 6-27D which indicates the relationship between the code errors and syndrome words.

However, as shown in the same diagram, we may come across a situation where the information contained in the syndrome word is the same for different code error patterns. This state of affairs is shown in Fig. 6-27D i) and ii). If the correction process is carried out in the standard manner under these conditions, the result will be a failure to correct. In the diagram under discussion, we may see that the patterns of the three code errors are identical and have occurred at the same time. The probability of such an error pattern actually occurring is, of course, extremely low, so low in fact that it is possible to ignore it for all practical purposes when considering error correction. However, from a theoretical point of view, it is not good practice to ignore such occurrences no matter how low their incidence.

For this reason, a crossword code will carry out the following functions based on the decisions of a decode algorithm:

1. Thoroughly investigate the combination of syndrome values when a code is suspected.
2. Carry out the correction process when the possibility of accurate correction is sufficiently high.
3. Carry out concealment procedures when the possibility of accurate correction is sufficiently low.

The general structure of the record blocks for the crossword code and its various divisions are shown in Table 6-4. The correct decode algorithm for Io (8, 4, 3,) is shown in Fig. 6-28.The right-

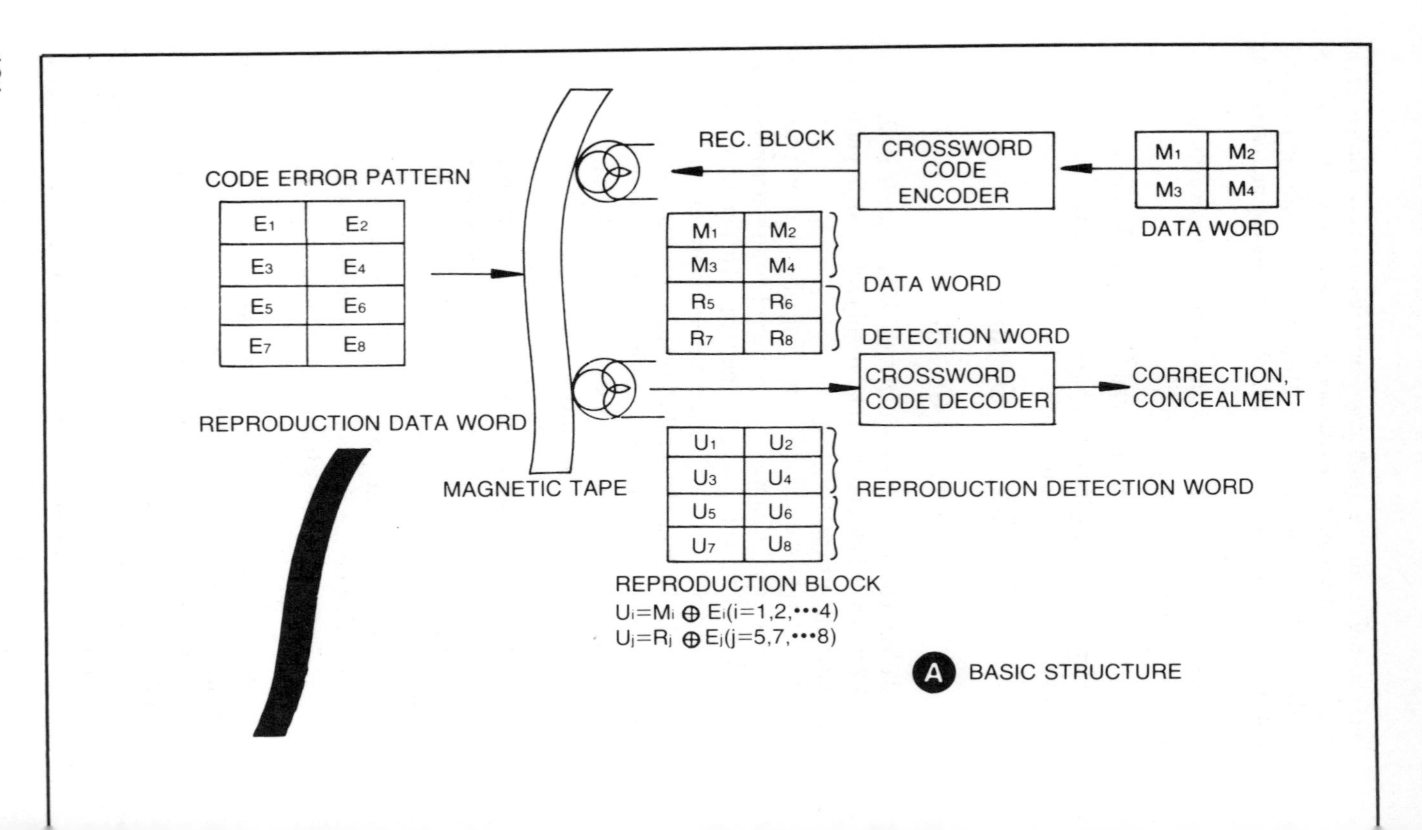

REC. BLOCK
CROSSWORD CODE ENCODER
M1 M2
M3 M4
DATA WORD
CODE ERROR PATTERN
E1 E2
E3 E4
E5 E6
E7 E8
M1 M2
M3 M4
R5 R6
R7 R8
DATA WORD
DETECTION WORD
CROSSWORD CODE DECODER
CORRECTION, CONCEALMENT
REPRODUCTION DATA WORD
MAGNETIC TAPE
U1 U2
U3 U4
U5 U6
U7 U8
REPRODUCTION DETECTION WORD
REPRODUCTION BLOCK
$U_i = M_i \oplus E_i (i=1,2,\cdots 4)$
$U_j = R_j \oplus E_j (j=5,7,\cdots 8)$
A BASIC STRUCTURE

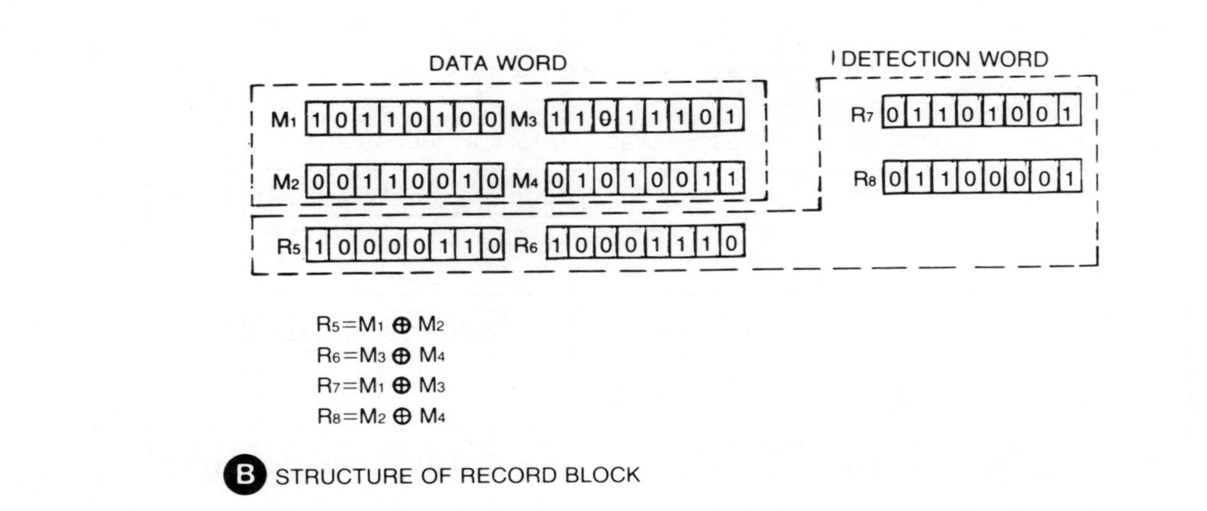

Fig. 6-27. An example of the structure of crossword code and its operation ($\oplus$ indicates modulo addition, the structure is I_0 [8,4,3]).

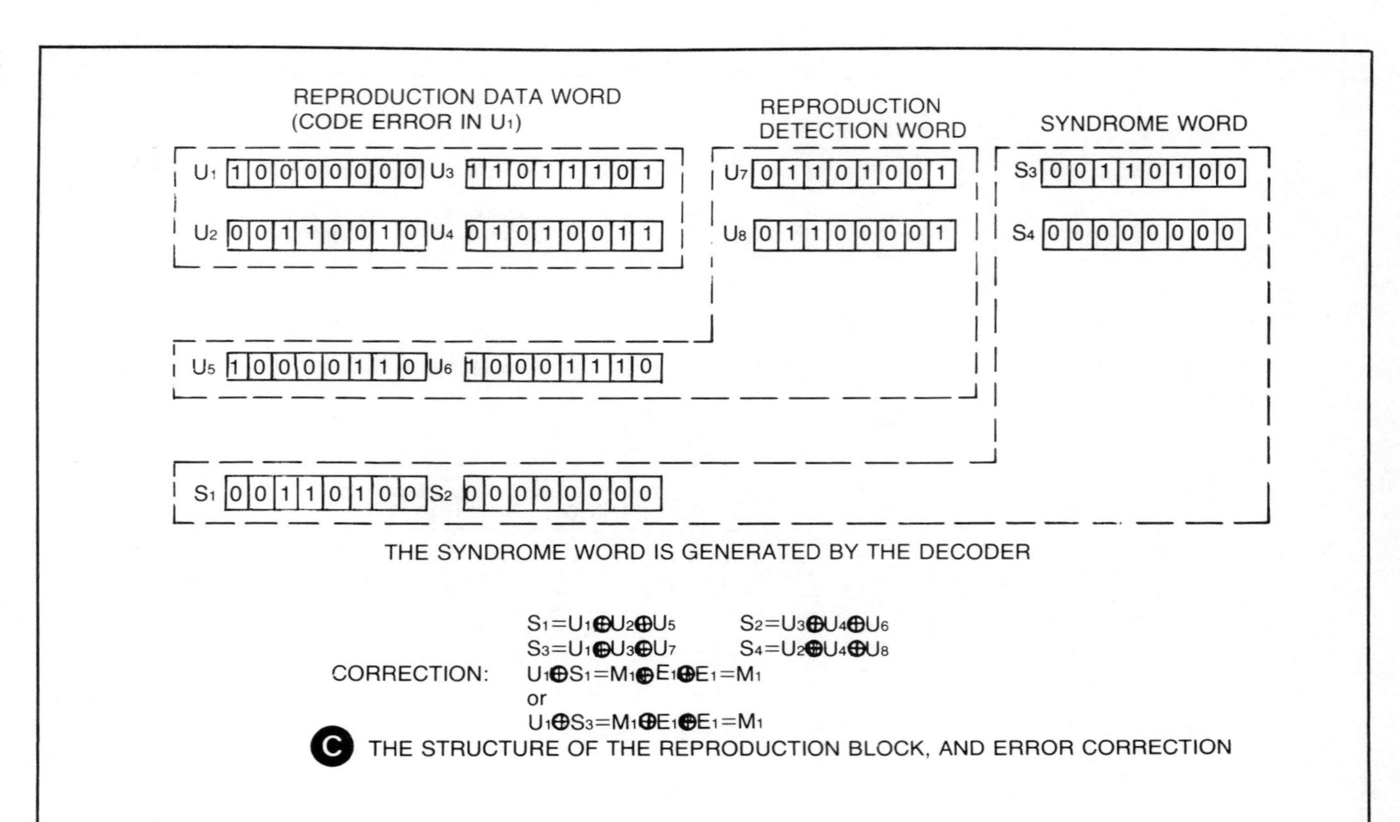

(C) THE STRUCTURE OF THE REPRODUCTION BLOCK, AND ERROR CORRECTION

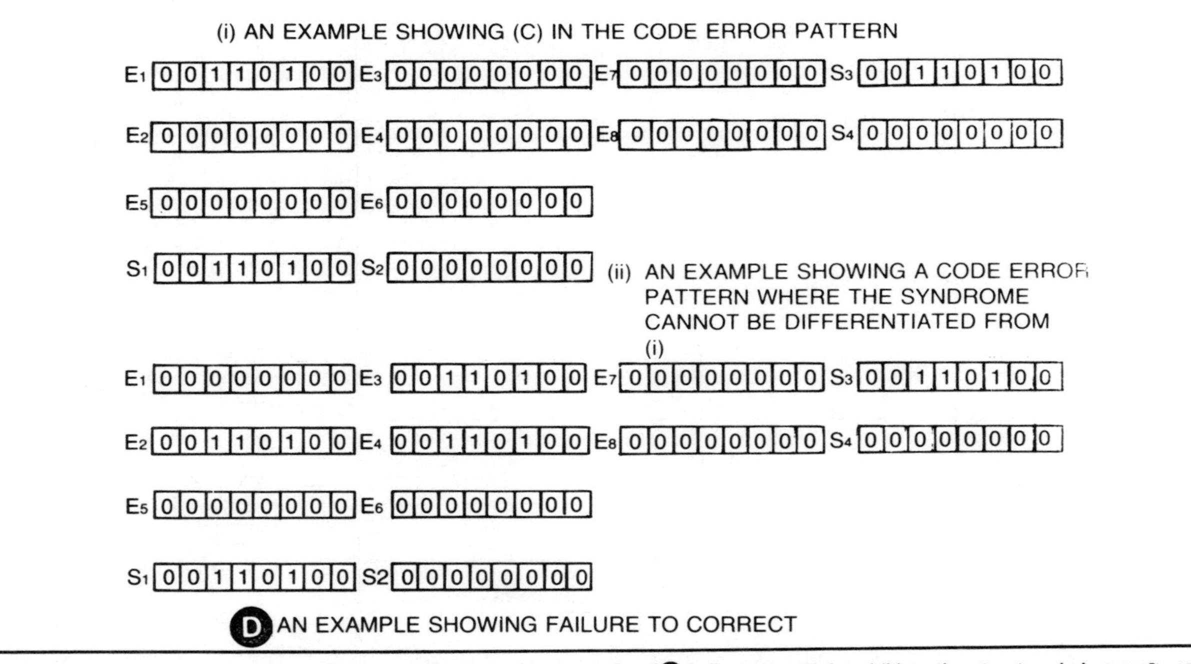

Fig. 6-27. An example of the structure of crossword code and its operation (⊕ Indicates modulo addition, the structure is Io 8,4,3]). (Continued from page 185.)

PROCESSING ORDER

Y = Yes
N = No

Start

$S_1=0$, $S_2=0$, $S_3=0$, $S_4=0$, $S_3=0$, $S_4=0$, $S_2=S_4$, $S_2=S_3$, $S_2=S_4$, $S_2=S_3+S_4$, $S_4=0$, $S_1=S_4$, $S_2=S_4$, $S_4=S_1+S_2$, $S_2=0$, $S_1=S_3$, $S_2=0$, $S_4=0$, $S_2=S_4$

PROCESSING ORDER	DECODE ORDER*				NUMBER OF ERROR WORDS									OCCURANCE RATIO (P : WORD ERROR RATIO)
	U_1	U_2	U_3	U_4	0	1	2	3	4	5	6	7	8	
(1)	0	0	0	0	1	2	1							$(1-p)^8+2p(1-p)^7+p^2(1-p)^6$
(2)	0	0	0	0		1	1							$p(1-p)^7+p^2(1-p)^6$
(3)	0	0	0	0		1								$p(1-p)^7$
(4)	0	0	0	1		1								$p(1-p)^7$
(5)	0	0	0	×			3	1						$3p^2(1-p)^6+p^3(1-p)^5$
(6)	0	0	1	0		1	1							$p(1-p)^7+p^2(1-p)^6$
(7)	0	0	0	1			1							$p^2(1-p)^6$
(8)	0	0	2	2			1							$p^2(1-p)^6$
(9)	0	0	×	0			3	1						$3p^2(1-p)^6+p^3(1-p)^5$
(10)	0	0	×	×				8	5	1				$8p^3(1-p)^5+5p^4(1-p)^4+p^5(1-p)^3$
(11)	0	1	0	0		1	1							$p(1-p)^7+p^2(1-p)^6$
(12)	0	0	0	1			1							$p^2(1-p)^6$
(13)	0	1	0	1			1							$p^2(1-p)^6$
(14)	0	×	0	0			3	1						$3p^2(1-p)^6+p^3(1-p)^5$
(15)	0	×	0	×				8	5	1				$8p^3(1-p)^5+5p^4(1-p)^4+p^5(1-p)^3$
(16)	1	0	0	0		1	1							$p(1-p)^7+p^2(1-p)^6$
(17)	1	0	0	0			1							$p^2(1-p)^6$
(18)	1	0	0	1			1							$p^2(1-p)^6$
(19)	1	0	0	×				3	1					$3p^3(1-p)^5+p^4(1-p)^4$

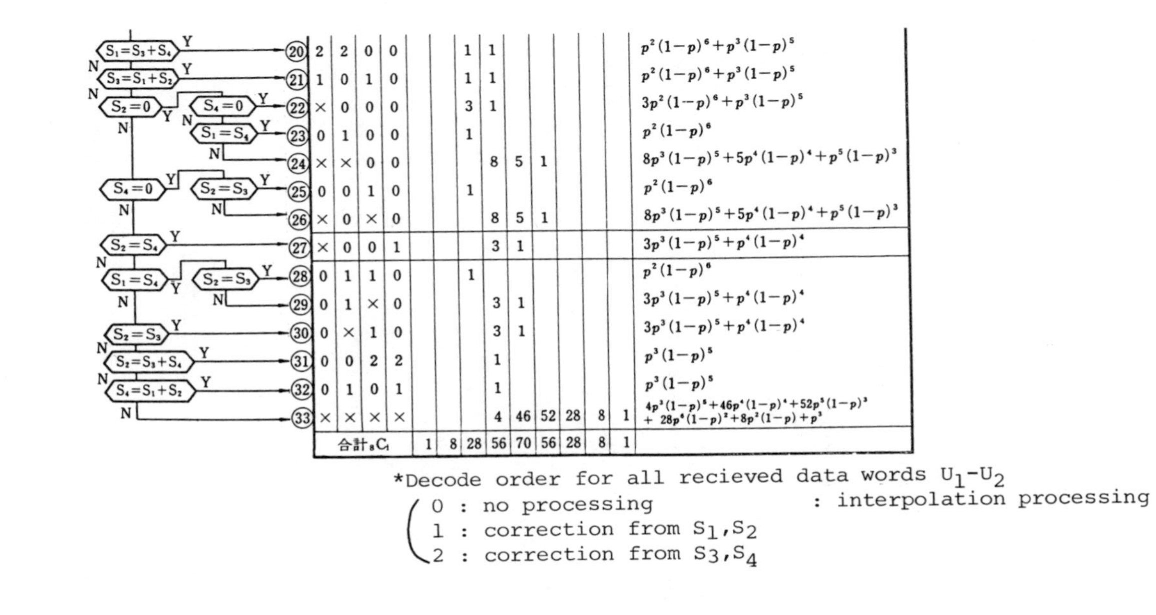

20	2	2	0	0			1	1						$p^2(1-p)^6+p^3(1-p)^5$
21	1	0	1	0			1	1						$p^2(1-p)^6+p^3(1-p)^5$
22	×	0	0	0			3	1						$3p^2(1-p)^6+p^3(1-p)^5$
23	0	1	0	0			1							$p^2(1-p)^6$
24	×	×	0	0				8	5	1				$8p^3(1-p)^5+5p^4(1-p)^4+p^5(1-p)^3$
25	0	0	1	0			1							$p^2(1-p)^6$
26	×	0	×	0				8	5	1				$8p^3(1-p)^5+5p^4(1-p)^4+p^5(1-p)^3$
27	×	0	0	1				3	1					$3p^3(1-p)^5+p^4(1-p)^4$
28	0	1	1	0			1							$p^2(1-p)^6$
29	0	1	×	0				3	1					$3p^3(1-p)^5+p^4(1-p)^4$
30	0	×	1	0				3	1					$3p^3(1-p)^5+p^4(1-p)^4$
31	0	0	2	2				1						$p^3(1-p)^5$
32	0	1	0	1				1						$p^3(1-p)^5$
33	×	×	×	×				4	46	52	28	8	1	$4p^3(1-p)^5+46p^4(1-p)^4+52p^5(1-p)^3 + 28p^6(1-p)^2+8p^7(1-p)+p^8$
合計 $_8C_i$					1	8	28	56	70	56	28	8	1	

*Decode order for all recieved data words U_1-U_2 : interpolation processing

0 : no processing
1 : correction from S_1, S_2
2 : correction from S_3, S_4

Fig. 6-28. Decode algorithm for crossword code Io (8,4,3).

Table 6-4. A General Demonstration of Crossword Code.

[P] [j]([N], [K], [L])

Code	Contents	
P	I	Compared to I_0, this is the words supplement to the detection words (created from the sum of the diagonal data words)
	C	Compared to C_0, this is the words supplement to the detection words (created from the sum of the other detection words)
j	number of supplementary detection words	
N	block size (unit : word)	
K	number of information words	
L	total number of rows	

Example	Code	Note
ex-1	$I_0(8,4,3)$ $M_1\ M_3\ R_7$ $M_2\ M_4\ R_8$ $R_5\ R_6$	$R_5 = M_1 \oplus M_2$ $R_6 = M_3 \oplus M_4$ $R_7 = M_1 \oplus M_3$ $R_8 = M_2 \oplus M_4$
ex-2	$I_1(9,4,3)$ $M_1\ M_3\ R_7$ $M_2\ M_4\ R_8$ $R_5\ R_6\ R_9$	add to 1. $R_9 = M_1 \oplus M_4$
ex-3	$C_1(9,4,3)$ $M_1\ M_3\ R_7$ $M_2\ M_4\ R_8$ $R_5\ R_6\ R_9$	add to 1. $R_9 = R_5 \oplus R_6$

hand column shows the various classifications of code errors, which are then shifted to various types of processing. The error correction performance with reference to random code errors is given in Table 6-5.

Other Error Correction Codes. There are a number of other error correction codes which are suitable for use in digital audio equipment. b-adjacent codes, for example, have had an extremely long association mathematically with computers. Originally, they were developed by D.C. Bossen in 1970. In addition to b-adjacent codes cross-interleave codes of various types employing heuristic techniques may also be used. In fact, they were originally developed along with crossword codes with the specific aim of application to digital audio equipment.

b-adjacent codes[15]

An example of error correction being carried out using a b-adjacent code is shown in Fig. 6-29A. If a b-adjacent code is used in

Table 6-5. Random Error Correction Capacity of Crossword Code Io (8,4,3).

Error word number	Number of occasions				Total	Probability
	correction possibility	1 word interpolation	2 words interpolation	4 words interpolation (block rejection)	$_8C_i$	
0	1				1	$(1-p)^8$
1	8				8	$p(1-p)^7$
2	16	12			28	$p^2(1-p)^6$
3	4	4	44	4	56	$p^3(1-p)^5$
4			24	46	70	$p^4(1-p)^4$
5			4	52	56	$p^5(1-p)^3$
6				28	28	$p^6(1-p)^2$
7				8	8	$p^7(1-p)^1$
8				1	1	p^8

p : Word error ratio.
Here we assume that the factor determining the non-existence of errors in an information word is the ability to correct.

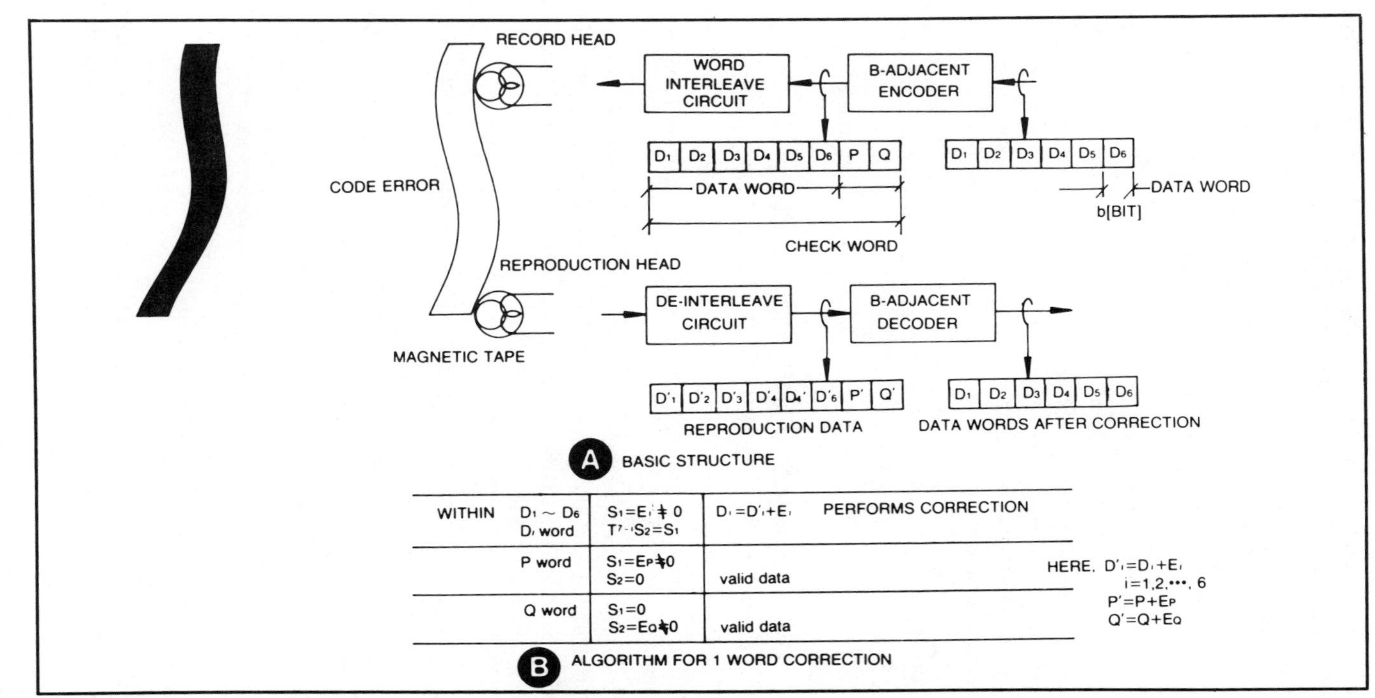

Fig. 6-29. Error correction (one word correction) using b-adjacent code.

digital audio equipment; firstly, code errors of b(bits) in length which are situated next to each other in one block may be corrected; secondly, the position of these errors may be confirmed by the techniques employed in other error detection codes. However, the most important feature of this code is its capacity to correct two code errors of b (bits) generated at the same time and lying to each other.

The construction diagram (Fig. 6-29A) takes the first point mentioned in the preceding paragraph as its hypothesis. Should a code error occur in one of the words shown, where each word is b (bits) long and where one complete block is made up of D_1 through D_6, P and Q, then the error may be corrected. In the example given, a word interleave circuit is included. This ensures that long continuous code errors caused by the use of recording tape are split up, so that continuous errors are distributed in such a way that each word in each block will contain a code error.

Detection bits P and Q may be calculated from the code as follows:

$$P = D_1 \oplus D_2 \oplus D_3 \oplus D_4 \oplus D_5 \oplus D_6$$

$$Q = T^6 D_1 \oplus T^5 D_2 \oplus T^4 D_3 \oplus T^3 D_4 \oplus T^2 D_5 \oplus T D_6$$

T is known as the generation matrix, and if, for example, b = 14, then the matrix may be used as shown in Section 9-4 of Table 8-1 in Chapter 8.

After P and Q have been calculated, the block is recorded, and reproduced as D_1—through D_6', P′, Q′. In the decoder for b-adjacent code, the following two syndromes, S_1 and S_2 (code error diagnos word) are then created.

$$S_1 = D_1' \oplus D_2' \oplus D_3' \oplus D_4' \oplus D_5' \oplus D_6' \oplus P.'$$

$$S_2 = Q' \oplus T^6 D_1' \oplus T^5 D_2' \oplus T^4 D_3' \oplus T^3 D_4' \oplus T^2 D_5' \oplus TD_6'$$

Then one word error correction is carried out using the algorithm shown in Fig. 6-29(B), relying on the two syndromes S_1 and S_2. S_1 becomes the code error pattern used in correction, and S_2 is used to discover the position of words which contain the code errors.

For example, if the data Di′ for the ith word is incorrect, by locating an i which satisfies the following equation,

$$T^{7-i} S_2 = S_1$$

it is possible to detect that the error data is in fact Di. Then complete correction may be carried out using the below equation:

$$\text{correct Di} = \text{Di}' + S_1.$$

Actually, some other type of code such as a CRCC is used in conjunction with a b-adjacent code. This code is highly effective where error correction has to be carried out frequently, and will be further explained with actual examples in Chapter 8.

Cross-interleave [16]

The basic design of an error correction system using cross-interleave is shown in Fig. 6-30A. In this explanation we shall consider the data word made up of one block as two words, for the sake of convenience. The correction process will be elucidated using the block structure shown in Fig. 6-30B.

The detection words P and Q in the diagram are created by half addition of the two chains (known as the P and Q chains). For example,

$$P_0 = Q_0 \oplus D_0 \oplus D_1$$
$$Q_{11} = D_{20} \oplus D_{17} \oplus P_4$$

In cross-interleave, the choice of the component parts of the Q chain (here we are using 1, 3 and 7) is extremely important. If the chain used has not been carefully evaluated, there will be a dramatic loss of error correction capacity. The normal evaluation methods for this chain are not based on theoretical mathematical calculations, but on practical tests involving computers.

For each block that is created in this way, interleave is carried out for the component words. The detection words from the CRCC are added, and the information is recorded as a new sub-block.

After reproduction, any code errors in the block units are detected. Subsequently, the data is returned to its original format by the de-interleave circuit shown in Fig. 6-30B. At this point, it is possible to confirm whether code errors have occurred in the word units.

Using the same diagram let us consider the situation where the correction of code error words is not possible. Let us assume that five words, D_{13}, Q_7, D_{15}, Q_9 and D_{18} contain error information. In this situation it is possible to correct all the errors using the P or Q chains, as shown below:

$D_{13} \leftarrow Q_6, D_{12}, D_{13}, P_6$ in the P chain
$Q_7 \leftarrow Q_7, D_{12}, D_9, D_0$ in the Q chain
$D_{15} \leftarrow Q_7$ (already corrected), D_{14}, D_{15}, P_7 in the P chain
$Q_9 \leftarrow Q_9, D_{16}, D_{13}$ (already corrected), P_2 in the Q chain

$D_{18} \leftarrow Q_9$ (already corrected) D_{18}, D_{19}, D_9 in the P chain. However, if we change the parameters slightly, and assume that a

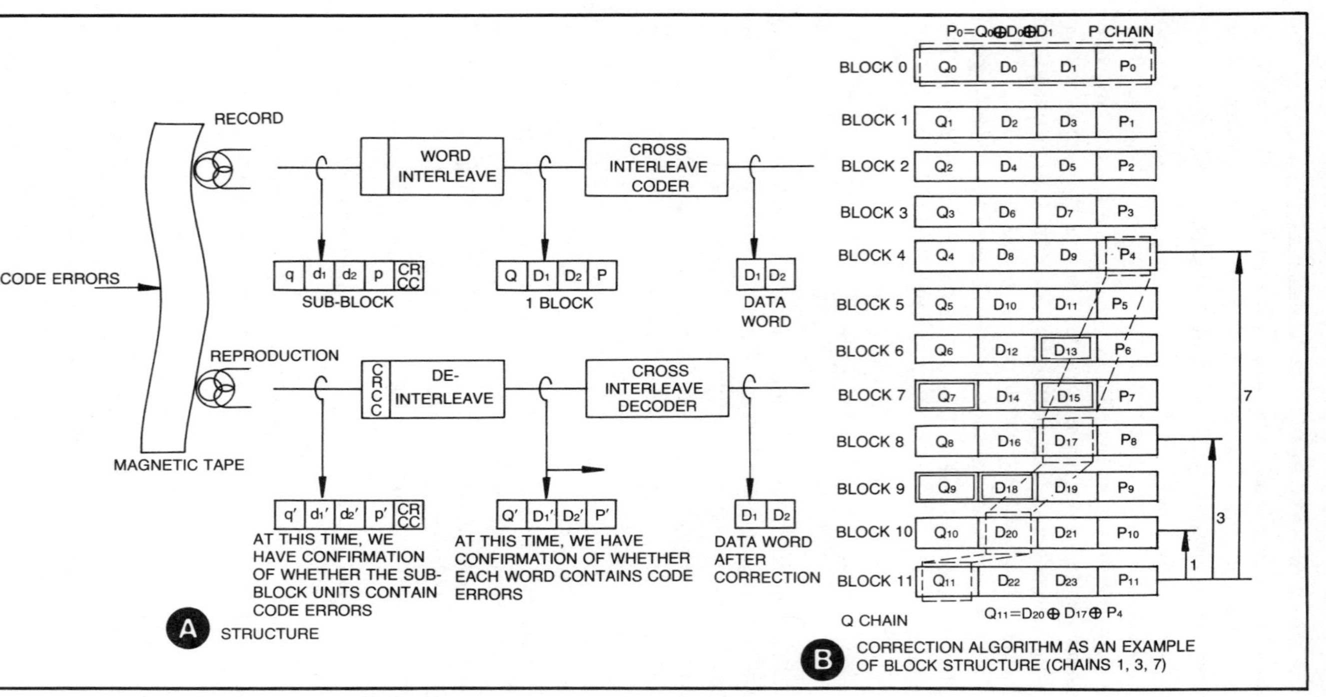

Fig. 6-30. Error correction using cross-interleave.

code error has also occurred in D_{12}, we can see that it would then be impossible to carry out complete error correction.

Thus, we may say that if six code errors occur simultaneously within the blocks preceding and following those associated with the P and Q chains, then a situation where correction becomes impossible will occur.

If we assume that the word error ratio is P_H, then the ratio indicating failure to correct will be $P_H{}^6$. This indicates that a very high ratio of error correction operations will be carried out.

STATISTICAL MODELS FOR CODE ERRORS

It is necessary to consider the nature and distribution of code errors in detail if code error compensation is to be carried out effectively. There are, however, numerous methods of investigation and appraisal which may be used, and it is of course desirable to select the most suitable one. This process of appraisal will, of course, require the necessary hardware, as well as a great deal of effort, time and money, so that the various error compensation systems can be properly evaluated. The ideal situation would, therefore, be one where problems not associated directly with the code error compensation systems under review could be eliminated. All the problems arising from the hardware, for example, take up a great deal of valuable time, which could be better spent assessing the actual compensation systems. In fact, the most convenient state of affairs would be an arrangement whereby a number of compensation systems could be appraised in a short time using the same set-up; in other words, simulation techniques with the appropriate software using computers.

To use simulation effectively statistical models have to be created which have the same qualitative and quantitative characteristics as the code errors caused in the real world.

The Stochastic Process for Statistical Models

The basic approach to handling statistical models for code errors is made through stochastic processes, especially the Markov Chain.

We can say that there are n types of phenomenon (event) which may occur, E_1 E_2 .... E_n, and because we will carry out trials m times in sequence from the first trial, we may say that various phenomena will occur during these trials. The event for which we have obtained results after m number of trials will be determined by the results of the previous trial, m—1; this chain of trials is known as the Markov chain. In a situation where the results of trial m depend solely on the results of trial m – 1 this method is known as the "simple" Markov chain.

In the pure Markov chain where event E_j occurs, the probability of the occurrence of E_j in the next trial is called the transition probability, which is shown by Pij.

We may show all the transition probabilities for the period covering all the events, E_1 to E_n, as follows:

$$\begin{bmatrix} P_{ii} & \cdots & P_{in} \\ \vdots & \ddots & \vdots \\ P_{ni} & \cdots & P_{nn} \end{bmatrix} \quad \text{Here,} \quad \sum_{j=1}^{n} Pij=1$$

This is known as the first order transition probability matrix.

Assuming that we can show the vectors for the probability of the occurrence of each event before the first trial as:

$$q^{(0)} = [q_1^{(0)} q_2^{(0)} \ldots\ldots q_n^{(0)}]$$

We can then extrapolate from the first event the probability of occurrence $q^{(1)}$ for each event as follows:

$$q^{(1)} = q^{(0)}P$$

The transition probability matrix is shown by the transition diagram, as in Fig. 6-31. This is known as a Shannon diagram.

The probability of occurrence $q^{(2)}$ for each event in the results from the second trial are shown as:

$$q^{(2)} = q^{(1)\,P} = q^{(0)}\, P \times P = q^{(0)}\, P^2$$

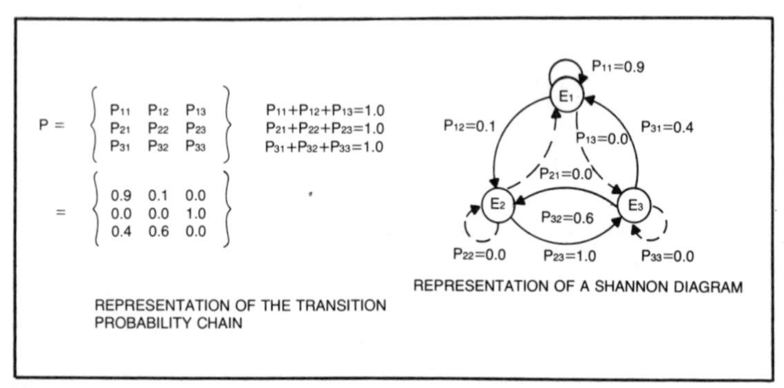

Fig. 6-31. A simple Markov Chain.

We can, in general, calculate the results from trial m as follows:

$$q^{(m)} = q^{(m-1)} P = ... = q^{(0)} P^m$$

P^m is known as the higher order transition probability matrix. If all the elements constituting the first order transition probability matrix P are correct, then P^m, which is the m order higher order transition probability, will converge on matrix T which contains elements of identical value in columnar direction, as m gets sequentially larger.

$$\lim_{m \to \infty} P^m = T = \begin{bmatrix} t_1 \ t_2 & \cdots & tn \\ \vdots & \ddots & \vdots \\ t_1 \ t_2 & \cdots & t_m \end{bmatrix}$$

The row vector $t = (t_1 \ t_2 .....t_m)$ is known as the inradiant probability vector, and if we choose the value of this vector as the initial probability $q^{(0)}$, we establish the relationships

$$t = tP = tP^2 = ...tP^m$$

This type of chain is known as a constant Markov chain and whatever point is chosen in the infinite chain, the probability of the occurrence of events E_1 to E_n will be shown as t_1 to t_n.

The Gilbert Model

A statistical model often used for continuous code errors occurring in transmission paths is known as the Gilbert Model, since it was developed by E.N. Gilbert in 1960.

As shown in Fig. 6-32 the Gilbert Model may be divided into two basic classifications of events: good and bad. These are depicted from transition probability of the events. This model employs the aforementioned simple Markov chain approach and is used as a model for the occurrence of code errors.

If we assume that the parameters P,ρ,Q and q are $S_{11} = p$, $S_{12} = P$, $S_{21} = Q$ and $S_{22} = q$ for the Gilbert model, the first order transition probability matrix for our Gilbert model may be shown as follows:

$$S = \begin{bmatrix} S_{11} & S_{12} \\ S_{21} & S_{22} \end{bmatrix}$$

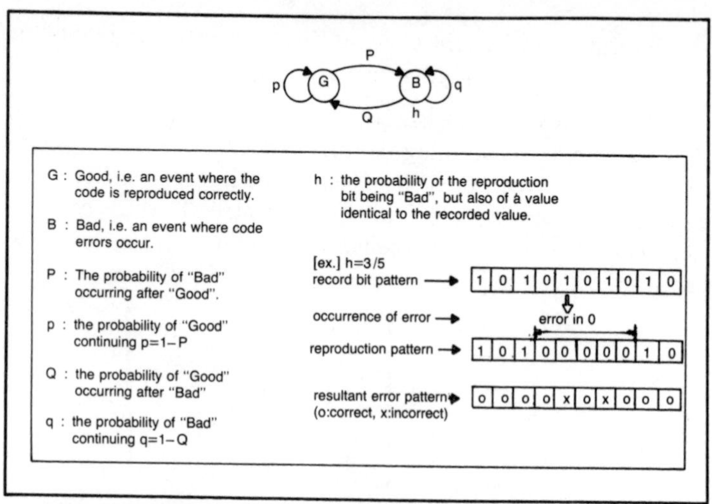

Fig. 6-32. Gilbert Model.

However, it may help to think of the bit chain represented by the Gilbert model as a constant Markov chain, because the amount of information being reproduced is very large. The component z_2 of the invariant probability vector

$$Z = (Z_1 Z_2) \text{ where } Z_2 = S_{12} / (S_{12} + S_{21})$$

is the probability of code errors occurring when the reproduction bit chains are being observed. This is, therefore, identical to the overall code error rate. Also we can define ρ in

$$\rho = 1 - S_{12} - S_{21} = 1 - P - Q$$

as the bit coefficient of correlation, and therefore, we can express the character, either burst or random, of the code errors. Generally speaking, where ρ is close to 1 the errors will tend to be burst errors, and where ρ is close to 0 the errors will be of a random nature.

Using the above information, we can then work out Pi(s) which is the probability of the occurrence of a continuous code error i bits long, as follows:

$$Pi(s) = Z_1 \, S_{12} \, S_{22}{}^{i-1} \, S_{21}$$

Estimations of parameters from measured code error data is then carried out by non-linear programs. This is to ensure that the squared error between the theoretical random code error distribution calculated using the above equation and the measured values is kept to a minimum.

Figure 6-33 shows the estimated parameters for a Gilbert model using code error data occurring in the middle section of a reel of magnetic tape.

However, h will be a different value depending on the method of record and reproduction. In general, it is known to be approximately 0.5.

We shall now consider the random distribution of the more complex code errors occurring at points close to each end of a reel of magnetic tape. This is shown in Fig. 6-34. In this situation, the normal Gilbert model is variously applicable for both long and short code errors, as shown by the dotted and solid lines on the diagram. However, it is difficult to use a single Gilbert model to express the whole picture.

Therefore, we shall next examine what happens when we align two independent Gilbert models to express error data. If we assume that we are expressing the first order transition probability of two

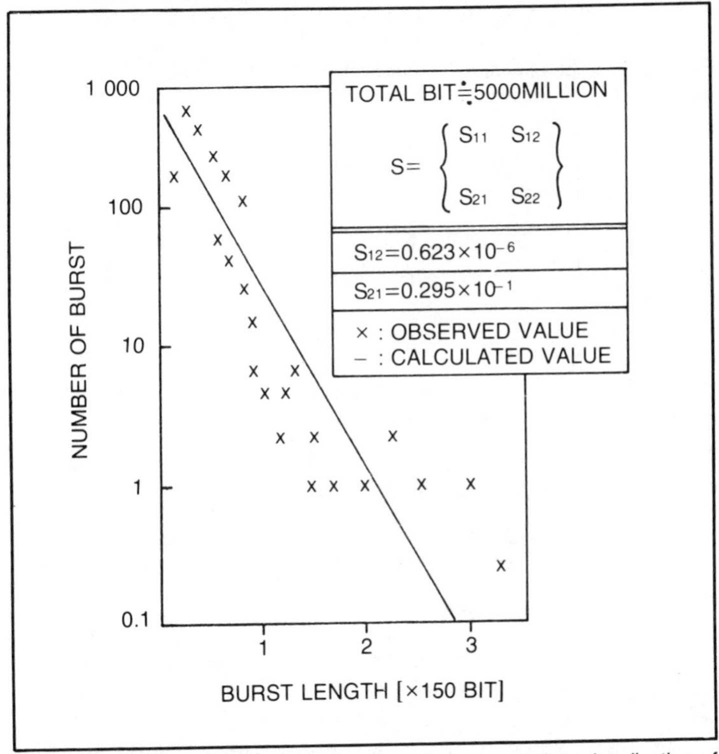

Fig. 6-33. Code errors in the central portion of a magnetic tape (application of random distribution and the Gilbert model) (transmission rate 1.764 Mb/s).

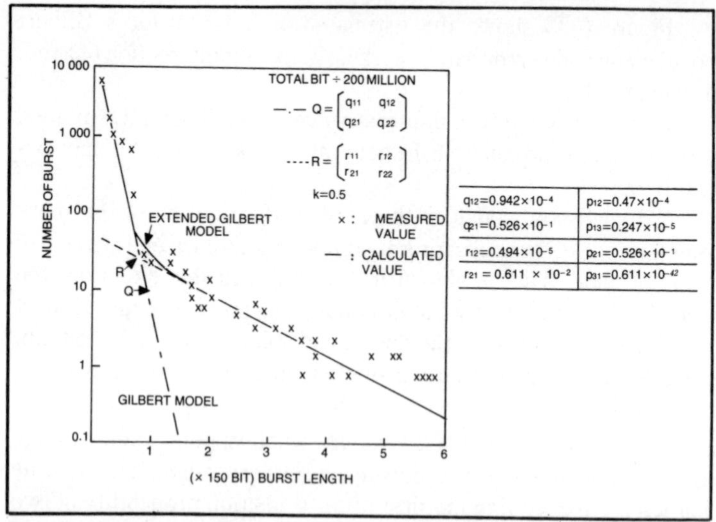

$q_{12}=0.942\times10^{-4}$	$p_{12}=0.47\times10^{-4}$
$q_{21}=0.526\times10^{-1}$	$p_{13}=0.247\times10^{-5}$
$r_{12}=0.494\times10^{-5}$	$p_{21}=0.526\times10^{-1}$
$r_{21}=0.611\times10^{-2}$	$p_{31}=0.611\times10^{-42}$

Fig. 6-34. Code errors occurring at the ends of the tape (application of random distribution and the extended Gilbert model) (transmission rate 1.764 Mb/s).

Gilbert models as Q and R, and that the passive probability vectors are X and Y, then:

$$Q = \begin{bmatrix} q_{11} & q_{12} \\ q_{21} & q_{22} \end{bmatrix} \qquad R = \begin{bmatrix} r_{11} & r_{11} \\ r_{21} & r_{22} \end{bmatrix}$$

$$X = \begin{bmatrix} \chi_1 & \chi_2 \end{bmatrix} \qquad Y = \begin{bmatrix} y_1 & y_2 \end{bmatrix}$$

and $P_{1\,(Q)}$ and $P_{i\,(R)}$, which are the respective probabilities of the occurrence of continuous code errors, may be expressed in the following equation:

$$P_i(Q)=\chi_1 q_{12} q_{22}{}^{i-1} q_{21} \quad P_i(R)=y_1 r_{12} r_{22}{}^{i-1} r_{21}$$

Probability P_i which is obtained by combining these calculations based on the combined constant $k(O \leqq k \leqq 1)$ may be calculated thus:

$$P_j = kP_{i\,(Q)} + (1-k)\,P_{i\,(R)}$$

Using the first term expressed in the above equation, and applying the second term independently to the long and short code errors respectively, it is possible to ensure comformity of the model to the random distribution of complex code errors. This is shown by the solid lines in Fig. 6-34.

A single model containing the equivalent theoretical random code error distribution is shown in Fig. 6-35. Probability P_i which is the occurrence of a continuous code error i bits long in the diagram is obtained thus:

$$P_i = t_1 P_{12} P_{22}^{\,i-1} P_{21} + t_1 P_{13} P_{33}^{\,i-1} P_{31}$$

The conditions where this is equivalent to the values for the solid lines for the two models are as follows:

$$P_{22}=1-q_{21},\ P_{21}=q_{21},\ P_{33}=1-r_{21},\ P_{31}=r_{21}$$

$$P_{12}=k\chi,\ q_{12}q_{21}r_{21}/\Delta$$

$$P_{13}=(1-k)y_1 r_{12} r_{21} q_{21}/\Delta$$

$$\text{Here } \Delta = q_{21}r_{21} - k\chi, q_{12}r_{21} - (1-k)y_1 q_{21} r_{12}$$

COMPUTER SIMULATION OF CODE ERROR COMPENSATION[(19)-(24)]

Simulation of code error compensation by computer, as shown in Fig. 6-36, may be divided into two sections. The first technique is the analytical appraisal of code error compensation performance by calculation of statistical probability (indicated as "1" in the dia-

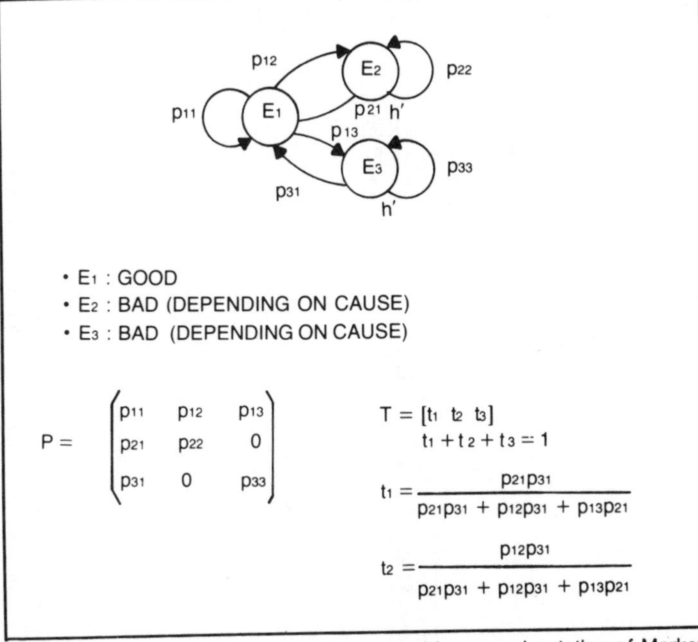

Fig. 6-35. Extended Gilbert Model (expressed in general notation of Markov chain).

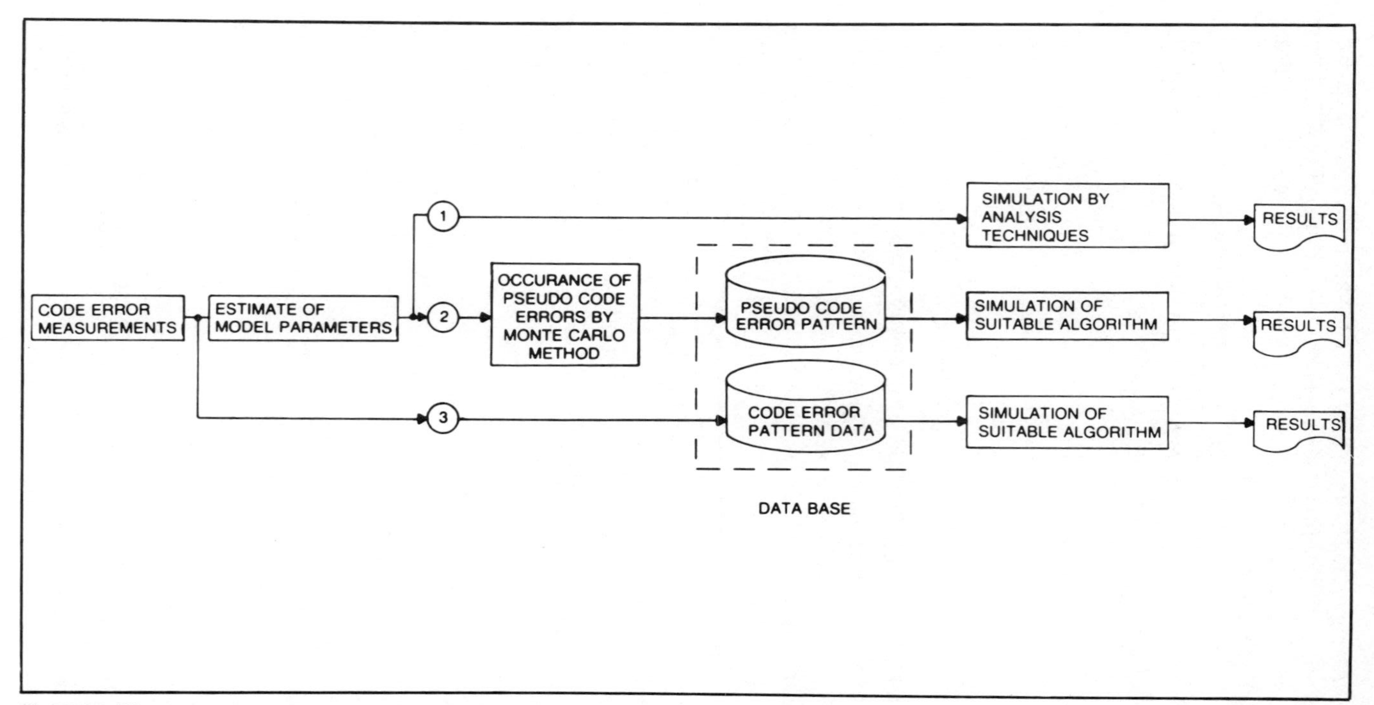

Fig. 3-36. Diagrammatic representation of computer simulation.

gram). The second technique is a method of replacing actual hardware operations with software (shown as "2" and "3".)

The former system is extremely advantageous for systems which may demand code error compensation performance in theoretical order, and its major advantage is that the operations required for simulation are completed relatively quickly. However, it is difficult to make this algorithm applicable to a non-standard code error. In this case, it is more effective to use the latter simulation method. In terms of the fundamental code error patterns used, the error detection and correction operations are carried out just as they would be by the actual hardware. Generally, rather long operation time is necessary, but an extremely accurate simulation may be achieved.

When using either of the techniques described, it is desirable to give sufficient thought to the structure of the code error compensation system under review, and also to consider the scope and accuracy of the simulation.

Analytical Techniques

These can be very effective. Even if, at the time of error compensation, error detection and correction are not possible, the data may still be depicted completely.

As shown in Fig. 6-37, this is achieved by using the code error occurrence patterns for several blocks (in this case, A and B, as shown in the diagram) and a constant Markov chain indicating the reproduction bit chain. In this type of situation, we can then calculate the occurrence probability for all cases by the simple probability calculation shown in the same diagram.

In real terms, evaluation of the code error correction performance of methods, such as pointer erasure (Fig. 6-26), are best accomplished, as shown in Fig. 6-38, by analytical techniques.

In the diagram, the bit coefficient of correlation alters over a very wide range, from 0.999 to 0.9, and calculations are made according to the results of these changes. It is also possible to read out the extent of the most efficient bit coefficient of correlation for code error compensation using this system. Thus, simulations based on analytical techniques may be used effectively and efficiently to evaluate a wide spread of code errors.

Simulation Methods Based On Hardware Operations

As indicated in Fig. 6-36 simulations based on actual hardware operation may also be divided into two broad categories, the differ-

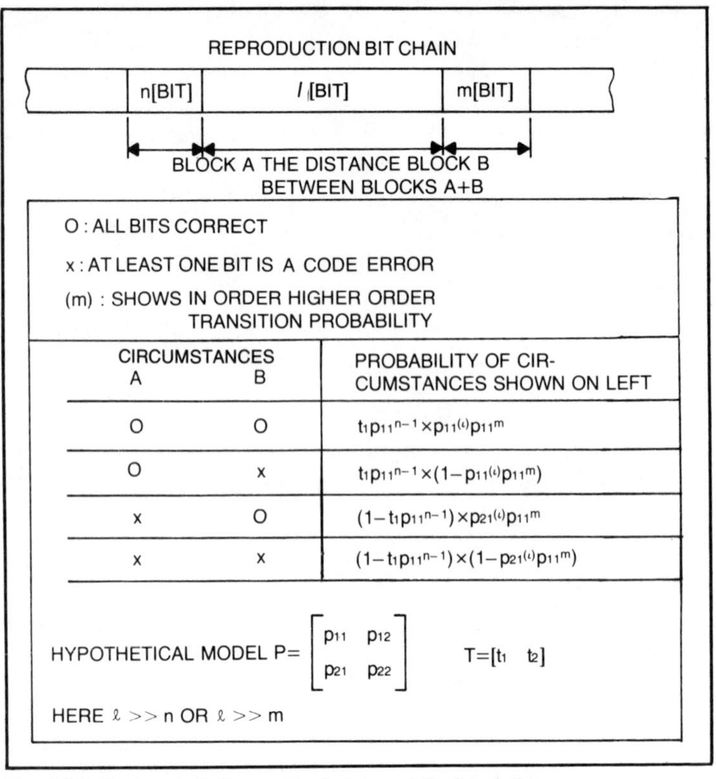

Fig. 6-37. Basic calculation methods for analytical techniques.

ence lying in the basic method of code error pattern generation. For one of these methods we infer the model parameters from measured data taken from actual code errors. From this we can produce the pseudo-code error pattern system shown as “2” in Fig. 6-36. The other section (“3”) uses raw code error measurement data.

Generally, the spread of code errors is quite random and relatively large, depending as it does on the recording medium and the record reproduction equipment used. This is particularly true when the equipment is being operated by a general user. There will be many occasions where long code errors occur, but, unfortunately, these cannot be forecasted from laboratory data. From this point of view, the former system is quite useful, because it can be used to change the form of the code errors. Furthermore, code error measurement is often carried out using the numerous bits of which each block is composed, because of the operational speed relationships in the computer used for the actual measurement. As a result, one would have to rely on the former system in cases where it is necessary to carry out bit unit simulation.

However, it is, of course, evident that from the point of view of conformity to actual code patterns a system using raw measured data is several degrees superior. Thus, when considering these two systems, the former is most suitable for investigating code error compensation systems, and the latter for evaluating the characteristics of the recording medium, or for inspecting the code error compensation performance inherent in production machines.

Figure 6-39 shows a code error pattern created by the Monte Carlo system and using a Gilbert model. (One row is about 3.6 seconds, the whole is approximately equivalent to 6 minutes of code errors.)

Figure 6-40A shows the code structure for a code error compensation system, while Fig. 6-40B shows an example of the evaluation of its compensation performance. The table uses the decode algorithm for the diagram and a pseudo-code error pattern.

This system is known as the double writing method, and the information is recorded either partially or totally using fixed time expansion. Then, the reproduction signal is examined using error

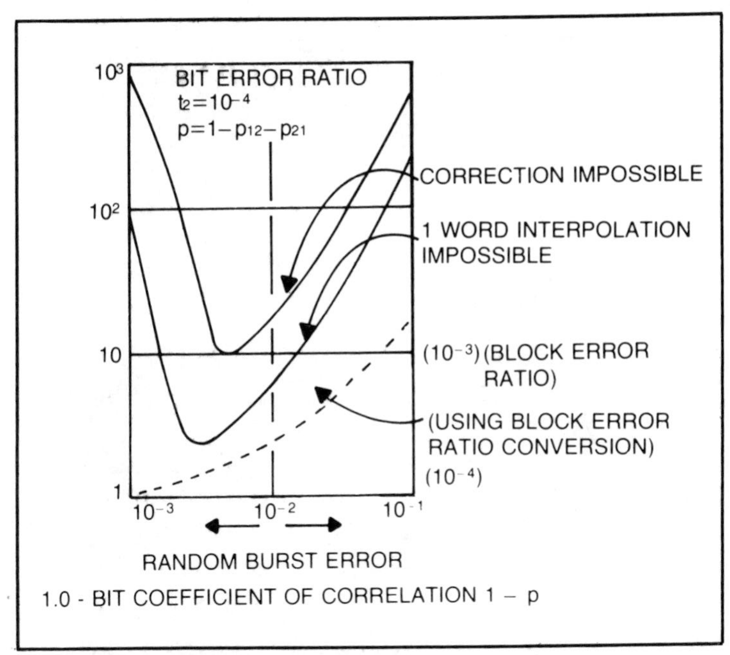

Fig. 6-38. Appraisal of error correction performance.

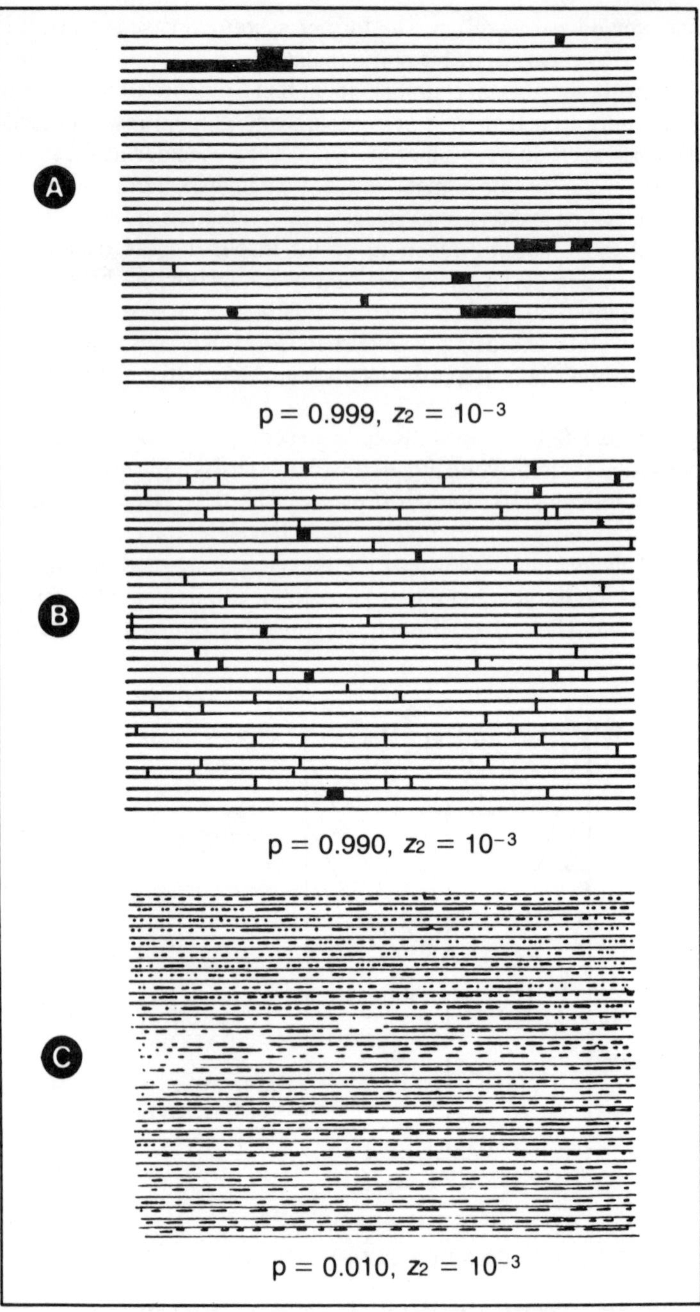

Fig. 6-39. Pseudo-code error pattern generated by the Monte Carlo Method.

detection to determine whether the re-transmitted signal conforms to the original signal or not. If the signal does not conform to the original, then the CRCC is used to discover which parts of the signal are correct. Then the contents of the correct parts of the signal is used for error correction using the code error compensation system. In the example given in Fig. 6-40 the reproduction signal is constructed only from the upper 12 bits of the information in the original signal, so the code errors occurring in the subordinate 4 bits are, for practical purposes, ignored.

As shown in Fig. 6-40B 1 to 9, simulation results for the number of occurrences may be obtained in all situations where errors may occur.

Thus, simulations based on hardware operation and evaluations of code error compensation systems may be used to investigate the operation of actual production machines. The techniques described above are, of course, widely used for checking digital audio equipment.

References

1. Shannon, C.E.: "A Mathematical Theory of Communications" BST J, pp. 379-423 (1948).
2. Imai, Iwadare, Miyakawa: "Lectures on Computers, 18 code theories" Shokodo.
3. Hashimoto, Fukuda, Doi: "Applications of code theory to PCM magnetic recorders" Shingaku Zendaiso, S9-14 (1977-3).
4. Doi, T.: "A new code for the detection and correction of dropout on magnetic tape," Nikkei Electronics, 20, pp. 94-113, (1978-5).
5. Iga et al: "A Measurement Method of Drop-outs of Audio PCM Recorder", ASJ Comf. No 3-2-12, Oct. '77.
6. Yamamoto, Inose: "Lectures on electronic calculators", "Data Communications" Saupo.
7. Sumoda et al: "A Listening Test of Dropout-Compensation of PCM Audio Recorder", ASJ Comf. No 3-2-13, Oct. '77.
8. Fukuda et al: "An evaluation method for burst error compensation in PCM recorders", IECE, S13-10, Mar. '78.
9. Fukuda et al: "Interpolation Method of Code Error on PCM Magnetic Recorder", ASJ Comf. No 3-5-14, May '78.
10. Fukuda et al: "For a Evaluation of Miscorrection on PCM Magnetic Recorder", ASJ Comf. No 3-P-14, Oct '78.
11. Doi: "A Class of Error Correcting Code for a Transmission System with High Error Rate", ASJ Comf. No 3-2-14, Oct. '77.

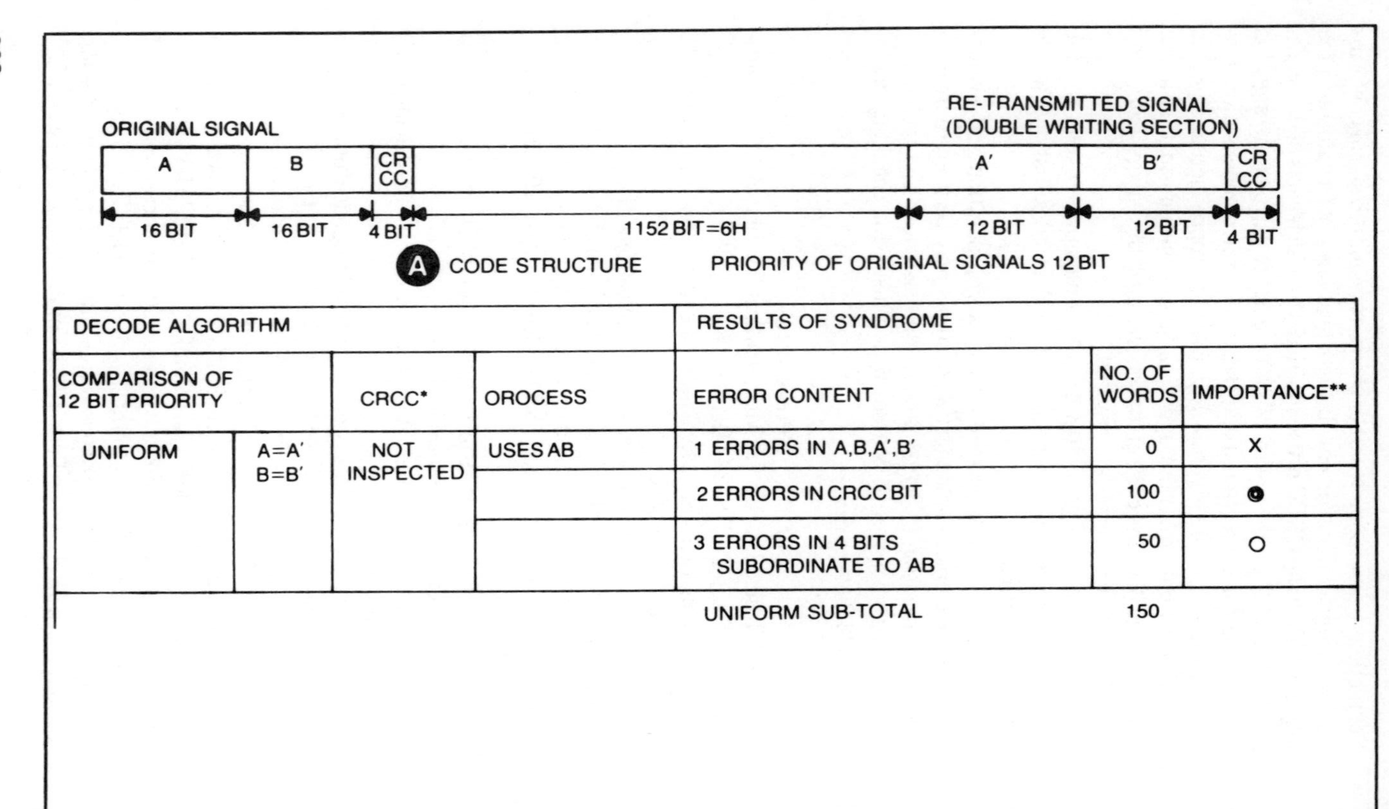

Ⓐ CODE STRUCTURE

DECODE ALGORITHM				RESULTS OF SYNDROME		
COMPARISON OF 12 BIT PRIORITY		CRCC*	OROCESS	ERROR CONTENT	NO. OF WORDS	IMPORTANCE**
UNIFORM	A=A′ B=B′	NOT INSPECTED	USES AB	1 ERRORS IN A,B,A′,B′	0	X
				2 ERRORS IN CRCC BIT	100	◉
				3 ERRORS IN 4 BITS SUBORDINATE TO AB	50	○
				UNIFORM SUB-TOTAL	150	

NON-UNIFORM	A≠A' OR B≠B'	AB : O A'B' : X	USES AB	4 ERRORS ONLY IN A'B'	1966	⊙
				5 ERRORS IN AB,A'B'	1	O
		AB : X A'B' : O	USES A'B' (SUBORDINATE 4 BITS USE AB)	6 ERRORS ONLY IN AB	2019	O
				7 ERRORS IN AB, A'B'	0	X
		AB : O A'B' : O	COMPENSATION	8	241	O
		AB : X A'B' : X		9	5	O

NON-UNIFORM SUB-TOTAL 4232

NUMBER OF ERROR WORDS IN 1 HOUR 4382

* CRCC : (O: ZERO SURPLUS, X : NON-ZERO SURPLUS)

** IMPORTANCE : (X : OCCURENCE OF NOISE, O: DIFFICULT TO DISTINGUISH AUDIBLY, ⊙: NO ERROR)

(B) AN APPRAISAL OF THE DECODE ALGORITHM AND ITS ERROR CORRECTION CAPACITY

Fig. 6-40. An example of simulation (double writing method) based on hardware operation.

12. Doi et al: "A New Error Correcting Code for PCM Recorder", IECE, S13-11, Mar. '78.

13. Doi: "On Decoding Algorithms of Cross Word Code", ASJ Comf. No 3-5-13, May, '78.

14. Doi: "Cross Word Code; A New Error Correcting Code for Digital Audio Systems", IECE, EA78-25, July, '78.

15. Fukuda et al: "Comparisons of Several Decoding Methods of Error Correcting Code on a New Format for Digital Audio Processors for Home-use VTR's", ASJ, No 3-5-19, June, '79.

16. Doi et al: "Error Correction by Improved Cross-Interleave Code", ASJ, No. 3-5-12, June, '79.

17. Gilbert, E.N.: "Capacity of a burst noise channel" BSTJ 39, 5, pp. 1253-1263 (1960-9).

18. Fukuda et al: "Drop-out Statistical Model on PCM Magnetic Sound Recording", ASJ, No 4-6-5, April, '77.

19. Doi et al: "Computer Simulations on Error Correcting of PCM Magnetic Tape Recorder", ASJ, No 4-6-4, April, '77.

20. Fukuda et al: "Drop-out Simulations (2) on PCM Magnetic Sound Recording", ASJ, No 3-2-11, Oct., '77.

21. Fukuda et al: "Drop-out Compensation on PCM Magnetic Recorder", IECE, EA78-34, July, '78.

22. Fukuda et al: "For a Method of Simplified Error Correction on a PCM Adaptor Connected with Home-use VTR", ASJ, No 3-P-15, Oct., '78.

23. Fukuda et al: "For a Method of Code Error Correction on a Stationary Head Type PCM Tape Recorder", ASJ, No 3-P-16, Oct., '78.

24. Fukuda G, Doi. T: "On dropout compensation of PCM systems: computer simulation method and a new error correction code (cross word code)" AES 60th Conv. No. 1254, (E-7) (1978-5).

Chapter 7

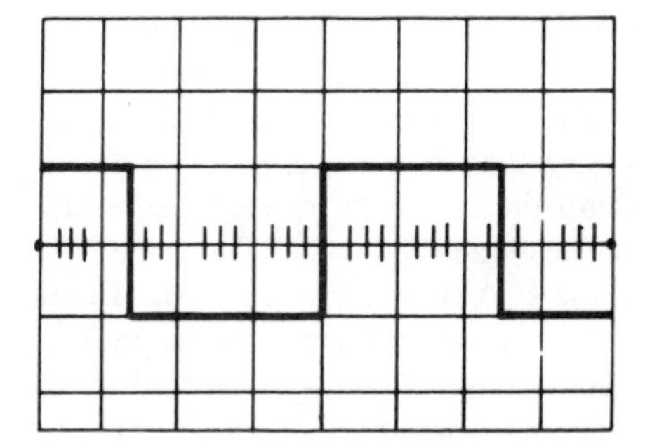

The Design of a PCM Tape Recorder

The design process for any type of equipment follows more or less the same procedure; the design engineers must select the circuits and parts of the equipment required on the basis of the performance expected of the system under construction.

A PCM tape recorder is designed in the same way; the qualities required are considered first, and are balanced against ultimate cost in order to form an effective design concept.

The major parameters for the appraisal of a PCM tape recorder are dynamic range, distortion, frequency characteristics and wow and flutter. In PCM tape recorders, unlike general analog tape recorders, it is not difficult to reduce wow and flutter, as the signal reproduced from the recording medium is initially in digital form. In brief, the digital signal including time discrepancies upon reproduction is momentarily stored in a digital memory, and is taken out by a very stable synchronizing signal generated from an equipment such as a crystal oscillator, which, in effect, lowers wow and flutter to a negligible level.

Dynamic range and distortion are determined by the number of quantization bits and linearity of the configuration of the A/D and D/A converters. Any improvements in these areas would, however, result in steep cost increase.

The maximum frequency is determined by the sampling frequency, and the frequency characteristics within a given band are effected by that of the low pass filters in the input and output stages as well as by the aperture time.

In this chapter an actual construction of a PCM tape recorder with a dynamic range of 80-95 dB, quantization bit of 14-16, and a frequency response of DC-20 kHz is considered.

Many types of PCM tape recorders using stationary or rotary head systems have been introduced. However, the circuitry before the A/D and after the D/A converter stages for any type of PCM recorder is fundamentally the same. The same holds true for the circuitry after the D/A converter for a DAD (Digital Audio Disc) player. Figure 7-1 shows the basic block diagram of a two-channel PCM tape recorder, which will be referred to throughout this chapter. The fold-out diagrams at the back of the book depict the whole circuitry of the PCM-1 digital audio processor.

It may be appropriate to make a few short comments here about the line amplifiers and dither generators which the explanations are omitted in this chapter. The line amplifier is not different from those generally used in other audio applications, and there are many textbooks available which explain them in depth. A dither generator is a device which amplifies the noise generated by a zener diode to an appropriate level, and is shown in the above mentioned diagram of PCM-1.

LOW-PASS FILTERS

Low pass filters are used in both the record and reproduction chains, and are likely to be represented by a single circuit construction for cost purposes. As a required characteristic, ripple of less than ± 0.2 dB at DC-20 kHz is essential due to the audio signal which has to be filtered twice, only during recording, and once during reproduction.

Furthermore, because attenuation of 80-90 dB is necessary for levels in excess of half the sampling frequency, a filter with suitable steep cut-off characteristics is designed, using the unified Tchebyscheff theory.

Figures 7-2 and 7-3 show the circuit and constants of 9th and 13th order Tchebyscheff filters with input and output impedance of 3.3 k ohms. Figures 7-4 and 7-5 show the frequency characteristics of each filter. It is necessary to adjust the constants upon manufacture, by altering the coils dc resistance and the suspended capacitance and so on.

Another point which must be noted is that current concentration is generated in the coil close to the filter input. Thus, the core

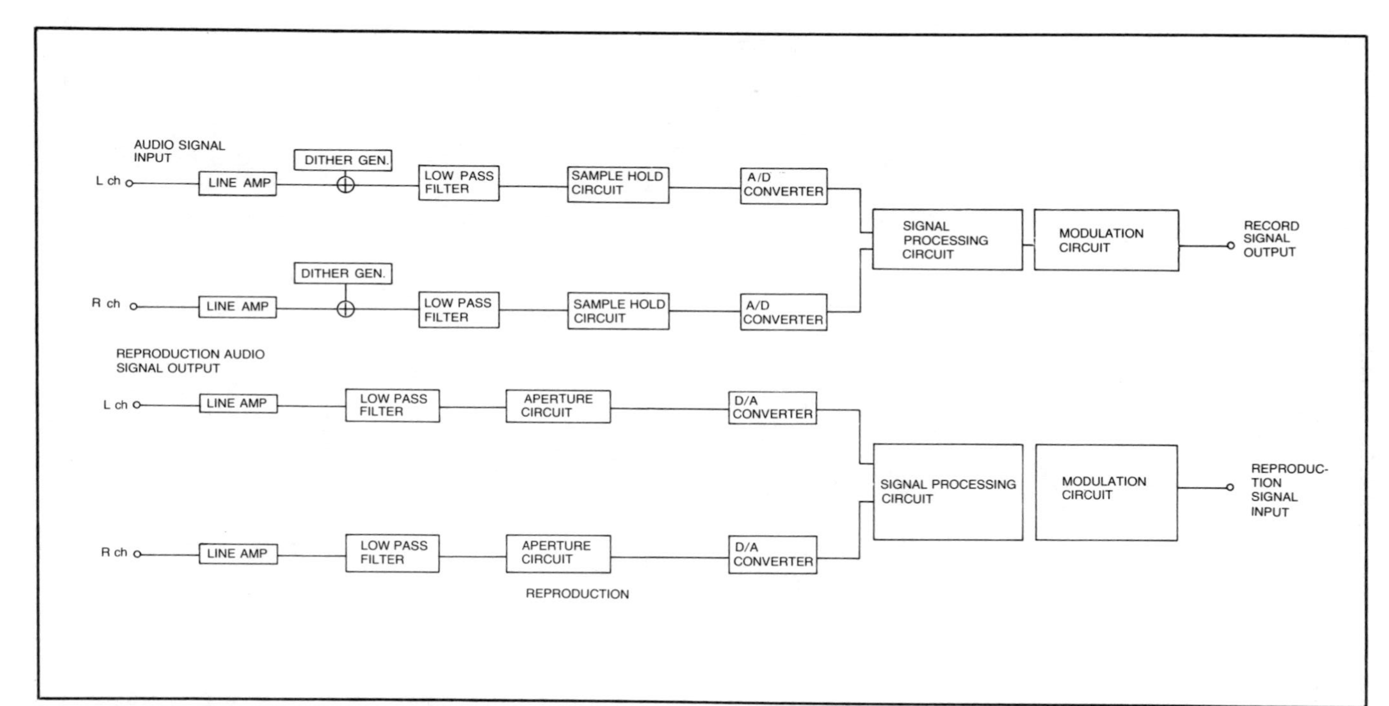

Fig. 7-1. Block diagram of a PCM tape recorder.

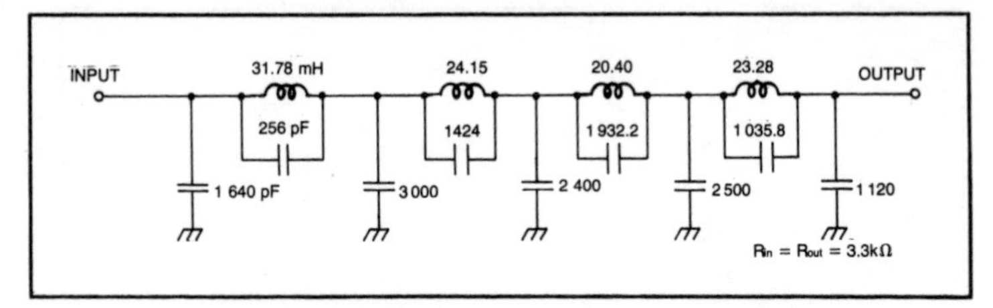

Fig. 7-2. 9th order Tchebyscheff filter circuit diagram and constants.

material must be chosen carefully so that non-linearity of the coil core material is not transmitted to the filter regardless of the input signal level. Non-linearity in the core material will result in distortion of the signal.

SAMPLE AND HOLD CIRCUIT

The sample and hold circuit is generally constructed from a capacitor, an analog switch, and buffer amplifiers, as shown in Fig. 3-16. The output voltage of an ideal sample and hold circuit conforms to the input voltage by the sample command, and maintains the above input voltage upon receiving the hold command. However, in actual circuits, errors are generated for a variety of reasons. We shall now look into the major types of error, the effect they have on the total characteristic of the circuit, and the measures taken against them are discussed.

Offset error

Offset errors are generated mostly in the input and output buffer amplifiers. Errors may also be generated at the analog switch of high speed buffer amplifiers employing transistor and diode bridges in the switch when off-balanced. However, such high speed devices are not necessary for PCM equipments for audio signal

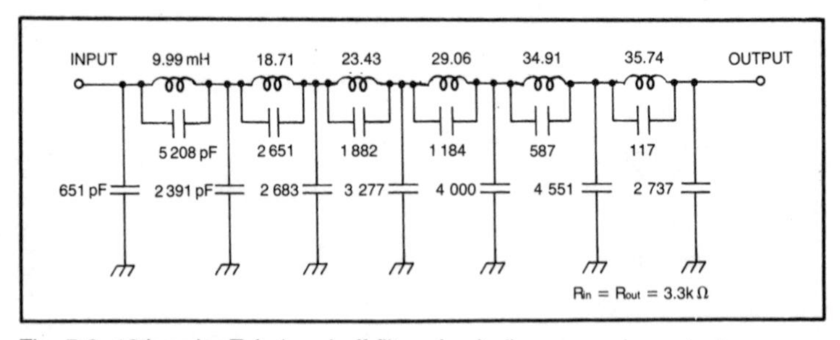

Fig. 7-3. 13th order Tchebyscheff filter circuit diagram and constants.

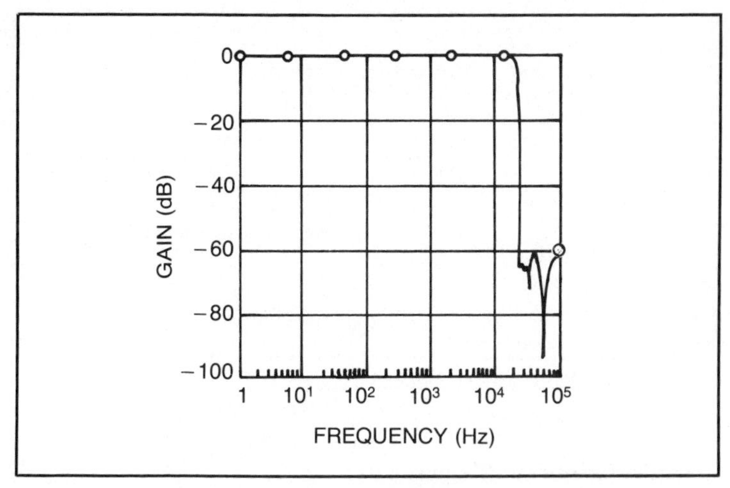

Fig. 7-4. Frequency characteristics of 9th order Tchebyscheff filter.

processing, and FETs are often used, in which case, error may be ignored.

The effect of offset in the output buffer amplifier may also be nullified by using a closed loop circuit, as shown in Fig. 7-6. In this case, however, the input buffer amplifier offset still remains, and

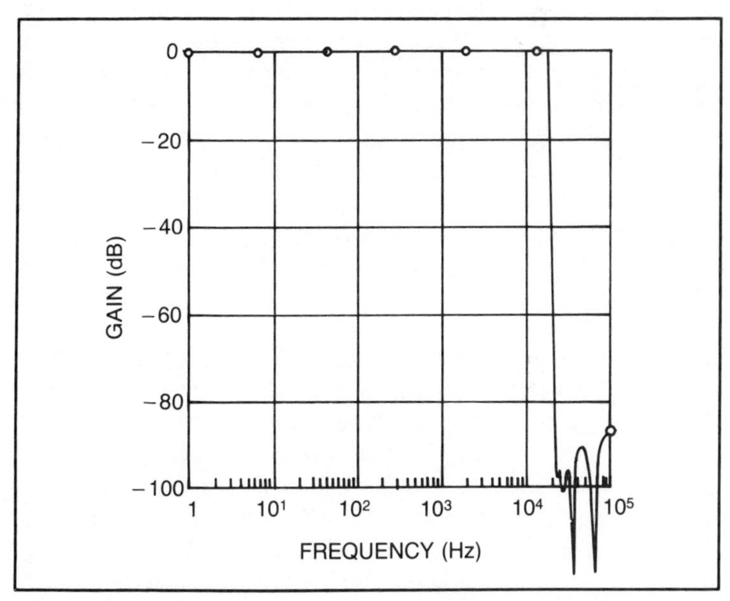

Fig. 7-5. Frequency characteristics of 13th order Tchebyscheff filter.

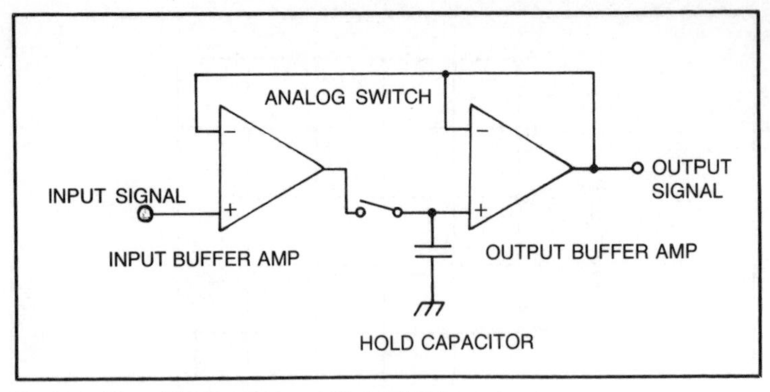

Fig. 7-6. Closed-loop sample and hold circuit.

becomes an offset error in the sample and hold circuit. Thus, low off-set is essential for input buffer amplifiers.

However, record and reproduction of a perfect dc signal is not necessary for audio signal record and reproduction, and even if there is an off-set of several tens of mV, there will be no problem provided that the dc components are disconnected upon reproduction.

Acquisition time

Acquisition time is the time necessary for the input voltage to conform to the output voltage upon receiving the sample command. The operational waveform of an ideal sample and hold circuit includes no time lag as shown in Fig. 3-12. However, the actual signal will accompany some discrepancy as shown in Fig. 7-7 due to buffer amplifier settling time, analog switch switching time, capacitor charging current, etc.

The acquisition time is essentially the same as the operational speed of the sample and hold circuit, and must be less than the sampling period.

In most PCM tape recorders, the acquisition time chosen is about 5 μs due to its relation to the A/D conversion speed. However, when constructing a sample and hold circuit which requires the same amount of acquisition time, the capacitor charging current requires more consideration than the operational speeds of the buffer amplifier and analog switch (FET). For example, if we use a standard operational amplifier as a buffer amplifier, the maximum output current would be approximately $\pm$ 10 mA, and the capacitance C, given by the below equation, required to conform to $\pm$ 10 V in 5 μs must be less than 2500 pF:

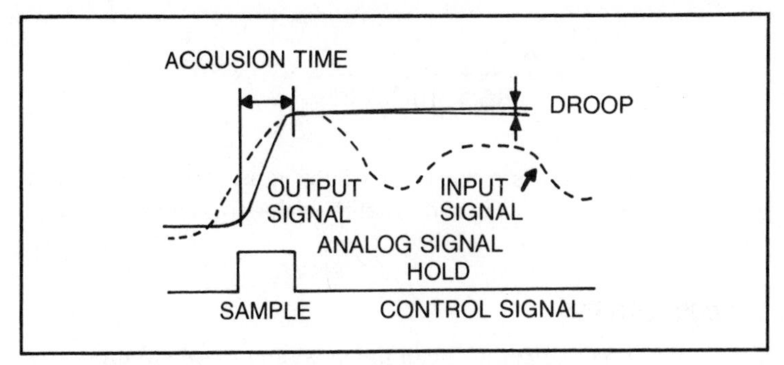

Fig. 7-7. Operationed waveform in a sample and hold circuit.

$$C = \frac{IT}{V} = \frac{10^{-2} \times 5 \times 10^{-6}}{20} = 2.5 \times 10^{-9} \text{ (F)}$$

Equation 7-1

Needless to say, the analog switch here must be one capable of holding a current of 10 mA at the ON position. From the point of view of acquisition time, the smaller the capcitance the better, whereas, from the point of view of droop, the opposite is the case.

Droop

The sample and hold circuit turns OFF the analog switch by a hold command, and the voltage stored in the capacitor is emitted. It is desirable that this voltage is kept at a constant value until the next sample command is sent. However, in an actual circuit it varies as shown in Fig. 7-7. This variation is due to the voltage change of the capacitor that results from the changes of the current of the output buffer amplifier and the leak current of the analog switch. This voltage variation must be less than the quantization steps, because otherwise, it would result in conversion errors in the A/D converter.

Assuming a system of 16 bit quantization with an input signal of ± 10 V, the quantization step E_o will be approximately 300 μV from the below equation:

$$E_o = \frac{20}{2^{16}-1} = 3 \times 10^{-4} \quad \text{(V)}$$

Equation 7-2

Taking A/D converter conversion time of 15 μs and capacitor volume of 2,500 pF, the permissible leakage and bias current Ie will be less than 50 nA from the below equation:

$$Ie = \frac{2.5 \times 10^{-9} \times 3 \times 10^{-4}}{15 \times 10^{-6}} = 5 \times 10^{-8} \text{ (A)}$$

Equation 7-3

This, however, is a marginal value; in an actual circuit it is better to allow for a margin and to set the value to about less than 10 nA.

A/D CONVERTERS[1]

The A/D converter is one of the most important circuits in the PCM tape recorder, as well as one of the most difficult parts to design. For an example, as can be seen from equation (3.5), a 16 bit A/D converter must be capable of breaking down the analog input signal into as many as 65,536 steps. Furthermore, the transition must be carried out in 10-20 μs.

Various types of A/D converter have been designed, but systems which can satisfy the demands for medium speed and high accuracy, as mentioned above, are limited. In this section, we shall attempt to explain the systems suitable for our applications and the details of their operation are explained.

Successive Approximation A/D Converters

A/D converters using successive approximation techniques are the most widely used high accuracy medium speed converters. The circuit is composed of a D/A converter, shift register, latch and analog comparator, as shown in the block diagram in Fig. 7-8.

D/A converter used here has the same number of bits as an A/D converter, and where the digital signal is in binary code and is ON for i bits, the output voltage Vi is:

$$Vi = V1 \frac{1}{2^{i-1}}$$

Equation 7-4

Here, V1 is the output voltage when the first bit is ON.

The operation of the A/D converter is as follows: first of all, the first bit of the D/A converter is turned ON by the shift register upon applying the sample value of the sample and hold circuit to the input terminals of the analog comparator. Here, the input sample value and the output voltage V_1 of the D/A converter are compared at the analog comparator, and if the former is larger, then the comparator output is tuned to "H" (high), and the latch is switched ON. Here, the first bit of the D/A converter is left ON. If, on the other hand, the

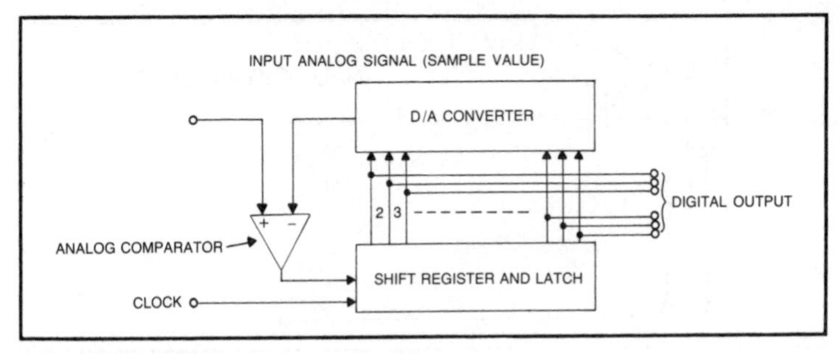

Fig. 7-8. Successive approximation A/D converter.

latter voltage is larger, the comparator output is tuned to "L" (low), and the first bit of the D/A converter is switched OFF. By the next clock, the shift register turns on the second bit of the D/A converter.

If the first bit of the D/A converter remains switched ON by the latch as a result of the previous comparison, then the output voltage of the D/A converter becomes $(3/2)V_1$, and is compared to the input sample value. If, on the other hand, the first bit is turned OFF as a result of the first bit comparison process, the output voltage of the D/A converter becomes $V_1/2$, and is compared at the comparator. As a result, the latch will be switched ON and OFF by the output of the comparator as it was for the first bit. After the third bit, following the same operation, the output voltage of the D/A converter will near the input voltage, as shown in Fig. 7-9. After the comparison for the last bit has been carried out, the output voltage of the D/A converter will be equivalent to the input voltage, or will result in error of less than half a quantization step. At this point, the latch output, that is, the input signal of the D/A converter, has been changed from a sample value to a digital code.

The largest problem, when designing 14-16 bit A/D converters using this system of successive comparison, lies in the selection of the D/A converter used. D/A converters are also used in the reproduction section of PCM tape recorders, but the D/A converter used here must be more superior to the ones used for reproduction in terms of conversion speed and accuracy.

For example, if we consider a 16 bit converter, the conversion speed of the D/A converter on the reproduction side is sufficient if it is slightly faster than the sampling period. However, using it in an A/D converter, it must operate much faster, at least 16 times in each sampling period. Furthermore, if we consider the operation time of the comparator, then a conversion speed of several hundreds

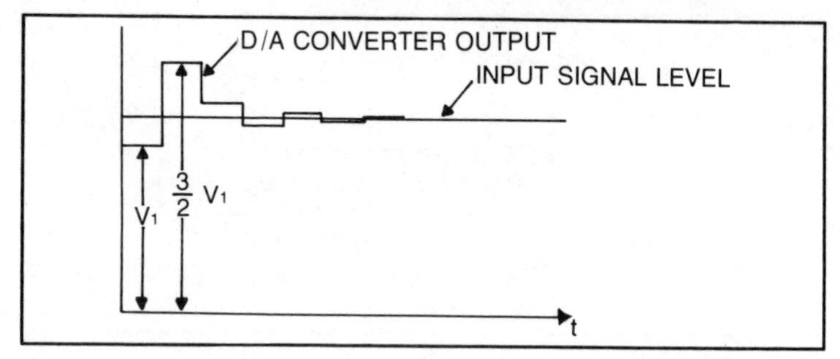

Fig. 7-9. The operation of a successive approximation A/D converter.

of nanoseconds is required for the D/A converter. There are not many high speed, high accuracy D/A converter systems. The D/A converter will be explained in more detail later in this chapter.

Integrated A/D Converters

The basic principle of an integrated A/D converter is as follows: A charge satisfying the input analog voltage is stored in the capacitor of the integrator. This is then discharged at a uniform rate, and the time required for the output voltage of the integrator to reach zero or a certain fixed value is measured. A digital counter is used for this measurement, in which the measured value automatically becomes digitally output corresponding to the input voltage.

This system has been widely used in measurement equipments, such as digital volt meters, for its high accuracy despite the low conversion speed.

Figure 7-10 shows an example of this type of converter. The circuit consists of an integrator, analog switches, comparators, a

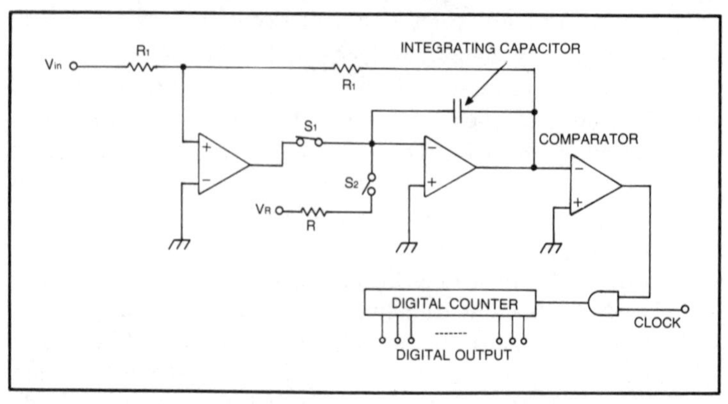

Fig. 7-10. Integrated A/D converter.

digital counter, etc. The analog switch S_2 is OFF and S_1 ON at the start of the conversion. Then, the integrating capacitor is charged until it equals the input analog voltage Vin, and the output of the integrator becomes –Vin. Then, S1 is turned OFF, and S_2 is turned ON in return. As a result, the input of the integrator is connected to the voltage source V_R through a resistor, and the voltage at the integrating capacitor terminals falls. At the same time, the digital counter starts operating.

When the voltage at the integrator terminals becomes zero, that is, when the output voltage of the integrator becomes zero, the clock within the digital counter is stopped by the comparator, and thus, the digital counter stops operating. Assuming that the digital counter has counted N times, then:

$$N = Vin \times C \times R \times fc/Vr \qquad \textbf{Equation 7-5}$$

Here, C is the volume of the integrating capacitor, and fc is the input clock frequency of the digital counter.

The above shows that the number of countings N is proportional to the input voltage Vin, and the counter output at this point will become the corresponding digital output. In this way, an integrated A/D converter may achieve a very high degree of accuracy as the input voltage changes with time. This also results in the advantage of requiring no sample and hold circuit, as shown in Fig. 7-10. This is because, at the start of the conversion, the input voltage is charged to the integrating capacitor, performing the same process as the sample and hold circuit otherwise would.

The major disadvantage of this system, however, is that the clock frequency must be set very high for effective conversions at high speed. For example, in a 16 bit system, the total number of countings N must equal the number of quantization steps, and therefore, N must be N ➔ 0~65536. In order to perform one conversion in 20 μs, the clock frequency fc must be:

$$fc = \frac{65536}{2\times 10^{-5}} = 3.28 \times 10^{9} \text{ (Hz)} \qquad \textbf{Equation 7-6}$$

Thus, frequency in excess of 3 GHz is required.

It is difficult to realize a digital counter that operates at such high frequencies and its peripheral circuitry. In general, clocks with a frequency of several tens of MHz are widely used, and as a result, a conversion speed is lowered to several ms.

A cascade arrangement of integrated A/D converter[3] shown in Fig. 7-11 is being designed in order to compensate for these deficiencies and to achieve a high speed conversion with a high level of accuracy using a low frequency clock.

The characteristic of this system is in the division of the integration at the standard voltage into two stages. The first part of the integration is carried out at high speed, while the second half is dealt with at a more slower speed. Two digital counters which correspond to these two levels of integration are operated during each corresponding period.

At the start of the conversion process, the analog switches S_2 and S_3 are turned OFF, while S_1 is turned ON. The integrating capacitor is charged to the input voltage Vin, and the output voltage of the integrator becomes –Vin (Fig. 7-12). Then, S_1 is switched OFF and S_2 is switched ON, and the digital counter D_1 starts operating.

The integrator output voltage nears zero at a fixed value. When this voltage becomes V_F, comparator 1 operates and sends a command to counter D_1. Upon receiving this command, counter D_1 turns S_2 OFF at the next clock pulse, and simultaneously, counter D_1 stops operating, S_3 turns ON, and counter D_2 starts operating. As a result, the output voltage of the integrator further nears zero; but this time, the ratio is 1/n compared to the time when S_2 was ON. When the output voltage of the integrator becomes zero, comparator 2 starts operating and stops the operation of counter D_2.

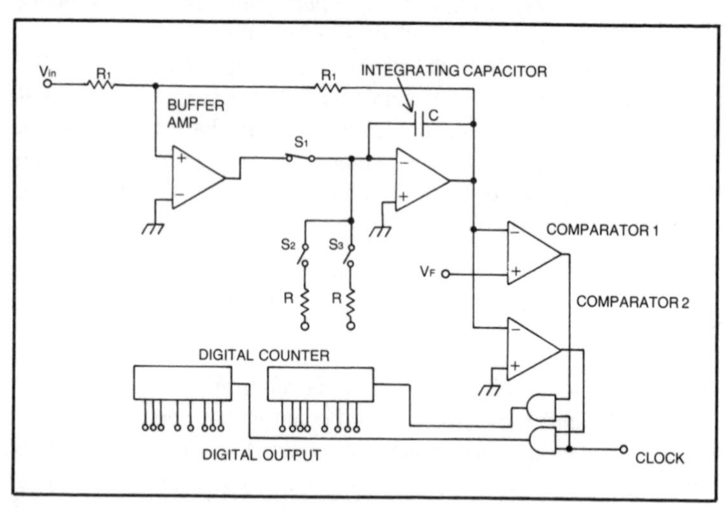

Fig. 7-11. Cascade integrated A/D converter.

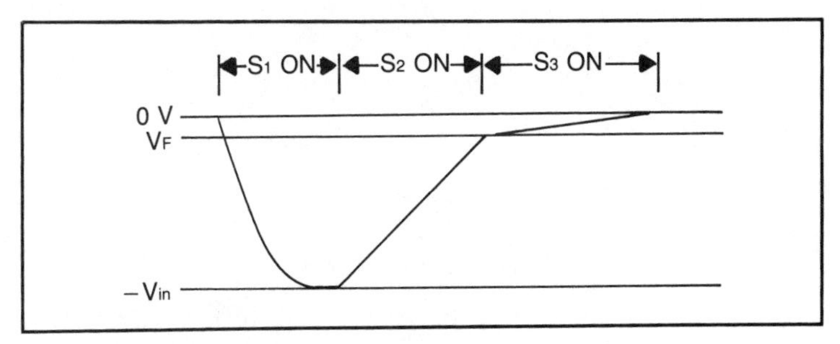

Fig. 7-12. Output voltage of the integrator.

Using this system for a 16 bit converter at n = 256, the integrator output voltage which changes during one counting period of counter D_1 becomes 256 times the voltage which changes during one counting period of counter D_2. If the input terminal voltage V_F of comparator 1 becomes equal to the integrator output voltage which changes during one counting period of counter D_1, then the number of countings for counter D_2 will be 0~255.

The number of countings, at which counter D_2 stops operating, therefore, corresponds to the subordinate 8 bits, while that of counter D_1 corresponds to the upper 8 bits, and thus a 16 bit A/D conversion is performed.

Using this system, the maximum number of counting for D_1 as well as for D_2 is 255, giving a total of 510 for both counters. In order to perform one conversion in 20 μs the clock frequency fc must be:

$$fc = \frac{510}{2 \times 10^{-5}} = 25.5 \times 10^{6} \text{ (Hz)}$$

Equation 7-7

which is 25.5 MHz.

With this order of clock frequency, it is relatively simple to construct the appropriate counters and peripheral circuits, as well as to produce them as a monolithic IC.

One of the characteristics of integrated A/D converters is that linearity is largely dependent on the dielectric loss of the integrating capacitor, and offset errors are caused by the effect of the integrator and the buffer amplifier. The offset error, as mentioned for the sample and hold circuit, does not cause serious problem from the point of view of record/reproduction of audio signals. However, when the dielectric loss of the integrating capacitor is large, as shown in Fig. 7-13, the integrator does not function properly, and linearity is thus ruined.

Since linearity has a direct effect on the distortion ratio of the

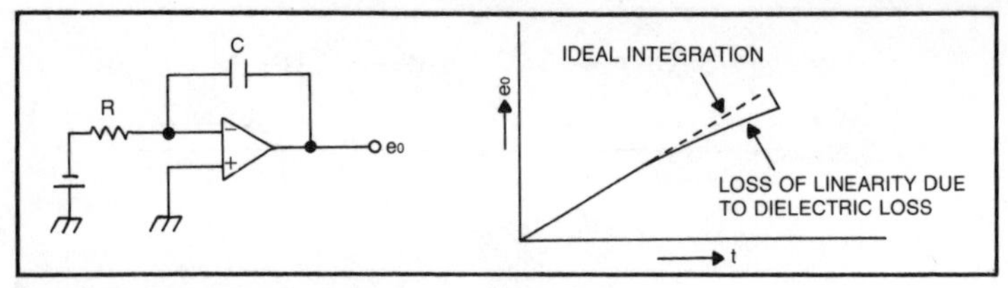

Fig. 7-13. Degraded linearity due to dielectric loss.

reproduction signal, it must be treated with care. Polystyrene, mica, polycarbonate[4], etc. are used as capacitors with low dielectric loss.

D/A CONVERTERS

Just as there are various types of A/D converters, there are also various types of D/A converters. Of these, however, only a few are suitable for use in PCM record and reproduction of audio signals, and the following explanations are limited to these few.

Weighted Resistance D/A Converters

With this type of D/A converter each bit of the input digital signal is converted into voltage or current according to its weighting, and is then added. When the digital signal is in binary code, the weighting for each bit changes in twofold, as shown in Equation 7-4.

From the above method, a D/A converter is constructed, as shown in Fig. 7-14, from analog switches, op amps, and resistances corresponding to each bit.

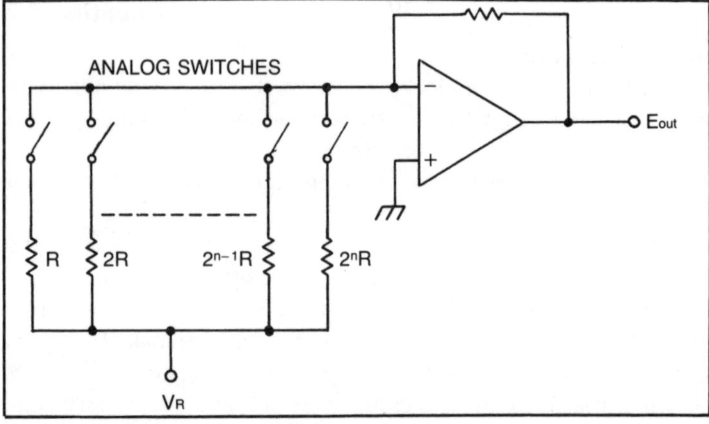

Fig. 7-14. D/A converter with weighted resistance.

This type of D/A converter has a simple, well-defined structure; however, it is necessary to have various types of resistance corresponding to the number of bits, and furthermore, the maximum and minimum resistance ratios are extremely large. Each analog switch is turned ON and OFF by corresponding to each bit of the incoming digital signal. For maximum accuracy of operation the value of R must be set at several k ohms, in order to avoid effects from ON resistance changes of the analog switch, etc. As a result, the larger the number of bits the higher the resistance value must be. For example, in a 16 bit system with resistance R at 1 k ohm, the maximum resistance $2^{16}R$ would be as large as = 65.5M ohms.

It is difficult to acquire high resistance values at a uniform accuracy, and in the case where the number of bits is large the circuit shown in Fig. 7-14 must be modified to that shown in Fig. 7-15.

This circuit shows a way to distribute the weighted resistance into each block, and to further add weight to the block units. In the case where four types of weighted resistance are used to construct a 16 bit converter, as shown in Fig. 7-15, the resistance ratio of the shunt connecting each block is 1:15, and the maximum value of the resistance ratio constructing each block is 1:8. There is no problem to determine the resistance value at this degree of resistance ratio; the problem in this case is with accuracy. The highest accuracy required among the resistances in Fig. 7-15 is that used in block A. From the view of linearity of the D/A converter, for an example, a discrepancy of 1% between this resistance and other resistances will result in an error of ± 0.5%.

The relative accuracy of resistance R must be less than 0.003% in order to maintain the linearity to less than 16 bit quantization step, and the resistance 2R in block A which requires a certain amount of accuracy must be less than 0.006%. The values for the required accuracy here are specified on the basis of relative errors occurring in one resistance only, and thus, when considering the possibility of errors occurring in all the resistances, the accuracy must be much higher. Here, the linearity of the D/A converter has been assumed to be less than a quantization step, but in reality it is desirable to aim at a value of less than ¼ of the quantization step, which makes the acquisition of the resistance accuracy even more difficult.

Thus, from the point of view of stability and accuracy of the resistance, it is difficult, at present, to construct the system as a

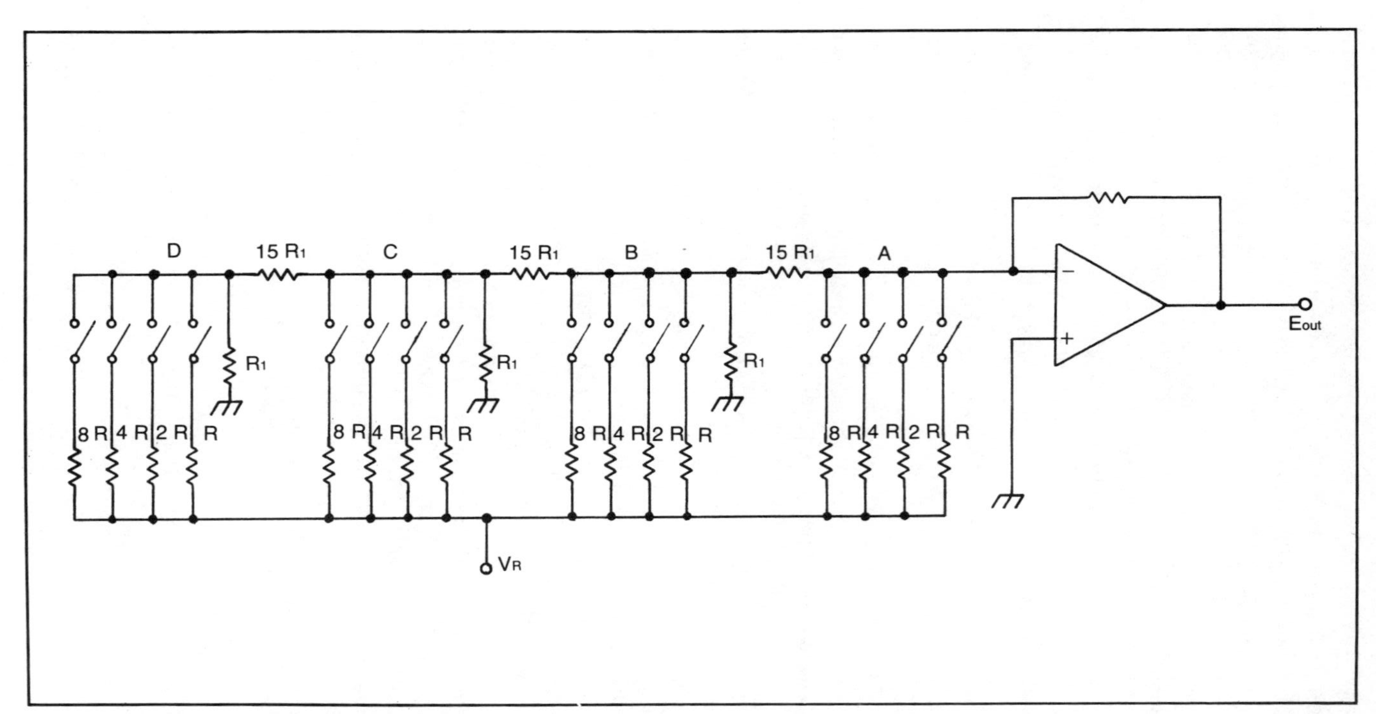

Fig. 7-15. D/A converter with block divided resistance.

monolithic IC, and, therefore, is mostly structured is a discrete form.

For an actual 16 bit D/A converter circuit, selected temperature-stable metal film resistors must be used as weighted resistance. In addition, very small values of resistance are arranged in series in order to achieve the required level of accuracy.

A weighted resistance D/A converter, as explained above, requires an extremely high degree of accuracy in the weighted resistances. However, the conversion speed is mostly determined by the analog switches, and therefore, it is relatively easy to achieve a high-speed system of several hundreds of nanoseconds. This type of D/A converter can, therefore, be used as a high speed, high accuracy D/A converter used in successive approximation A/D converters mentioned earlier in this chapter.

DEM (Dynamic Element Matching) D/A Converters

DEM D/A converter is a system designed by Plassche[2], and is basically a current adding device, as shown in Fig. 7-16. As a method to acquire highly accurate current source, the DEM D/A converter uses switching circuit as well as resistance to switch ON/OFF the current.

The current source used in a current adding device produces I, I/2, I/4,..... as shown in Fig. 7-16. And thus, if a divider that accurately divides the current in half is available, a D/A converter is easily made by constructing the dividers in multi-stage, as shown in Fig. 7-17.

As a method to accurately divide the current in half, a circuit shown in Fig. 7-18 is considered. In this circuit, current I is roughly divided into two parts by the resistance. High accuracy is not

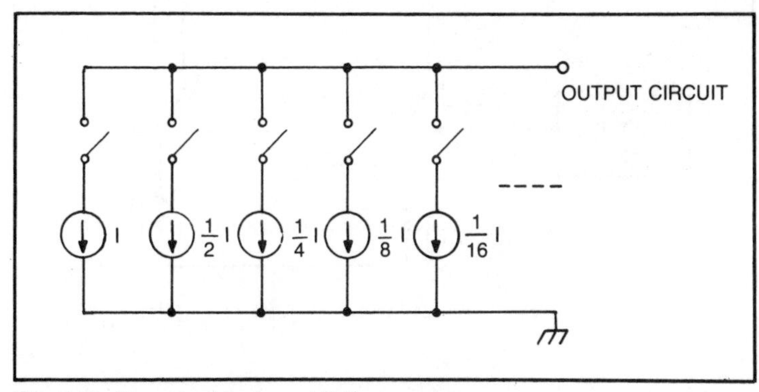

Fig. 7-16. Current adder format D/A converter.

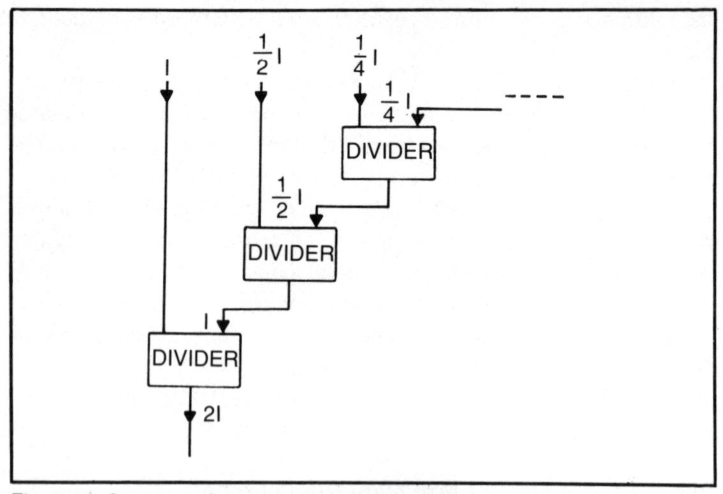

Fig. 7-17. Structure of current source using dividers.

required in this type of current dividing, and therefore, divided currents I_1, I_2 contain error of ΔI as shown in the below equations:

$$I_1 = \frac{1}{2} - \Delta I \quad \textbf{Equation 7-8}$$

$$I_2 = \frac{1}{2} + \Delta I \quad \textbf{Equation 7-9}$$

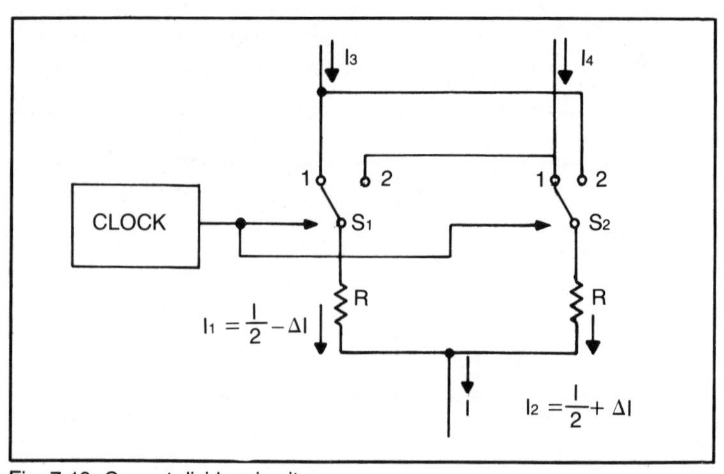

Fig. 7-18. Current divider circuit.

The divided currents are then supplied to the output terminals alternately by the switches S_1, S_2. In brief, switches S_1 and S_2 are linked, and are simultaneously in contact with either terminal 1 or terminal 2. When these switches are operating in conjunction with a timing of a clock shown in Fig. 7-19, output currents I_3, I_4 will both contain a ripple of ΔI. If, however, the clock frequency is sufficiently high, the output currents can be made dc with a simple CR ripple filter.

Given clock period T which turns ON/OFF switches S_1 and S_2 and time t_1, t_2 during which S_1 and S_2 are switched to terminals 1 and 2, and given that the below equations hold true:

$$t_1 = \frac{T}{2} + \Delta T \qquad \textbf{Equation 7-10}$$

$$t_2 = \frac{T}{2} - \Delta T \qquad \textbf{Equation 7-11}$$

then the output currents I_3, I_4 may be calculated as follows:

$$I_3 = (t_1 I_1 + t_2 I_2) / T$$

$$= \left[\left(\frac{T}{2} + \Delta T\right)\left(\frac{I}{2} - \Delta I\right) + \left(\frac{T}{2} - \Delta T\right)\left(\frac{I}{2} - \Delta I\right)\right] / T = \frac{I}{2} - \frac{2\Delta I\Delta}{T}$$

Equation 7-12

$$I_4 = (t_1 I_2 - t_2 I_1) / T$$

$$= \left[\left(\frac{T}{2} + \Delta T\right)\left(\frac{I}{2} + \Delta I\right) + \left(\frac{T}{2} - \Delta T\right)\left(\frac{I}{2} - \Delta I\right)\right] / T = \frac{I}{2} + \frac{2\Delta I\Delta T}{T}$$

Equation 7-13

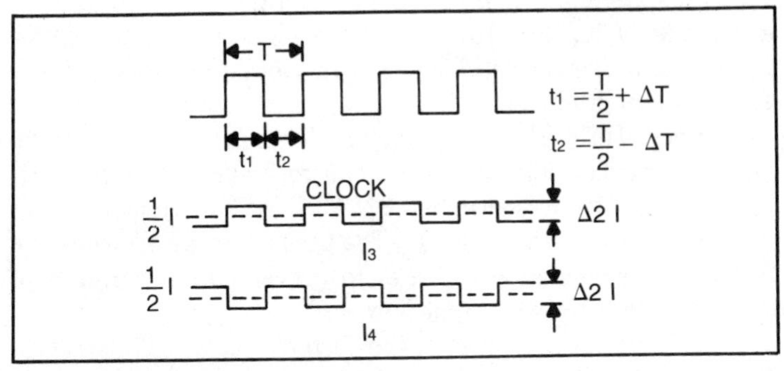

Fig. 7-19. Output current waveform.

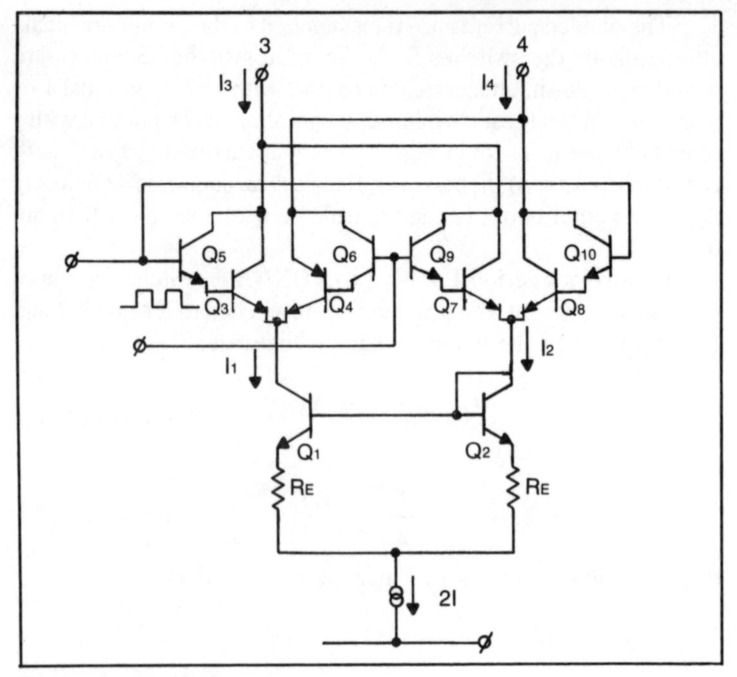

Fig. 7-20. Actual dividing circuit.

And the error related to the perfect division of the current in half will be:

$$\frac{\Delta I_{3,4}}{I_{3,4}} = 4 \bullet \frac{\Delta T}{T} \bullet \frac{\Delta I}{I} \qquad \textbf{Equation 7-14}$$

The clock timing accuracy may be made to about 0.01% without difficulty. It is also possible to set $\Delta I/I$ to 1% by the resistance. Using this system, it is then possible to acquire current dividing accuracy in the region of 10^{-5} to 10^{-6}, as well as enough accuracy for the 16 bit operation with the care of clock frequency and the occurrence of jitter.

An actual example of this circuit is shown in Fig. 7-20; after the current is roughly divided by current mirror from transistors Q_1, Q_2 and resistance R_E, it is switched ON/OFF by the switching circuit composed of transistors Q3~Q_{10}. The transistors are arranged in a Darlington configuration in order to prevent the occurrence of errors by the external current flow.

In a DEM D/A converter, the current obtained in the above explained way is then added by the current switches corresponding to each bit, as shown in Fig. 7-16. The current switches used here are identical to those used in a weighted resistance D/A converter, and thus a high conversion speed of several hundred nanoseconds may easily be obtained. It is also possible to use this type of converter as the D/A converter required for successive approximation A/D converters.

The characteristic of this system is that high accuracy elements are not required for high accuracy operation, while ripple filters are required for each current source. It is, therefore, apt for monolithic construction provided that the ripple filter is excluded and capacitor for the ripple filter attached.

This is also possible by setting the clock at an extremely high frequency and making the required volume of the capacitance small, and thus acquiring the necessary capacitance within the IC chip. It is, however, difficult to improve temporal accuracy, or in other words, to decrease the level of $\Delta T/T$, by increasing the clock frequency due to operation time of the current dividing circuit.

Current Adder D/A Converters

The basic structure of a current adder D/A converter is shown in Fig. 7-16. It is difficult to achieve high accuracy current source using the well-known R-2R ladder resistance arrangement, and DEM type has often been proposed. However, monolithic system using the DEM type module has the drawback of requiring many associated components.

Digital data of n bits fed into a D/A converter corresponds to 2^n steps of current or voltage. Thus, as shown in Fig. 7-21, 2^n equal current sources are arranged, and the current switch is controlled by adding the current sources indicated by the digital data of n [bit].

It would be ideal to have equal current values for all the current sources, but in practice, there will be some variation. Assuming that the distribution of these variations is regular and the average cur-

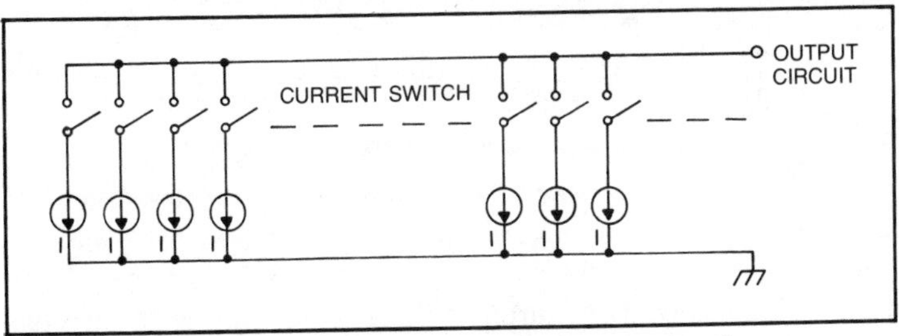

Fig. 7-21. Current adder D/A converter using the same current source.

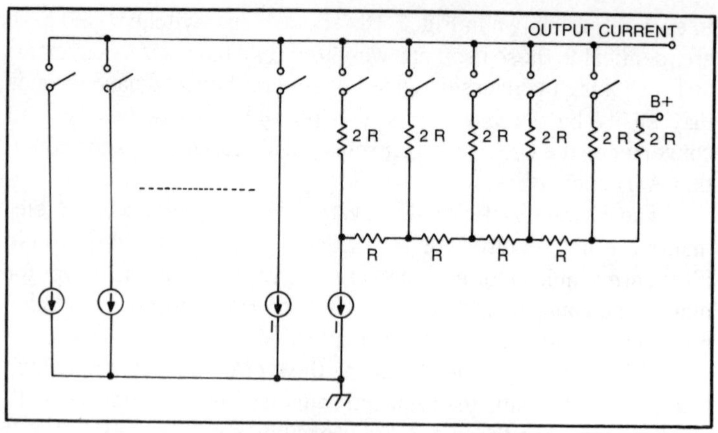

Fig. 7-22. Subordinate bits constructed by R-2R ladder.

rent value is I, current value at which m number of current sources is added will be mI, and the standard deviation of the variation σm will be:

$$\sigma m = \sqrt{m}\,\sigma\Delta \qquad \textbf{Equation 7-15}$$

Here, $\sigma\Delta$ is assumed to be the standard deviation of each current sources.

Thus, the ratio for the added current value mI will be:

$$\frac{\sigma m}{mI} = \frac{1}{\sqrt{m}} \cdot \frac{\sigma\Delta}{I} \qquad \textbf{Equation 7-16}$$

When using a 16 bit system, for example, assuming a variation $\sigma\Delta/I$ of 1% and substituting $m = 2^{16}$ into Equation 7-16, we acquire:

$$\frac{\sigma m}{mI} = \frac{0.01}{\sqrt{2^{16}}} \doteqdot 3.9 \times 10^{-15} \qquad \textbf{Equation 7-17}$$

This is insufficient for the accuracy required for 16 bits, but the accuracy for the changes in each stage would be:

$$\frac{\sigma m}{mI} \doteqdot \frac{0.01}{2^{16}} = 1.5 \times 10^{-7} \qquad \textbf{Equation 7-18}$$

which is sufficient for the 16 bit accuracy.

This system requires a large number of current sources and current switches, so there is no real alternative to monolithic construction. Even in the monolithic form, however, there are still too many components required for this circuit as it stands. An actual circuit therefore, is constructed as shown in Fig. 7-22, in which the subordinate bits are made from a resistance ladder arrangement. As an accuracy of around 5% may also be obtained from the scattered resistances within the IC, it is sufficient for the subordinate 4 to 6 bits.

The accuracy of the added current will be slightly reduced by reducing the number of current sources, but the accuracy shown in Equation 7-18 still provides sufficient margin for satisfactory operation.

Conversion speed is determined by the speeds of the current switch connected to each current sources, and a speed of around several hundred nanoseconds may be achieved, as with weighted resistance or DEM D/A converters. This type can also be used in successive approximation A/D converters.

SIGNAL PROCESSING CIRCUITS

After the analog signal has been converted into a digital signal by the A/D converter, digital signal processing corresponding to the recording system (rotary head, stationary head or DAD) is carried out. At this stage of the signal processing, the inspection bits for error detection and correction are added, and interleaved with the existing signal.

There are many different methods of error correction and detection, as described in Chapter 6, and it is necessary to choose a proper system depending on the code error rate (or the bit error rate of the recording medium), recording system, and the required performance.

Here, an actual circuit construction for parity check and CRCC, which are often used in digital audio devices and were explained in the previous chapter as practical error detection codes, as well as a simple interpolation circuit are explained. The crossword code is taken as an example of an error correction code, of which the actual circuits construction is explained.

Parity Check

When parity bit P is added to information of k [bit], the structure must satisfy the following equations:

$$b_1 + b_2 + b_3 + \ldots\ldots\ldots b_k + P = 0 \qquad \textbf{Equation 7-19}$$

$$b_1 + b_2 + b_3 + \ldots\ldots\ldots b_k + P' = 1 \qquad \textbf{Equation 7-20}$$

Here, $b_1, b_2, b_3 \ldots\ldots b_k$ represent the information bits, and the plus sign indicates modulo 2 addition. The parity shown in Equation 7-19 is known as even parity, while that shown in Equation 7-20 is known as odd parity.

When designing the above as a circuit, a modulo 2 addition circuit is required. Construction of a mod 2 addition circuit using gate circuits is shown in Fig. 7-23, and is known as an exclusive OR. The circuit for producing parity bits using the above mentioned circuit is shown in Fig. 7-24, at K = 7 as an example.

Parity check circuit at the reproduction side requires all the mod 2 addition for the information bits and the parity bits. Result of zero for the even parity indicates no code errors (to be precise, the same result may be obtained with error in excess of 2 bits), while 1 is obtained for the odd parity for the same result. Thus, the circuit used for parity check, as shown in Fig. 7-25, requires mod 2 addition for all the bits, including the parity bits.

CRCC Circuit

A division circuit is used to add detection bits of n−k[bit] to information bits of k[bit]. Taking k = 7 as an example, a circuit in which information bit polynomial M(x) multiplied by x^{n-k} is divided by generation polynomial G(x) is considered, assuming that:

$$M(x) = x^6 + x^5 + x^4 + x^3 + 1 \qquad \textbf{Equation 7-21}$$

$$G(x) = x^4 + x^3 + 1 \qquad \textbf{Equation 7-22}$$

This division circuit is constructed by an adder circuit with the shift register, as shown in Fig. 7-26. This circuit operates as follows: first of all, shift registers one to five are re-set to zero before operation starts. 1 is fed to the input of shift register 1 by the shift clock upon 1 corresponding to x^{11} is fed into the input. The input is

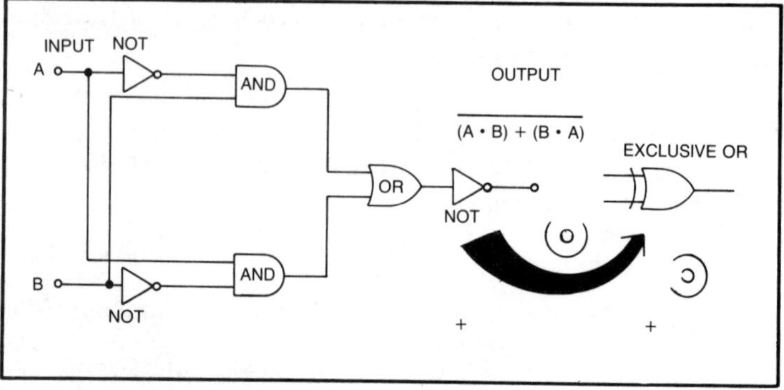

Fig. 7-23. Modulo 2 addition circuit (Exclusive-OR).

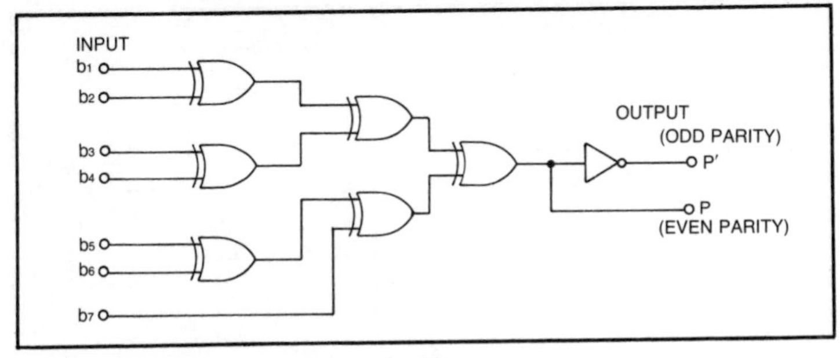

Fig. 7-24. Circuit for generating parity bits.

fed to each shift register by the shift clock in order. However, the output will remain at zero until the third shift is carried out. Then the output of shift register 4 fed by the fourth shift clock pulse will be 1, and is simultaneously fed into the adder circuit, while the input of each shift register changes in order to correspond with the generation polynomial. Thus, the output of each shift register indicates $x^{11} + x^{10} + x^9 + x^8$, while the input of each is $(x^{10} + x^9 + x^8) + (x^9 + x^7)$, and the first operation will be carried out in the next shift clock.

In this way, the division continues until the last data is absorbed into shift register 4, and the contents of the shift register will indicate the excess.

The detection circuit at the reproduction side is a division circuit using the same generation polynomial G(x) used for recording; the operation is also the same.

Code Error Compensation Circuits

There are many types of code error compensation, ranging from simple muting to very complex methods of using linear predic-

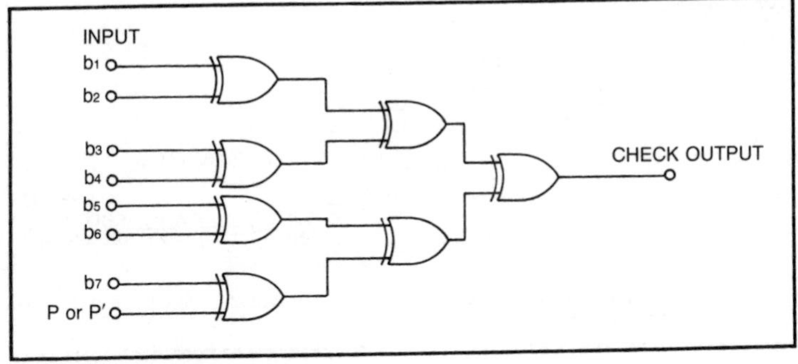

Fig. 7-25. Parity check circuit.

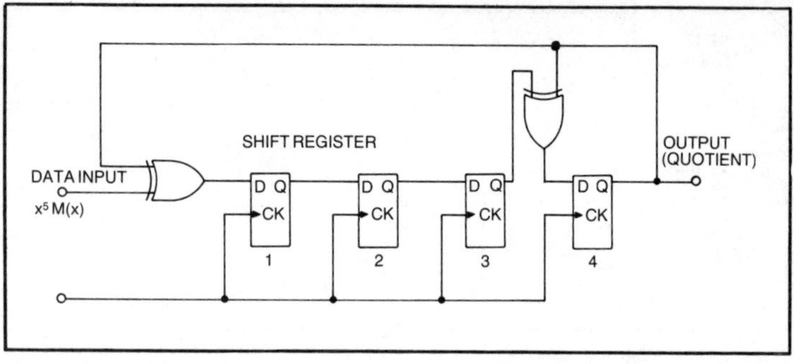

Fig. 7-26. Division circuit.

tion. Here, an effective linear interpolation method which the circuit is constructed easily is considered.

With linear interpolation, the average value of the words preceding and following the error word are calculated, and is substituted for the error word. This circuit can be constructed by a full adder, as shown in Fig. 7-27. When there is no error in the input word, it is fed out by the ON/OFF switch after passing through the first stage of the shift registers. On the other hand, when a word contains a bit error the output word of the full adder circuit at the first stage of the shift register will equal to ½ of the added value of the preceding and following words. Here, the output word is linearly interpolated by switching over to the output side of the full adder circuit.

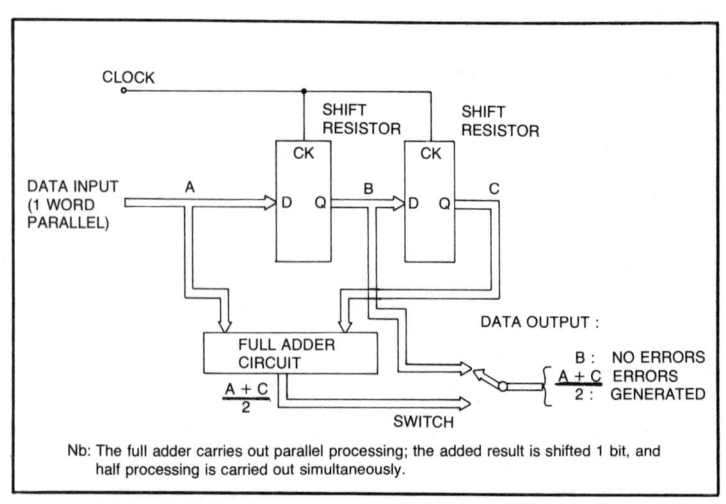

Fig. 7-27. Block diagram for a interpolation circuit.

Crossword Code

As mentioned in the previous chapter, the crossword code is one of the most effective error correction methods used in digital audio equipments. The encoding circuit is composed of a digital memory which stores the incoming words in order in block units, and of a mod 2 adder circuit which creates the horizontal and vertical parities. Figure 7-28 shows the crossword code and interleave structure of an actual PCM tape recorder. Here, six information and detection words make up one block, and the interleave is carried out over 35 blocks. Interleave is carried out to change the order of the data to be sent down the transmission path in order for long burst code errors to be dealt with effectively. The data is returned to its original order upon decoding, and long burst errors are made random in nature, thus simplifying the correction process. Figure 7-28 shows an example in which a code which is capable of correcting burst error of only up to four words becomes capable of correcting burst error up to 140 words by the effect of interleaving.

Figure 7-29 shows the encoding and decoding paths, while an example of an encoding circuit is shown in Fig. 7-30. Information words A, B are fed out from rows 1 and 3, respectively, and each bit of these words are added using mod 2 addition and fed out from row 2. After three information words have been transmitted, the horizontal parities are memorized in the shift register, and encoding for one block is completed upon stopping the input of the information words and feeding out the horizontal parities in order.

Figure 7-31 shows the decode algorithm for a theoretical distinction circuit. Process addresses 2 to 5 are used for burst errors, while that from 6 to 25 are used for random errors. This decode algorithm uses only one part of the coding capacity. It would be necessary to include addresses up to 162 to demonstrate the total capacity of the system, but, taking into consideration the hardware complexity, here it is kept only up to 26.

MODULATION CIRCUITS

By modulation circuits we are here referring to the signal conversion circuits used for recording the digital data onto the relevant record medium. As explained in Chapter 3, there are many types of modulation systems which are used for recording digital signals onto magnetic tapes. For PCM processor connected to a VTR, it is also necessary to add the requisite horizontal and vertical sync signals.

As a practical modulation method the optical DAD (Digital

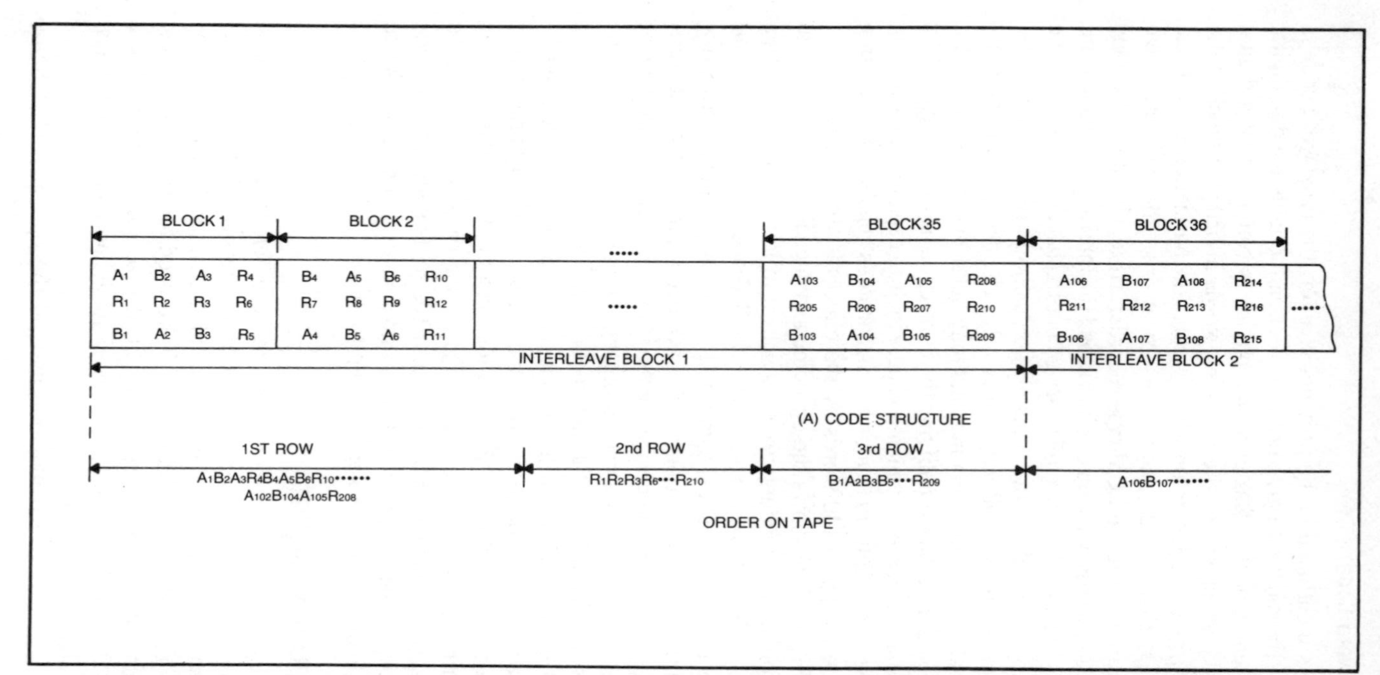

Fig. 7-28. Code and interleave structure for a PCM tape recorder (By changing the order of the words written on the magnetic tape-interleave-long burst errors can be corrected).

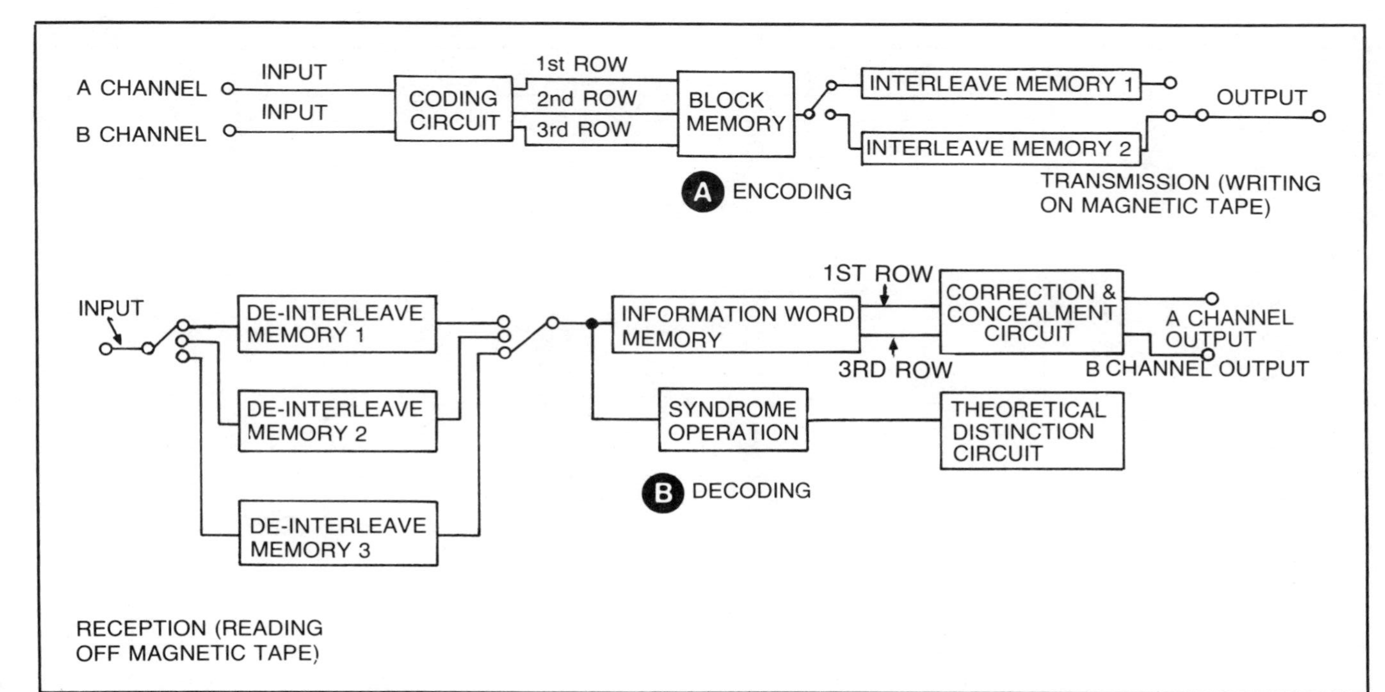

Fig. 7-29. Block diagram for coding and decoding.

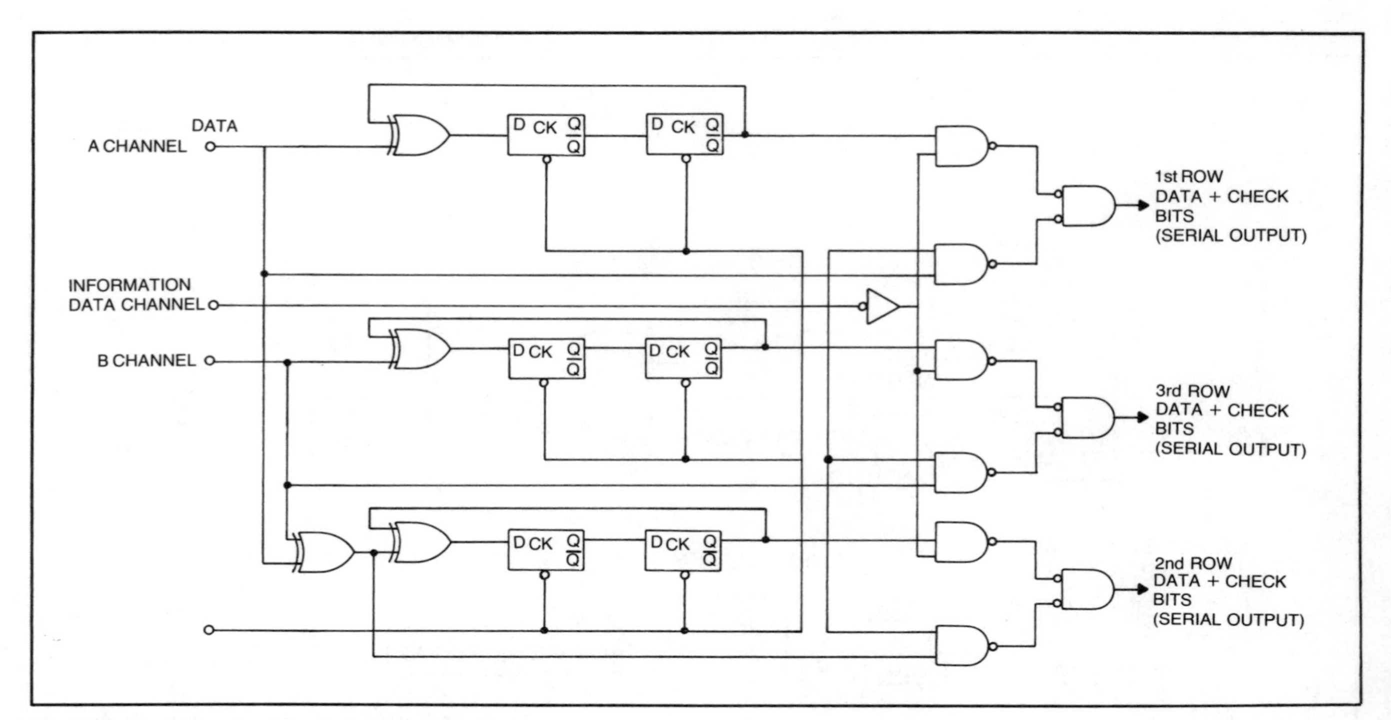

Fig. 7-30. Encoding circuit for the crossword code.

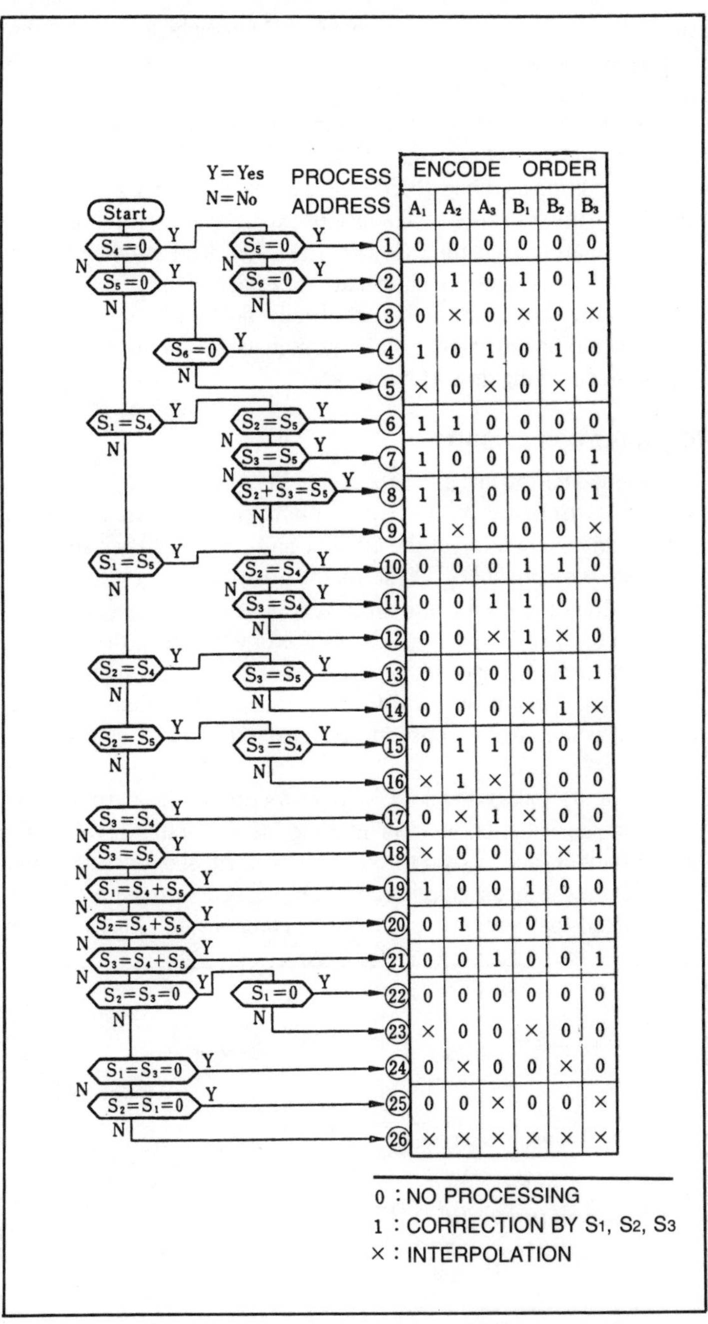

PROCESS ADDRESS	ENCODE ORDER					
	A_1	A_2	A_3	B_1	B_2	B_3
1	0	0	0	0	0	0
2	0	1	0	1	0	1
3	0	×	0	×	0	×
4	1	0	1	0	1	0
5	×	0	×	0	×	0
6	1	1	0	0	0	0
7	1	0	0	0	0	1
8	1	1	0	0	0	1
9	1	×	0	0	0	×
10	0	0	0	1	1	0
11	0	0	1	1	0	0
12	0	0	×	1	×	0
13	0	0	0	0	1	1
14	0	0	0	×	1	×
15	0	1	1	0	0	0
16	×	1	×	0	0	0
17	0	×	1	×	0	0
18	×	0	0	0	×	1
19	1	0	0	1	0	0
20	0	1	0	0	1	0
21	0	0	1	0	0	1
22	0	0	0	0	0	0
23	×	0	0	×	0	0
24	0	×	0	0	×	0
25	0	0	×	0	0	×
26	×	×	×	×	×	×

0 : NO PROCESSING
1 : CORRECTION BY S_1, S_2, S_3
× : INTERPOLATION

Fig. 7-31. Crossword code decode algorithm for a PCM tape recorder.

Audio Disc) utilizes modulation by 3PM in which 0 and 1 are corresponded to the presence or absence of pit for the recording. In this type of recording, a laser beam is modulated into signals of various strength by a laser photo-sensitive crystal, and the process is carried out by sensitising a photo-resistor.

For stationary head PCM tape recorders, 0 and 1 are corresponded to the presence or absence of magnetic inversions after the 3PM modulation, as is the case with the DAD. A circuit diagram and associated waveforms are shown in Fig. 7-32, and the record head is driven by the current drive. In digital audio processor connected to a VTR, the horizontal and vertical sync signals are applied to the NRZ digital signal using an adder circuit.

DEMODULATION CIRCUITS

Demodulation circuits are designed to convert the reproduction signal from the recording medium into the original digital signal, and to separate out the bit sync and the frame sync signals.

With the optical DAD system, the positive or negative electrical signal is extracted from photo-receptive elements (phototransistors, etc.) depending on the presence or absence of the pits, and is passed through a waveform shaper to obtain digital signal identical to that at the time of recording. Bit sync signal is simultaneously acquired by feeding the reproduced signal into the PLL (Phase Lock Loop).

On a stationary head PCM tape recorder, the output signal from the reproduction head is amplified to a suitable level by the head amplifier, and passed through a waveform shaper. The output signal from the head amplifier is then passed through a filter so constructed that any peak shift can be compensated for by a delay

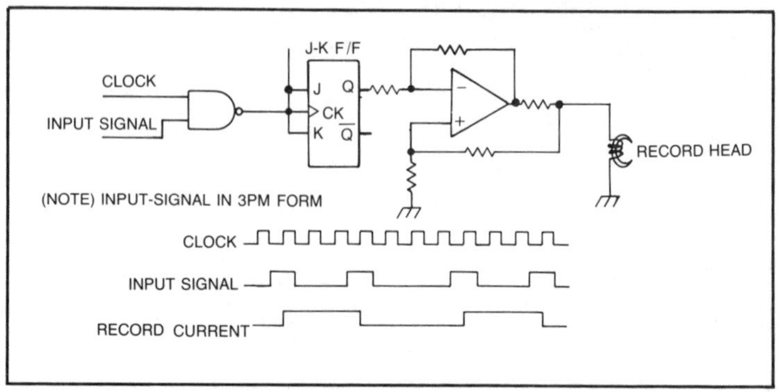

Fig. 7-32. A record circuit for a fixed head PCM tape recorder.

line, and is fed into an integrator sampling circuit. Differentiation circuit may also be used for the sampling circuit, as it is strong against noise, and the acquired signal possesses the equivalent current to that flowed through the head at the time of recording. (Applicable only for method in which the reproduction head responds to the flux changes.)

Sync separation circuits used in PCM processor connected to a VTR is the same in principle to that used in TV receivers; in order to improve stability, however, the time constants for AGC and clamp circuits are set different.

Fig. 7-33. Sony CX-899 A/D and CX-890 D/A LSI converters.

References

1. Gorden, B.M.:IEEE Trans., CAS-25, 7 (1978-7).
2. Plassche, R.J.: IEEE Trans., SC-11, 6 (1976-12).
3. "Analogue-to-Digital Conversion Techniques", Motorola Application Note AN-471 (1974-10).
4. Shingakkai-hen: Electronic Communications Handbook, Ohm Co.

Chapter 8

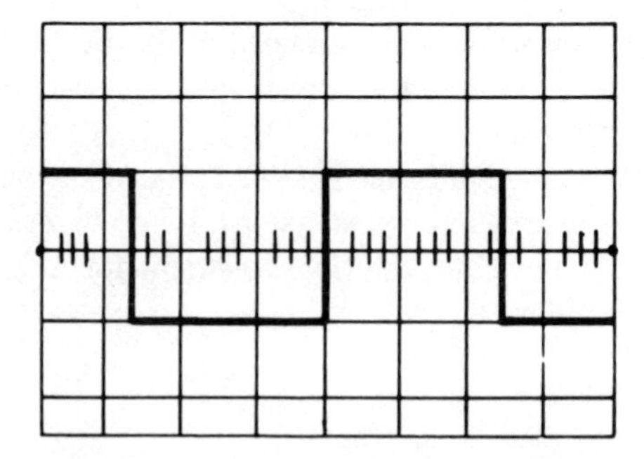

Standardization for Domestic PCM Tape Recorders

Of the rotary head PCM tape recorders shown in Table 4-2, the most important type for the purposes of standardization is no. 9, C format. This chapter is devoted to the explanation of C format, and will therefore have a specialist content.

The following points are a summary of the basic policy necessary as a preamble to more detailed discussions on C format:

1. A standard NTSC video signal is used (the standard TV signal for the U.S.A. and Japan).
2. It is possible to use a VTR as an adaptor. Decks designed as one unit with an integral rotary head system should use the same format.
3. It is necessary to develop sufficient facilities for future use as a domestic super hi-fi tape recorder.
4. A sufficient variety of product must be maintained. Differentials in performance and price ranging from products price-competitive with compact cassette recorders to prestige models.
5. There must be some guarantee against performance variations in domestic VTRs and tapes, and against time-related deterioration, so far as the PCM side is concerned. In other words, PCM equipment must be designed so that reliability of performance can be maintained for the VTR.

Table 8-1 gives an overview of C format standardization, and each paragraph details the relevant content and basis. The number of channels is specified as two, because this equipment will be used as a domestic tape recorder. However, as mentioned above in point 4, the actual electronic characteristics of a PCM tape recorder based on the C format definition will vary slightly to provide the necessary range of choice. The theoretical limits are shown below in Table 8-2.

DATA DENSITY FOR A DOMESTIC VTR

C format is aimed primarily at use with domestic VTRs, and it is therefore necessary to investigate the data density which can be recorded on a domestic VTR. Data density is commonly defined as time density (bit transmission rate, bit/s) and as spatial density on the tape surface (BPI=bit/inch or bit/mm^2). We will now examine the results of experiments carried out on the former premise.

The data shown in Fig. 8-1 is based on figures obtained from experiments carried out using a domestic VTR (Betamax, SL-8500), which has to date had the largest world distribution. In the graph, the bit transmission rate is plotted along the horizontal axis, and the block error ratio along the vertical axis. Assuming that each block unit occurs in one horizontal scan period (H), then the error rate for one block in every 1000 blocks will be $P_H=10^{-3}$. Since there are approximately 870,000 information lines every minute (this will be explained later), then $P_H=10^{-3}$ will correspond to about 870 block errors in one minute.

There are many reasons for an increase in the error ratio when the bit transmission rate is increased. A lack of frequency response in the VTR, a deterioration in the S/N ratio, and an increase in the effect of jitter will all lead to a failure to transmit the information accurately.

The range of dispersion in Fig. 8-1 reaches one to two figures, but this is due to tape characteristics and the diversity of VTR adjustments. The results for other domestic VTRs are in general the same, and since the dispersion is quite large, we can generally ignore the difference between the various VTRs themselves. If we consider the dispersion carefully, we can assume that the upper limit for data density is 3 Mb/s.

RELATIONSHIP WITH HORIZONTAL AND VERTICAL SYNC

As mentioned in Chapter 4, horizontal and vertical sync pulses must be added when the PCM signal is converted into a pseudo-video signal. In addition, provision for periods of no signal must be added, coincident with the head switching interval if a rotary head VTR is to be used. A total of 9 H is necessary for the vertical sync

Table 8-1. Standardization for Domestic PCM Tape Recorders (C Format).

No.	Items	Contents
1	Number of channel	2
2	Number of bits	14 bits/slots per channel
3	Quantization	linear quantization, 2's complement
4	Modulation	NRZ
5	Sampling frequency f_s	44.056 kHz
6	Max. bit transmission rate	2.643 Mb/s
7	Pseudo image signal	standard NTSC video signal (except above)
7-1	Data disposition in horizontal scan period	

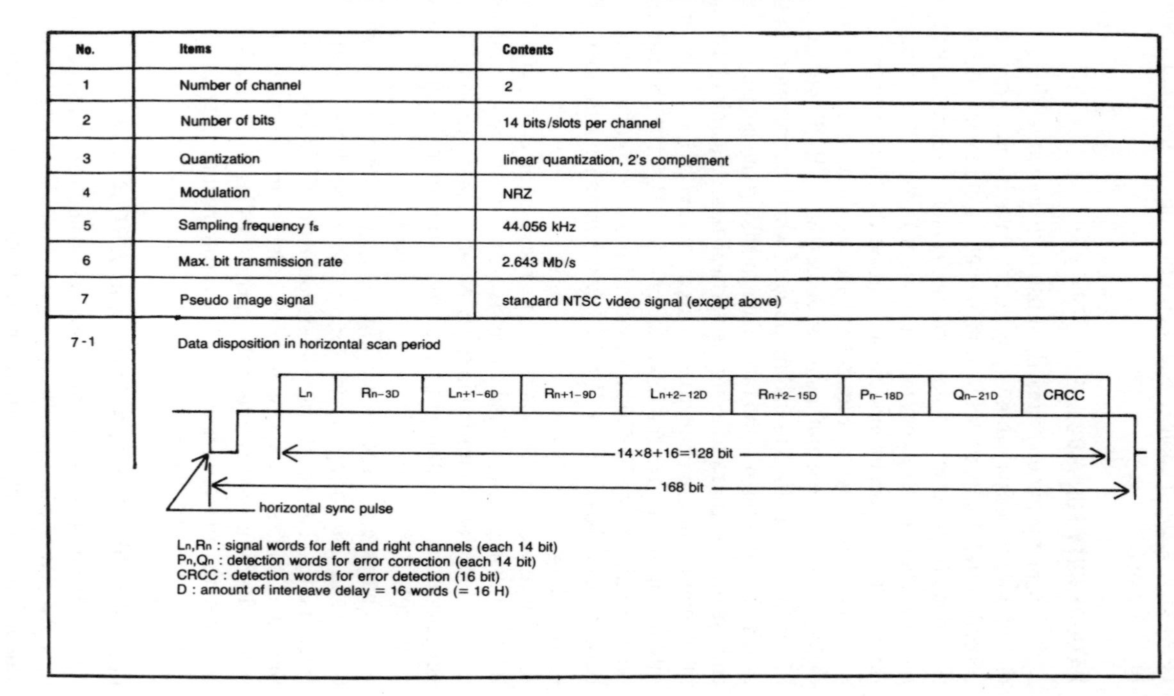

L_n, R_n : signal words for left and right channels (each 14 bit)
P_n, Q_n : detection words for error correction (each 14 bit)
CRCC : detection words for error detection (16 bit)
D : amount of interleave delay = 16 words (= 16 H)

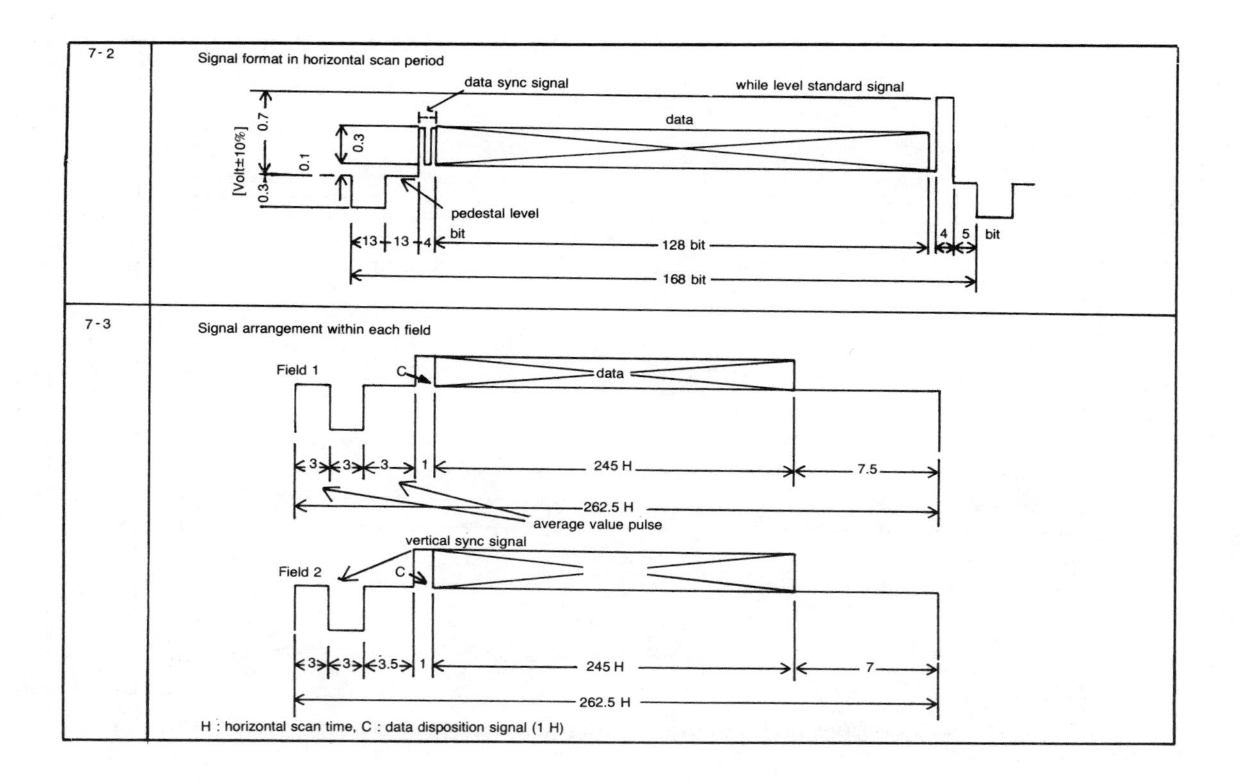
7-2
Signal format in horizontal scan period
data sync signal
while level standard signal
data
[Volt±10%]
0.7
0.3
0.1
0.3
pedestal level
bit
13
13
4
128 bit
4
5
bit
168 bit
7-3
Signal arrangement within each field
Field 1
C
data
3
3
3
1
245 H
7.5
262.5 H
average value pulse
vertical sync signal
Field 2
C
3
3
3.5
1
245 H
7
262.5 H
H : horizontal scan time, C : data disposition signal (1 H)

Table 8-1. Standardization for Domestic PCM Tape Recorders (C Format). (Continued from page 247.)

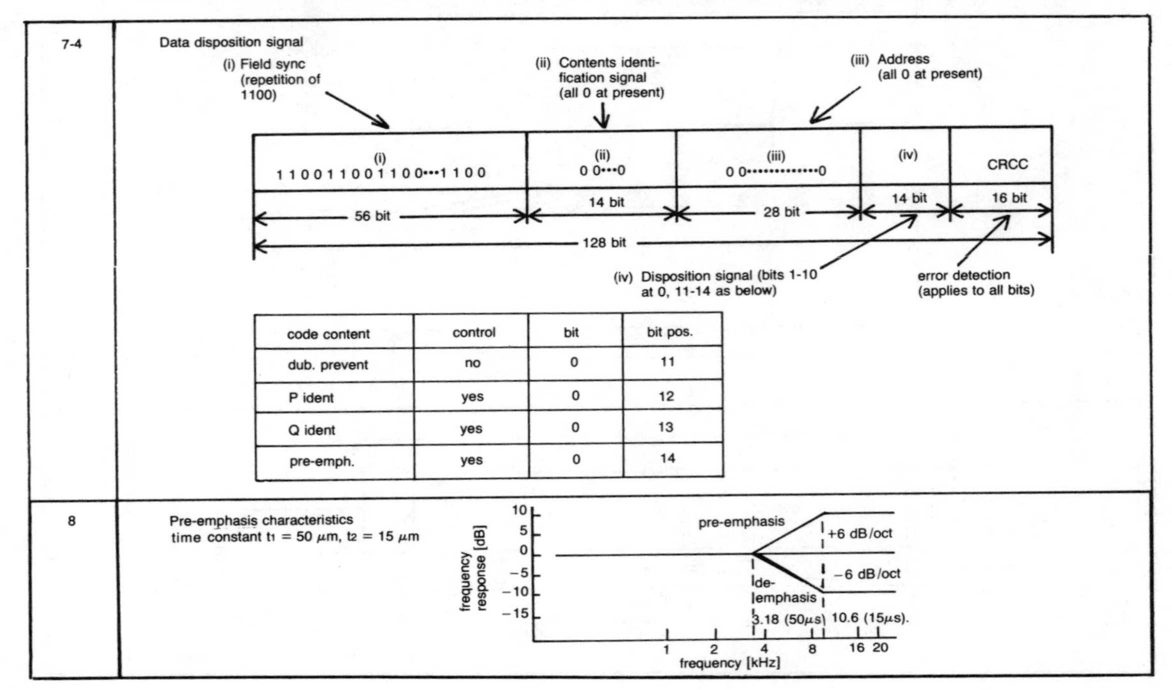

code content	control	bit	bit pos.
dub. prevent	no	0	11
P ident	yes	0	12
Q ident	yes	0	13
pre-emph.	yes	0	14

9	Error detection and correction signal : A combination of CRCC, interleave, erasure, adjacent codes, crossword codes, etc. with a redundancy of 34.4% (excluding sync periods)
9-1	Error detection code : CRCC, generation polynominal $G(X)=X^{16}+X^{12}+X^{5}+1$, shift resister with all 1 reset
9-2	Interleave : Simply delay interleave for 16 word blocks (see 8.5, 8.7)
9-3	Erasure decoding : Using the CRCC as pointer, a one word error anywhere in the block can be corrected using the erasure method.
9-4	Adjacent codes : Using the CRCC as a pointer, a two word error anywhere in the block can be corrected using an adjacent code. $P_n = L_n \oplus R_n \oplus L_{n+1} \oplus R_{n+1} \oplus L_{n+2} \oplus R_{n+2}$ $Q_n = T^6 \cdot L_n \oplus T^5 \oplus \cdot R_n \oplus T^4 \cdot L_{n+1} \oplus T^3 \cdot R_{n+1} \oplus T^2 \cdot L_{n+2} \oplus T \cdot R_{n+2}$ nb. $\oplus$ INDICATES ADDITION OF 2 FOR EACH BIT CORRESPONDING TO A WORD $T = \begin{bmatrix} 0&0&0&0&0&0&0&0&0&0&0&0&0&1 \\ 1&0&0&0&0&0&0&0&0&0&0&0&0&0 \\ 0&1&0&0&0&0&0&0&0&0&0&0&0&0 \\ 0&0&1&0&0&0&0&0&0&0&0&0&0&0 \\ 0&0&0&1&0&0&0&0&0&0&0&0&0&0 \\ 0&0&0&0&1&0&0&0&0&0&0&0&0&0 \\ 0&0&0&0&0&1&0&0&0&0&0&0&0&0 \\ 0&0&0&0&0&0&1&0&0&0&0&0&0&0 \\ 0&0&0&0&0&0&0&1&0&0&0&0&0&1 \\ 0&0&0&0&0&0&0&0&1&0&0&0&0&0 \\ 0&0&0&0&0&0&0&0&0&1&0&0&0&0 \\ 0&0&0&0&0&0&0&0&0&0&1&0&0&0 \\ 0&0&0&0&0&0&0&0&0&0&0&1&0&0 \\ 0&0&0&0&0&0&0&0&0&0&0&0&1&0 \end{bmatrix}$
9-5	Crossword code : Using the CRCC and P_n, a very high degree of correction for random errors can be achieved.

Table 8-2. Theoretical Performance Limits.

1.	Frequency and bandwidth	Dc-20 kHz
2.	Dynamic range	85.5 dB (noise bandwidth 20 kHz)
3.	Distortion	0.003% (maximum signal level)
4.	Wow and flutter	0 (to accuracy of crystal used)

pulse and the average value pulse period (H:horizontal scan and return period. See Table 8-3). The approved value for the head switching period is 4 H, and for every V (V:vertical scan and return period) a 1 H data control signal must be added. Thus, for every 1 V there are 14 H where the PCM signal cannot be recorded. The number of H where a PCM signal can be recorded in 1V is shown in the equation below:

$$H_{PCM} \leqq 248 = \left(\frac{525}{2}\right) - 14 \qquad \textbf{Equation 8-1}$$

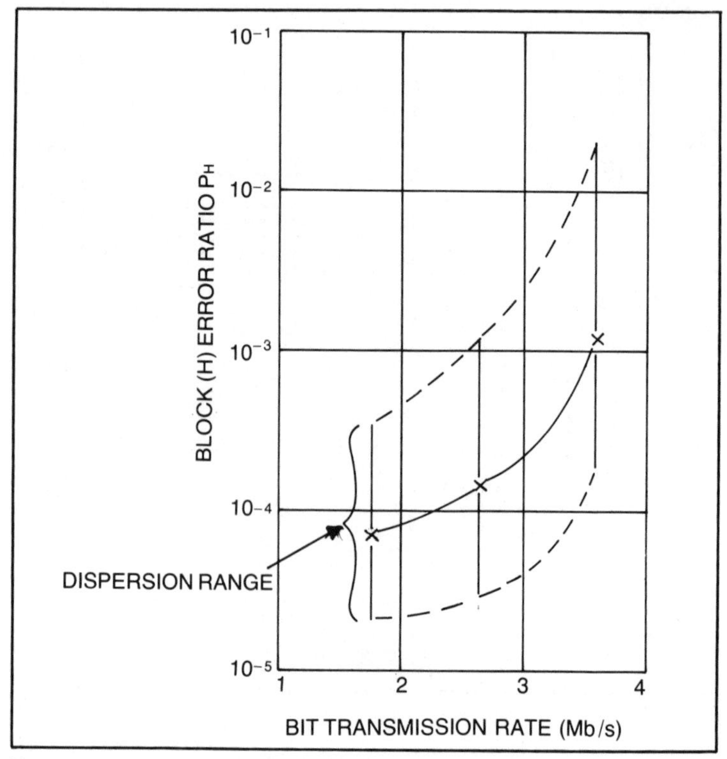

Fig. 8-1. Bit transmission rate and error ratio (SL-8500).

Assuming that we use vertical sync pulses as described above for the sync on a PCM signal, it is desirable that we make W_H (the number of words in 1 H) an even number. (This is because we are using two channels). Figure 8-2 shows the relationship between the bit transmission rate and the amount of redundancy where W_H is 4, 6, 8. If we calculate that one word contains either 14 or 16 bits, and do not, for the moment, include the sync pulses or the head switching interval, the redundancy is as shown in the diagram.

30% or more redundancy must be included for code error correction purposes (see Chapter 6). Therefore, in order to keep the bit transmission rate below 3 Mb/s, it is necessary to calculate that W_H=6 where one word is made up of 14 bits. See Fig. 8-3.

SELECTION OF A SAMPLING FREQUENCY

The number of Vs recorded in one second is f_V, and the number of Hs per V is H_{PCM}, while the number of words in one H is, for each channel, $W_H/2$. The number of words recorded in one second can be calculated using Equation 8-2.

$$fs = \frac{W_H}{2} H_{PCM\,fv} = 3\, H_{PCM\,fv}$$ **Equation 8-2**

The lower limit for f_s can be determined from the bandwidth of the signal to be recorded and the characteristics of the filter used. Generally speaking, the upper limit for human hearing is 20 kHz, and it is desirable to fix the sampling frequency higher than 44 kHz

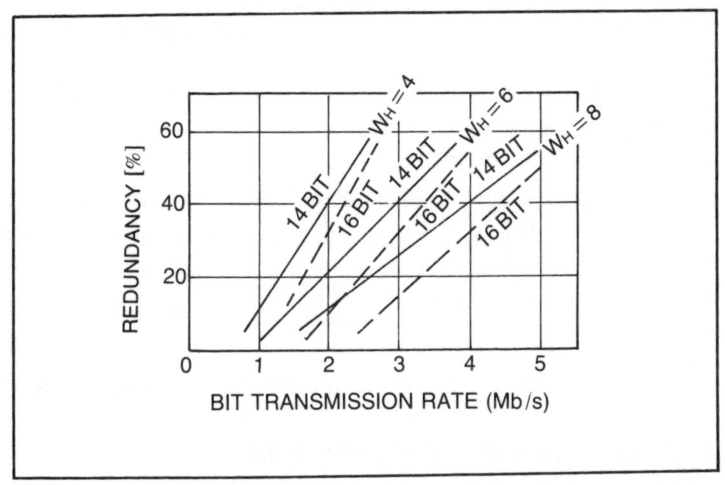

Fig. 8-2. Number of words (WH) in one H and bit transmission rate.

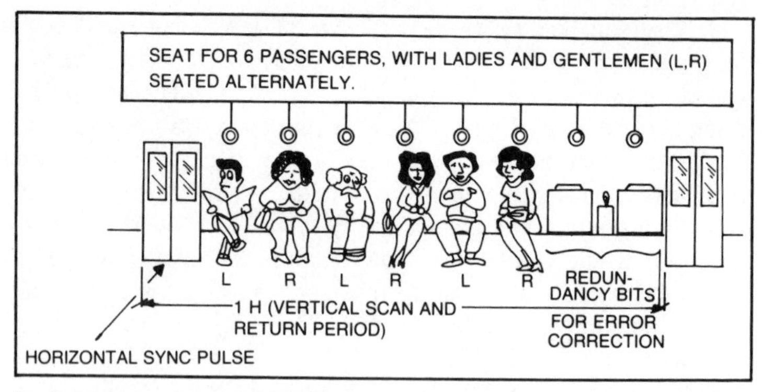

Fig. 8-3. Using a domestic VTR (Betamax, VHS) 6 words can be mounted on 1H (3 words each for left and right channels).

to prevent the price of the filters from soaring dramatically. Equation 8-4 can be used to calculate the lower limits of H_{PCM}:

$$f_s = 3H_{PCM\,fv} \geqq 44 \text{ (kHz)} \qquad \textbf{Equation 8-3}$$

$$H_{PCM} \geqq 244.698 \qquad \textbf{Equation 8-4}$$

The upper limit of H_{PCM} has already been decided in Equation 8-1, and it is, of course, desirable to ensure that H_{PCM} is an integer by definition. Thus, the possible values of H_{PCM} are limited to four: 245, 246, 247 and 248. We must, however, choose the lowest common denominator for the system master clock frequency f_M, using the PCM related frequencies f_s and the bit clock frequency as well as video related frequencies, such as f_V and f_H (see Table 8-3). To put this another way: in this type of system we must have a master clock f_M in the form of a crystal oscillator. All the other frequencies needed must then be constructed from the master clock frequency.

From the above-mentioned choices available for H_{PCM}, if we choose H_{PCM}=245, then $f_M \doteqdot 7.05$ MHz(1), and if we choose another value, $f_M > 200$ MHz. In the latter case, not only would the crystal become very expensive, but the counter itself would not conform to normal TTL. From Equation 8-2 we can calculate f_s using Equation 8-5:

$$f_s = 3H_{PCM} f_V \doteqdot 44.056 \text{ (kHz)} \qquad \textbf{Equation 8-5}$$

DATA DISPOSITION AND SIGNAL WAVEFORM

Section 7-1 in Table 8-1 shows the data arrangement for 1 H.

Table 8-3. The NTSC Video Signal.

Vertical (V) sync frequency	f_V=59.940 Hz
Horizontal (H) sync frequency	f_H=15.734k Hz
color subcarrier frequency	f_c=3.5764 MHz
Number of fields (V) in one picture	2V
Number of H in 1 V	525/2=262.5H
Size of vertical sync pulse	3H
Period of vertical sync pulse	6H (3H preceeding & following vertical sync)
V : vertical scan and return period (field period)	1/f_V=16.683ms
H : horizontal scan and return period	1/f_H=63.556μs

To summarize, in 1 H there are 3 information words each for the right and left channels, giving a total of six, and there is also one word for error detection. The actual sequence of information words is different from their original order, because they are "shuffled" using a simple delay interleave technique (see Fig. 8-4). Using this

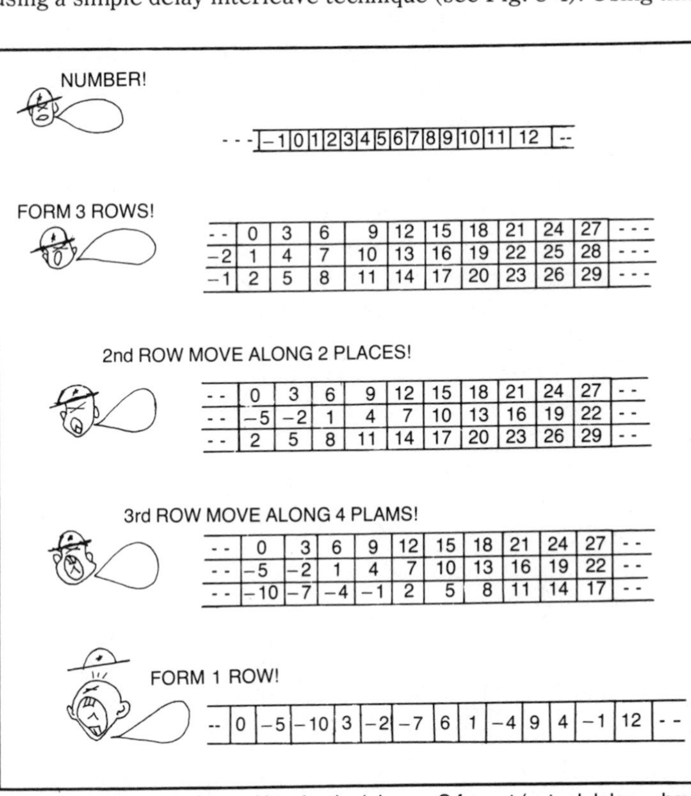

Fig. 8-4. Interleave performed by simple delay on C format (actual delay = how many places moved = 16 words).

type of interleave, error correction and concealment is simplified because a continuous code error on the tape is thus converted into random errors when the "shuffled" data is returned to its original order. (This is explained in detail in Chapter 6.)

The PCM signal levels used are 0.4 V for 1, and 0.1 V for 0. (That is, values based on pedestal level or video black level. See 7-2 in Table 8-1.) Peak white on a video signal is 0.7 V, so only about 57% of the available signal is actually being used. The reason for making these levels so low is that the effect of non-linear emphasis, used in the Betamax, is avoided. The zero level is raised to 0.1 V in order to set the PCM signal between white level and black level for VHS equipment to avoid clipping.

There are various types of AGC (Automatic Gain Control) used on VTRs; for example, one type uses the horizontal sync signal, while others use the peak video signal. Because on C format the PCM signal level is never more than 0.4 V, this would be recognized as the video signal peak and the AGC would come into operation, and thus the gain would be increased. (Mistaken operation.) To avoid this state of affairs, a signal corresponding to peak white (white reference, equivalent to 14 bits) is fed to the equipment during H flyback.

Just before the data commences, a 4 bit (1010) data sync signal is added. This is done so that automatic adjustment of the data threshold level (the level used to distinguish between 1 and 0) and phase (temporal bit position) can be carried out.

Section 7-3 in Table 8-1 shows the data arrangement in 1 V. As shown in the diagram, odd and even fields (V) are separated only by 0.5 H. Each V has 245 H of data area, and just before this there is 1 H of data control signal. The composition of the data control is as shown in section 7-4, and it consists of four words.

(i) Field (V) sync signal (56 bits, repetition of 11001100.....)

(ii) Contents identification signal (14 bits, indicates contents of recorded tape).

(iii) Address signal (28 bits, used for electronic editor and random access.)

(iv) Control signal (14 bits, dubbing prohibited. This will prevent digital dubbing of recorded tapes by disconnecting the digital output terminals. The control signal also indicates presence or absence of the detection words P and Q, or emphasis.)

The detection words P and Q have been explained in detail in the section on error correction, and even equipment which does not use them is still fairly resistant to errors and uses error compensation. It is possible that this sort of system will be used at the cheapest end of the market. The emphasis characteristics are shown in section 8 of Table 8-1. CRCC* for error detection is included in the data control signal, and it makes invalid code errors generated.

CODE ERROR CONCEALMENT AND CORRECTION

The code error concealment and correction techniques used in C format can be divided into the combination of the below three groups:

a) simple delay interleave
b) CRCC
c) adjacent codes
   Using only part of c), the following method is possible:
d) error correction using the erasure method
   It is also possible to use the highly effective
e) crossword code system
   This is a system for highly effective error correction using the surplus pattern from CRCC (used in error detection) and also using c) and d). Code error concealment and correction is explained below.

Encoding

A simple delay interleave encoder is shown in Fig. 8-5. The encoder itself is constructed from an adjacent encoder, a delay memory and a CRCC encoder. If we call the data blocks before the interleave B_0, B_1,......., then these blocks will be arranged in the order of the original information. If we take B_0 as an example, corresponding to the information word consists of L_0, R_0, L_1, R_1, L_2, R_2, the detection words P_0 and Q_0 are constructed by the adjacent encoder as shown below:

$$P_0 = L_0 \oplus R_0 \oplus L_1 \oplus R_1 \oplus L_2 \oplus R_2$$ **Equation 8-6**

$$Q_0 = T^6L_0 \oplus T^5R_0 \oplus T^4L_1 \oplus T^3R_1 \oplus T^2L_2 \oplus TR_2$$ **Equation 8-7**

The $\oplus$ sign indicates mod 2 addition, and as shown in Fig. 8-5, the addition is carried out using an exclusive-or circuit. P_0 is a simple parity word and, as shown in Fig. 8-6, Q_0 is generated using a matrix addition circuit.

The contents of matrix T in Equation 8-7 is given in Section 9-4 Table 8-1. Essentially it is a one bit shift system, but only every

*see Chapter 6

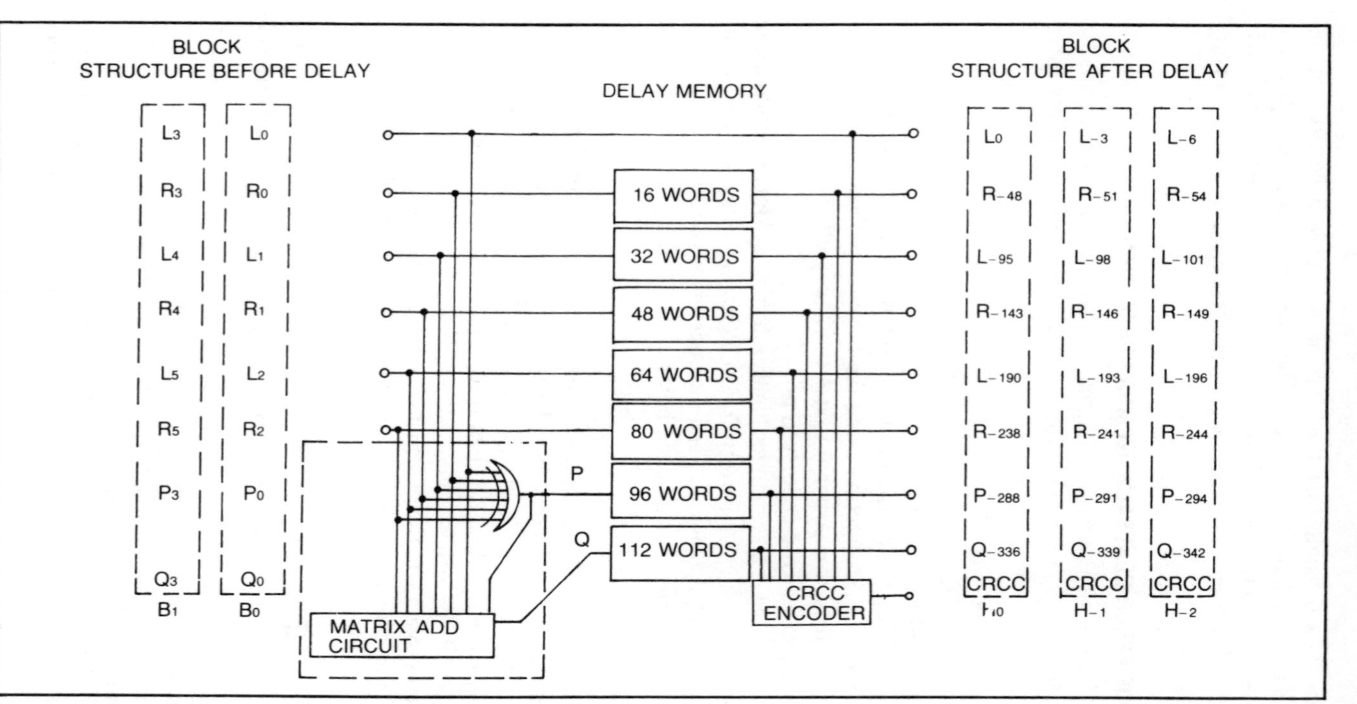

Fig. 8-5. Encoder.

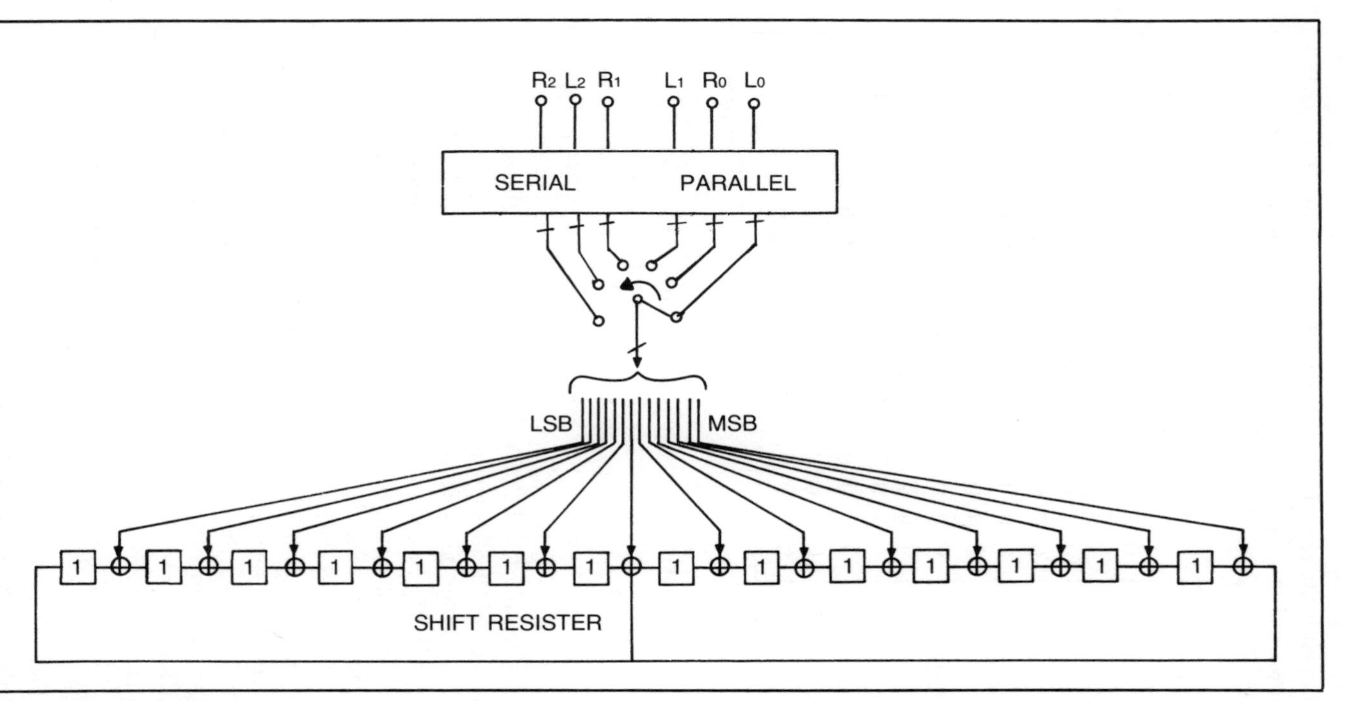

Fig. 8-6. Matrix addition circuit.

ninth bit is the sum of two bits (mod 2). The addition for matrix T is shown in Equation 8-8, where the first bit of the word is a_1, the second bit a_2 and so on.

$$\begin{bmatrix} a_{14} \\ a_1 \\ a_2 \\ a_3 \\ a_4 \\ a_5 \\ a_6 \\ a_7 \\ a_8 \oplus a_{14} \\ a_9 \\ a_{10} \\ a_{11} \\ a_{12} \\ a_{13} \end{bmatrix} = \begin{bmatrix} 0&0&0&0&0&0&0&0&0&0&0&0&0&1 \\ 1&0&0&0&0&0&0&0&0&0&0&0&0&0 \\ 0&1&0&0&0&0&0&0&0&0&0&0&0&0 \\ 0&0&1&0&0&0&0&0&0&0&0&0&0&0 \\ 0&0&0&1&0&0&0&0&0&0&0&0&0&0 \\ 0&0&0&0&1&0&0&0&0&0&0&0&0&0 \\ 0&0&0&0&0&1&0&0&0&0&0&0&0&0 \\ 0&0&0&0&0&0&1&0&0&0&0&0&0&0 \\ 0&0&0&0&0&0&0&1&0&0&0&0&0&1 \\ 0&0&0&0&0&0&0&0&1&0&0&0&0&0 \\ 0&0&0&0&0&0&0&0&0&1&0&0&0&0 \\ 0&0&0&0&0&0&0&0&0&0&1&0&0&0 \\ 0&0&0&0&0&0&0&0&0&0&0&1&0&0 \\ 0&0&0&0&0&0&0&0&0&0&0&0&1&0 \end{bmatrix} \begin{bmatrix} a_1 \\ a_2 \\ a_3 \\ a_4 \\ a_5 \\ a_6 \\ a_7 \\ a_8 \\ a_9 \\ a_{10} \\ a_{11} \\ a_{12} \\ a_{13} \\ a_{14} \end{bmatrix}$$

Equation 8-8

Putting matrix T into operation is equivalent to one shift of the shift register shown in Fig. 8-6. If the shift is repeated while supplying all the information words L_0, R_0,........ and so on, we can see that eventually Q_0 will remain in the form shown in Equation 8-7.

Each block with its adjacent code passes through the delay memory which adds a different amount of delay to each word. The CRCC is then added (Fig. 8-5). After delay, these blocks are then called H_0, H_1,..... and so on. These are the blocks stored on each 1 H after the delay has been carried out, and the data is recorded on tape in this order. (See Section 7-1, Table 8-1.)

Figure 8-7 shows the data arrangement on tape. Although H_{16} is shown as following H_0, H_1 to H_{15} have merely been omitted for convenience' sake. Before delay, the blocks were arranged diagonally as shown in the diagram. The data arrangement before and after delay is inter-related, since B_0, B_{16}, B_{32}........etc. is the data arrangement before delay, and for example, B_0 comprises the diagonal arrangement through H_0, H_{16}, H_{32}......etc.

The major advantage of this code is that it is possible to use many types of encoding while preserving complete interchangeability. Thus it is possible to adapt this system to any grade of product from a low cost, low efficiency encoder right up to an

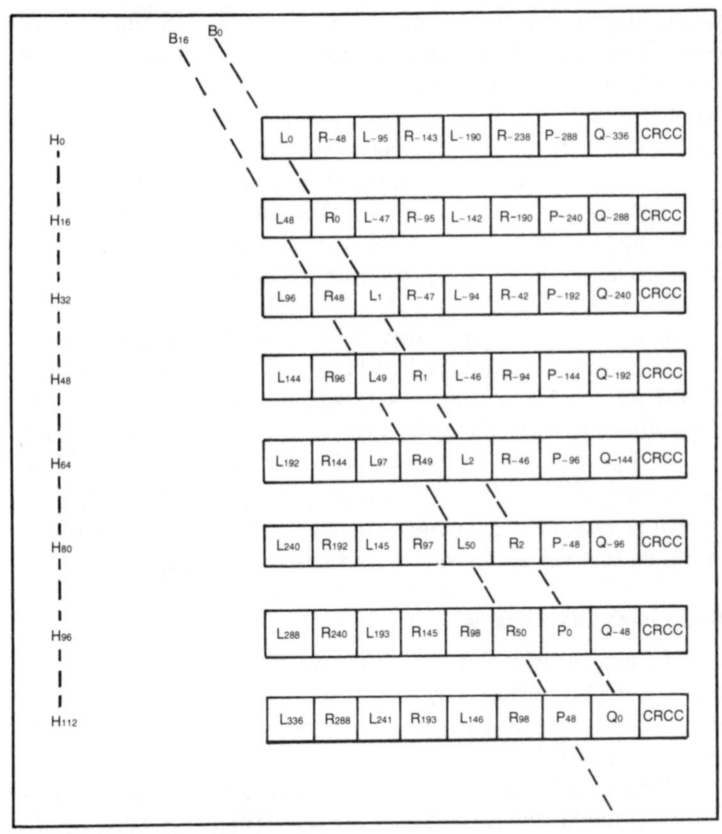

Fig. 8-7. Data arrangement on tape.

extremely sophisticated, very expensive encoder. Next we shall consider various, encoding systems.

Methods for Code Error Concealment Only

As mentioned in Chapter 6, provided that the position of the error words is sufficiently spaced, then it is generally speaking very difficult to detect where errors have been concealed in a musical signal. As shown in Fig. 8-7, the individual words are sufficiently dispersed along the tape using only a simple delay interleave. However, if a very strong 16 bit error detection code (CRCC) is added to each H, then the operation of detecting and concealing errors will be improved significantly.

For example, let us assume that all the blocks H_0 to H_{31} (4096 bits continuously) are lost, this will result in one error every three words after the data is de-interleaved (see Fig. 8-8). Similarly, if the 64H comprising H_0 to H_{63} are lost (8192 bits), there will be two error words for every three words, and error concealment will still be possible. If the length of the error exceeds 64 H, so that continuous errors longer than 5 words appear after de-interleave, then concealment noise will be noticeable when concealment is used.

Therefore, a PCM tape recorder using only concealment techniques can be acceptable for domestic use. This has, in fact, been proved by the PCM-1 processor as shown in no. 7 in Table 4-2 Chapter 4 (B format).

In this case, the error correction words P and Q are not used, and the control signal for the P and Q identification code (Table 8-1, 7-4) must stand at 1. Conversely, since it is set at 1, complete compatibility with all the various types of decoding described below can still be maintained, and the position of P and Q (total 28 bits) can be used as required. If these 28 bits are apportioned out amongst the data, it is possible to have the equivalent of 18 bits quantization per word, so that the electronic characteristics are improved.

Basic Decoding Systems

Figure 8-9 shows the simplest type of error correction decoding circuit. This ignores the detection word Q, and uses only the parity detection word P and carries out error correction by the erasure method.

First of all, the CRCC decoder checks for the presence of errors in the blocks composing each line. Then after applying the

× L_1 L_2 × L_4 L_5 × L_7 L_8 × L_{10} L_{11} × ......
× R_1 R_2 × R_4 R_5 × R_7 R_8 × R_{10} R_{11} × ......

If all the blocks H_0 to H_{31} (4,096 bits) are lost, then only one word in three is an error.

× × L_2 × × L_5 × × L_8 × × L_{11} × × ....
× × R_2 × × R_5 × × R_8 × × R_{11} × × ....

If all the blocks H_0 to H_{63} (8,192 bits) are lost, then one word in three remains correct.

Fig. 8-8. Code error concealment performance.

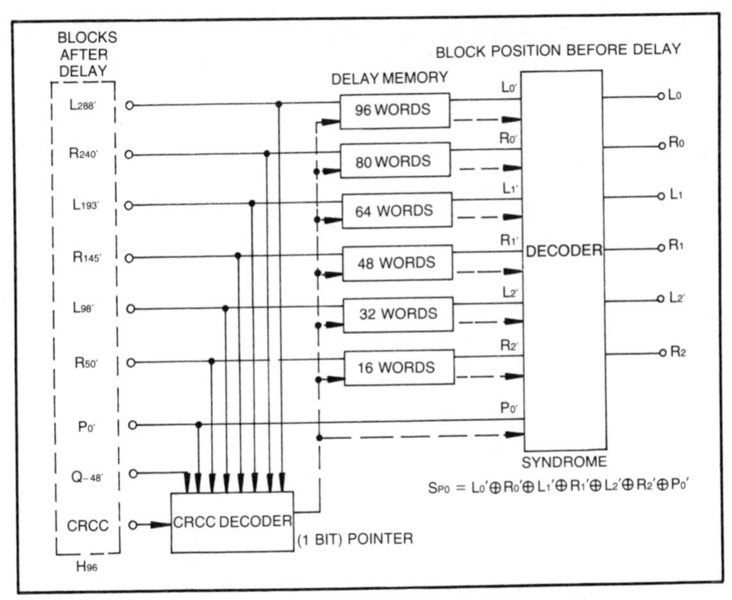

Fig. 8-9. Basic decoding system (memory contents 336 words).

reverse delay to that applied by the encoder, the data is reassembled in blocks identical to those before encoding (before delay).

To detect the presence of errors, one bit of information (pointer) is used, and the same delay is applied to every word, and these are then fed to the P decoder. Here, syndrome shown below is calculated:

$$S_{P0} = L_1' \oplus R_0' \oplus L_1' \oplus R_1' \oplus L_2' \oplus R_2' \oplus P_0' \quad \textbf{Equation 8-9}$$

The sign $'$ indicates the possibility of code errors having occurred during the record-reproduction process. If there are no errors, then,

$$S_{P0}=0 \quad \textbf{Equation 8-10}$$

on the basis of the definition given in Equation 8-6.

The possibility of correction is limited to one word only. If we assume that only the pointer corresponding to L_0' shows the existence of an error, $S_{P0} \neq 0$, and error correction is carried out using the following equation:

$$L_0' \oplus S_{P0} = L_0 \quad \textbf{Equation 8-11}$$

so that the error pointer for L_0' corresponds to S_{P0}. This method of error correction is well-known as the erasure method (see Chapter 6.

It is not possible to correct an error of two words or more, and concealment techniques must be used as explained in the previous section. As shown in Fig. 8-7, before delay the blocks of words are separated by 16 H, and furthermore, it is possible to correct a continuous 16 H (2048 bit) code error completely with this system.

Adjacent Decoding Systems

A decoder for an adjacent encoder is shown in Fig. 8-10. In an adjacent encoder the syndrome shown in Equation 8-9 and the syndrome shown in the following equation are calculated:

$$S_{Q0} = T^6L_0' \oplus T^5R_0' \oplus T^4L_1' \oplus T^3R_1' \oplus T^2L_2' \oplus TR_2' \oplus Q_0'$$

Equation 8-12

For example, if L_0' and L_1' contain errors, and the error pattern* are E_{L0}, E_{L1}, then the following equation can be derived from Equations 8-9 and 8-12. (The syndrome will be zero when there are no errors present; if errors are present, they are described only by the error pattern.)

*The error pattern is the situation where the error bit for a playback word is 1, and a non-error is 0.

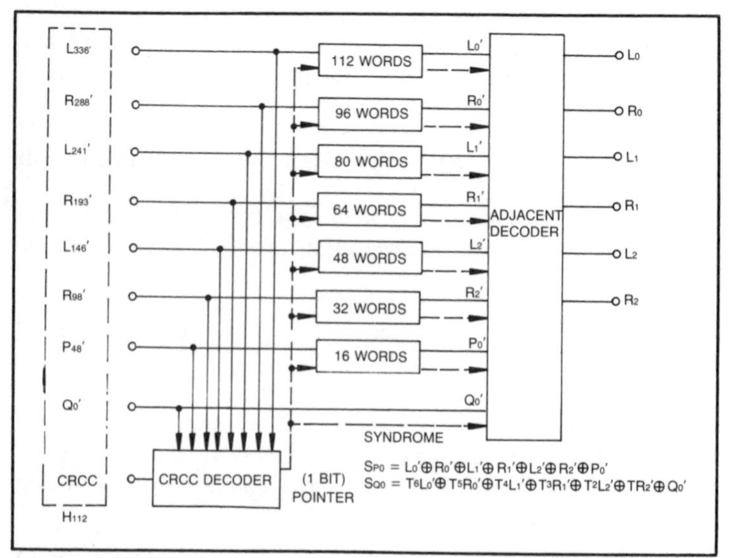

Fig. 8-10. Adjacent decoding system (memory contents 448 words).

$$S_{P0} = E_{L0} \oplus E_{L1}$$ **Equation 8-13**

$$S_{Q0} = T^6 E_{L0} \oplus T^4 E_{L1}$$ **Equation 8-14**

After these equations have been solved, the error pattern can be resolved using the following equations, and error correction is complete.

$$E_{L0} = (1 \oplus T^2)^{-1} \ (S_1 \oplus T^{-4} S_2)$$ **Equation 8-15**

$$E_{L1} = S_{P0} \oplus E_{L0}$$ **Equation 8-16**

With reference to actual hardware, it is necessary to have the information relating to the reverse matrix described in Equation 8-15 stored in a ROM. Using this type of decoder system, it is possible to correct two word errors anywhere in each block. This means that a continuous burst error of up to 32 H (4096 bits) can be corrected.

Crossword Decoding Systems[2]

For example, if the three words L_0', L_{48}', and R_{48}' are the only errors, then the CRCC pointer will indicate errors in the three blocks H_0, H_{16} and H_{32}. Correction would not then be possible with the adjacent decoder shown in Fig. 8-10. All the words (L_0, R_{-48}, L_{-95}, R_{-143}, L_{-190}, R_{-238}, L_{48}, R_0, L_{-47}, R_{-95}, L_{-142}, R_{-190}, L_{96}, R_{48}, L_1, R_{-47}, L_{-94}, R_{-42}) would, in this case, have to be interpolated. (See Fig. 8-7.)

However, these are one word errors in terms of blocks before delay,

$$B_0 = (L_0^X, R_0, L_1, R_1, L_2, R_2, P_0, Q_0)$$ **Equation 8-17**

although they are two word errors in terms of the sectional block,

$$B_{48} = (L_{48}^X, R_{48}^X, L_{49}, R_{49}, L_{50}, R_{50}, P_{48}, Q_{48})$$ **Equation 8-18**

Thus, if the detection of the error position is carried out with sufficient accuracy, then it must be possible to correct the errors using an adjacent code. (The error words are indicated by a ^{X} sign.)

If a crossword decoder, as shown in Fig. 8-11, is used, these errors can be completely corrected. The following equations and diagram indicate the basic principles. First of all the crossword

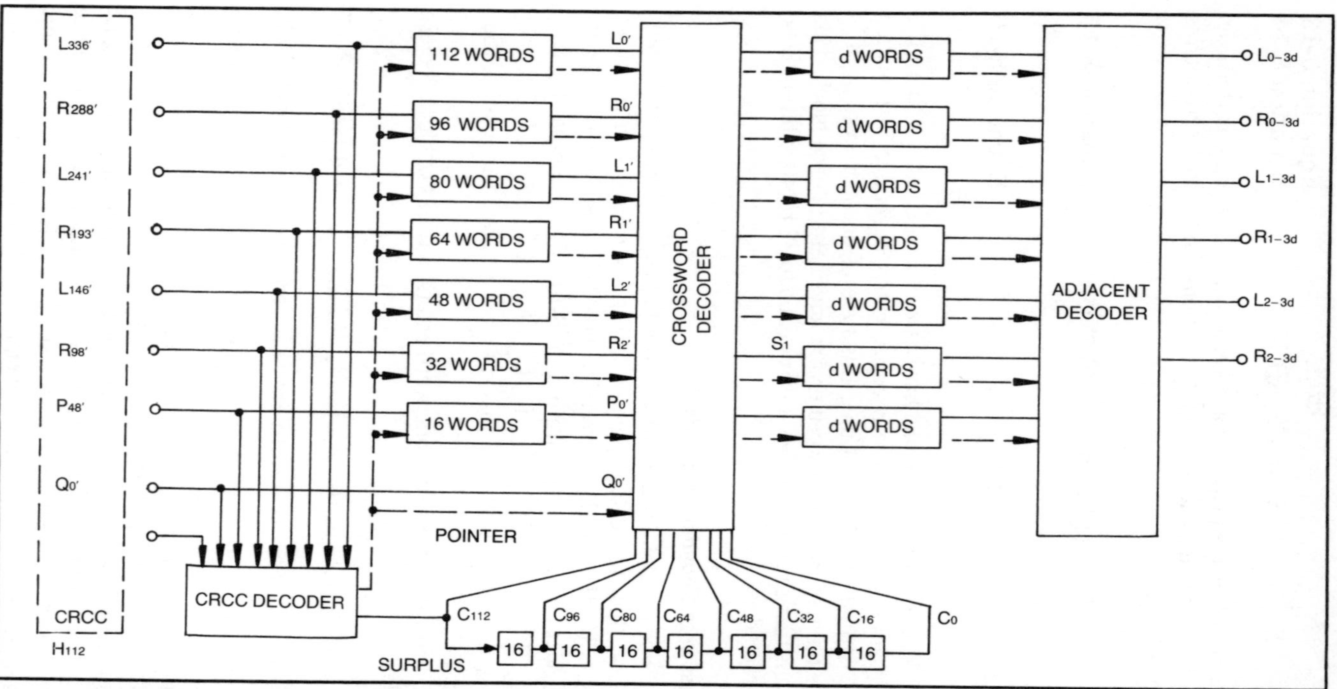

Fig. 8-11. Crossword decoding system (memory contents 560 to 1,456 words).

decoder calculates S_{P0}, as given in Equation 8-9. Then the following eight types of supplementary syndromes are used in conjunction with the CRCC decoder.

$$S_1=Res(\chi^{114}S_{P0}) \quad \textbf{Equation 8-19}$$
$$S_2=Res(\chi^{100}S_{P0}) \quad \textbf{Equation 8-20}$$
$$S_3=Res(\chi^{86}S_{P0}) \quad \textbf{Equation 8-21}$$
$$S_4=Res(\chi^{72}S_{P0}) \quad \textbf{Equation 8-22}$$
$$S_5=Res(\chi^{58}S_{P0}) \quad \textbf{Equation 8-23}$$
$$S_6=Res(\chi^{44}S_{P0}) \quad \textbf{Equation 8-24}$$
$$S_7=Res(\chi^{30}S_{P0}) \quad \textbf{Equation 8-25}$$
$$S_8=Res(\chi^{16}S_{P0}) \quad \textbf{Equation 8-26}$$

Here, Res ( ) indicates the remainder after the signals ( ) have been divided by a fixed generation polynomial. When χ^i includes the remainder, this indicates that the CRCC decoder shift register has shifted the surplus i times. Below it is shown that χ^i corresponds to the position of the word within the block. Because the CRCC is made up of all the blocks after delay (the H blocks), if only L_0^X is an error, we can assume that its error pattern is E_{LO}, and then the overall error pattern for H_0 will be:

$$\underbrace{E_{LO}}_{14\text{ bit}} \quad \underbrace{0000...........0000}_{114\text{ bit}} \quad \textbf{Equation 8-27}$$

$$\longleftarrow 128\text{ bit} \longrightarrow$$

If we assume that, at the time of CRCC decoding, the remainder is C_0, then

$$C_0=Res(\chi^{114}E_{L0}) \quad \textbf{Equation 8-28}$$

That is, χ^i corresponds to the number of zeros after the error pattern.
It is clear from Equation 8-9 that when L_0^X is the only error,

$$S_{P0}=E_{L0} \quad \textbf{Equation 8-29}$$

and from Equation 8-19 we can arrive at Equation 8-30,

$$S_1=Pes(\chi^{114}S_{P0})=Pes(\chi^{114}E_{L0})=C_0 \quad \textbf{Equation 8-30}$$

Conversely, once we have established these equations, we can conclude that L_0^X is the only error in B_n as proved by Equation 8-17, and the only error in sectional block H_0. The probability of a mistaken conclusion is less than 10^{-8} (in the case of standard error ratios).

Thus, $L_0{}^X$ can be corrected ignoring the pointer. After correction, if we clear all the pointers relating to H_0, we can correct $L_{48}{}^X$, $R_{48}{}^X$ and so on using the adjacent decoder. Figure 8-12 shows the correction algorithm.

If there are further different errors and we conclude that $L_{48}{}^X$, $R_{48}{}^X$, etc. are therefore not correctable, so long as we use a high quality crossword decoder, we can still recognize the location of one word errors in any position regardless of the CRCC pointers.[2].

A Comparison of Correction Performance

Figure 8-13 shows the results of a computer simulation of error correction performance.[5] Here, the vertical axis shows the frequency of occurrence of uncorrectable code errors (occurence/time), while the horizontal axis shows the bit co-efficient of correlation $\rho \cdot \rho$ indicates the nature of the code errors: the higher values indicate burst (continuous) errors, and the lower values indicate the

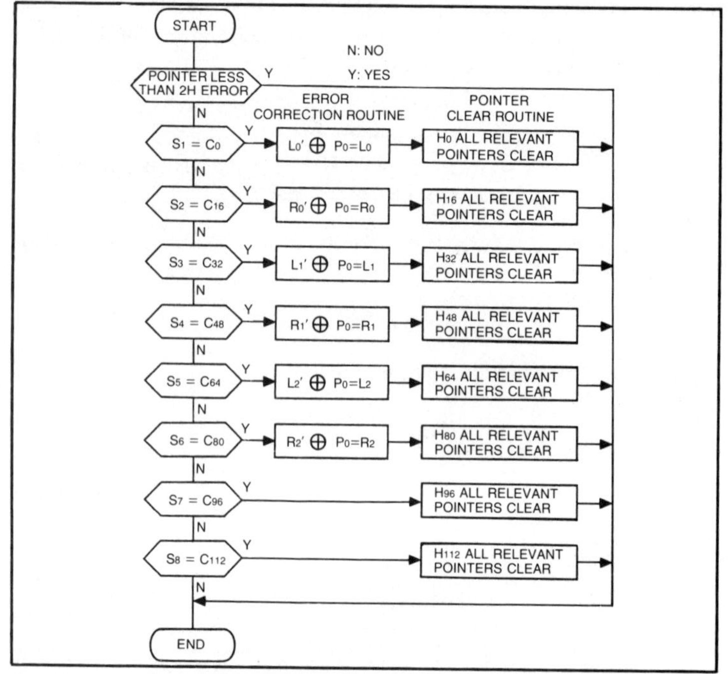

Fig. 8-12. Crossword decode algorithm.

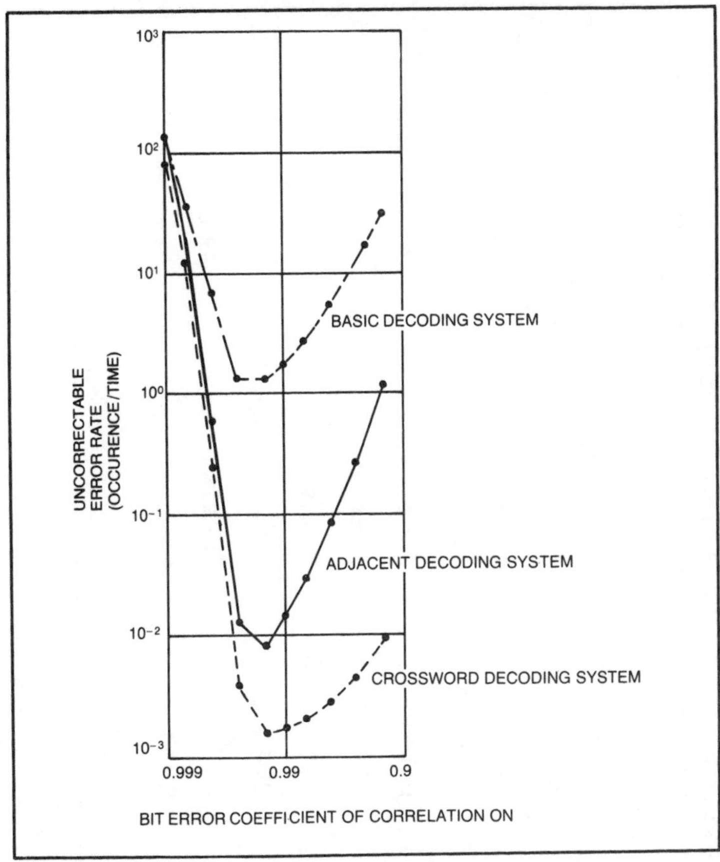

Fig. 8-13. Simulation results for error correction capacity.

more frequent random errors (see Chapter 7). The actual measured period of ρ is between 0.9 and 0.99, but this may vary widely depending on the state of the tape and the amount of jitter. The bit error rate is 10^{-4}.

If an adjacent decoder is used, there is an increase of about three figures in correction performance compared to basic decoding systems. If a crossword decoder is used, then an increase of between one to two digits may be expected.

EQUIPMENT PRODUCED CONFORMING TO EIAJ CONSUMER FORMAT

A brief overview of the PCM equipment produced by various companies conforming to a common standard is shown in Table 8-4. At present, prices are rather high, and this type of equipment is

Table 8-4. Standardized Consumer Digital Processors (as of 1983).

No. Manufacturer	Model Name	PCM Type	Quantization
1. Hitachi	PCM-V300	VCR-processor	14 bit linear
2. Sansui	PC-X1	Processor	14 bit linear
3. Sharp	RX-3	Processor	14 bit linear
4. Nakamichi (Sony)	DMP-100	Processor	14/16 bit linear
5. Sony	PCM-F1	Processor	14/16 bit linear
6. Sony	PCM-701ES	Processor	14/16 bit linear
7. Technics	SV-P100	VCR-processor	14 bit linear
8. Technics	SV-100	Processor	14 bit linear
9. Technics	SV-110	Processor	14 bit linear

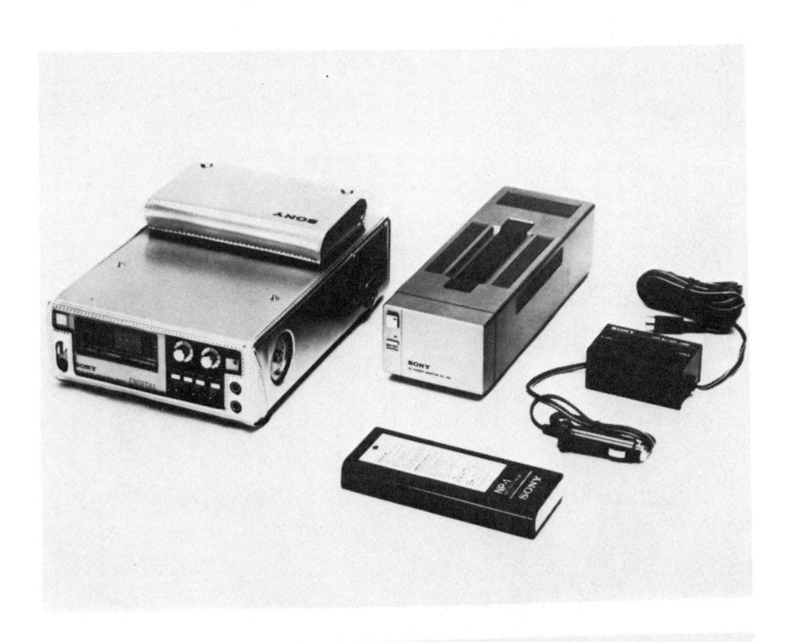

Figs. 8-14 and 8-15. Sony's consumer 16 bit PCM-F1 and PCM-701ES digital audio processors.

therefore not within the reach of the general consumer market. The equipment shown can be broadly divided into two categories. The major difference between these two categories lies in the A/D and D/A converters ; most use 14 bit linear quantization, while the Sony is a dual 14 bit/16 bit linear quantization system. The prices of the

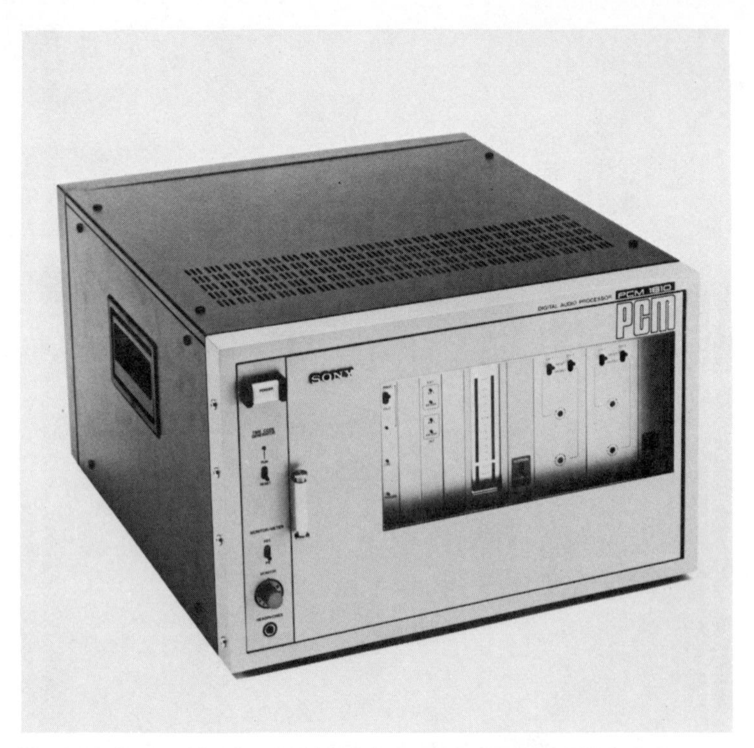

Fig. 8-16. Sony professional 16 bit digital audio processor (PCM-1610).

systems depend very largely on the prices of the A/D converters used.

However, as mentioned in Chapter 7, once new types of converter are developed, we can expect a big reduction in price. In addition, the cost of digital circuits themselves will also fall as the development of LSI speeds up, and we will see the same sort of phenomenon as occurred with digital watches and electronic calculators. Using semi-conductor technology it is now possible to build all the above-mentioned error correction circuitry into a one-chip LSI. The processors shown in Table 8-4 are, generally speaking, constructed from one or two LSIs, A/D and D/A co converters, and one chip of RAM, and the day when this type of package will cost below 400 dollars is not so very far away. Figures 8-14 and 8-15 consist of photographs of the Sony products described in Table 8-4.

References

1. Doi, T., Y. Tsuchiya, and A. Iga: "On several standards for

converting PCM signals into video signals" J.AES, 26, 9, pp. 641-649 (1978-8).

2. Doi. T.: "Crossword code for interleave adjacent codes", Nichi-on Gakkai Koron, 3-4-14 (1979-6).

3. Ishida, Nishi, Ishida, Shibutani, Sate: "Domestic PCM adaptors with error correction capability", Nichi-on Gakkai Koron, 3-5-16.

4. Ishida, Nishi, Inagi, Kunii: "An investigation into error correction codes for domestic PCM recording systems" Nichi-on Gakkai Koron 3-5-18.

5. Fukuda, Tsuchiya, Odaka, Doi: "A comparison of various decoding systems for error correction codes to be used in a new standard format for VTR based PCM recorders", Nichi-on Gakkai Koron, 3-5-19.

Chapter 9

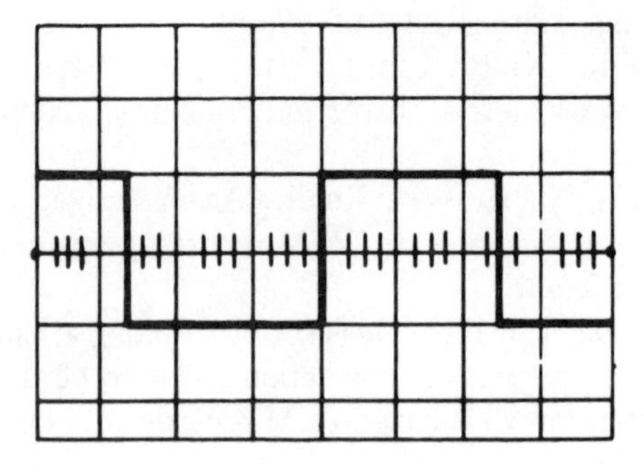

A Look at the Future

The first step towards the digitization of the audio field, as mentioned before, will be the increasing changeover to using digital recorders instead of analog recorders as secondary sound sources. PCM tape recorders will be used in the process of putting together master tapes for record production, and we can see even now the beginnings of an advance towards an improvement of sound quality and a simplification of the direct cutting system. In May 1978, the PCM-1600 and PCM-1 digital audio processors were used as the masters for an FM broadcast, and in September of the same year, a special 1.544 Mb/s PCM circuit was supplied for a stereo broadcast in the Tokyo, Nagoya and Osaka areas. Thus, digital technology has been absorbed into one corner of the FM stereo broadcast area. The great enthusiasm for the excellent quality of these broadcasts was expressed by a large number of listeners in the form of letters to the broadcasting companies, and such was the impression made that they have probably never forgotten the impact of those first broadcasts.

THE FUTURE OF DIGITAL AUDIO

The most interesting point to arise from this is the problem of how exactly the increasing use of digital audio equipment will affect the audio field in general. When PCM equipment is used as the sound source for the above-mentioned two software systems, as shown in Fig. 9-1, then the only differences between FM sound quality and record quality will be those of transmission. That is to

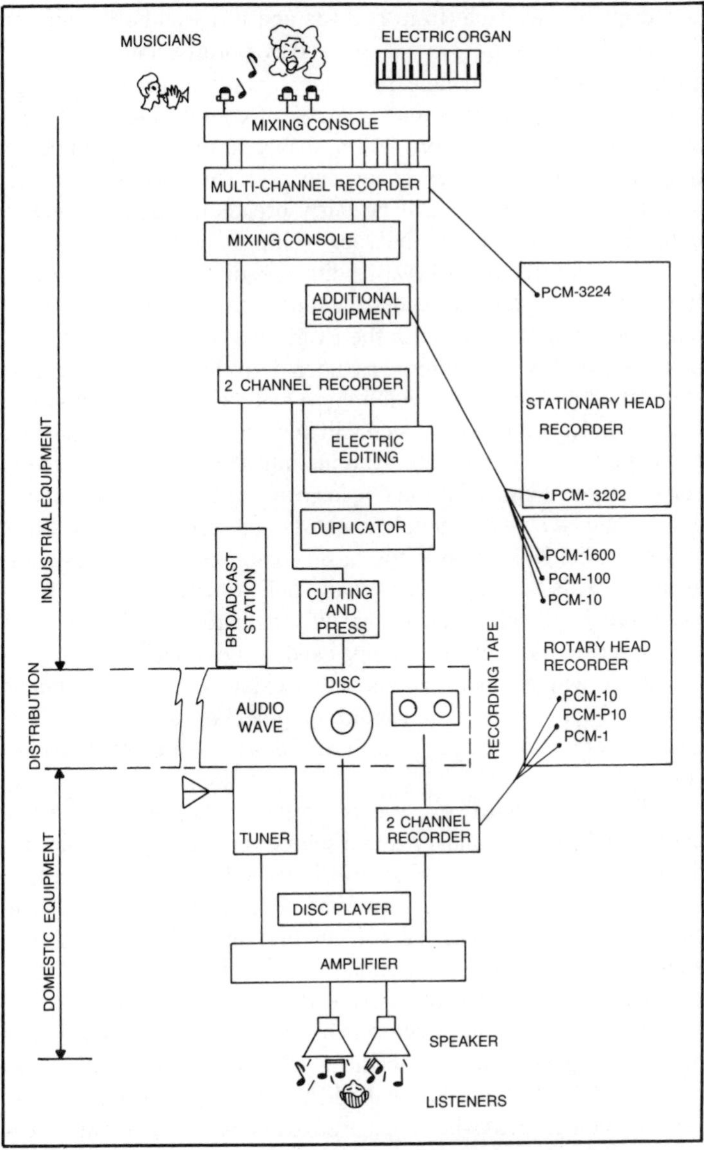

Fig. 9-1. The audio signal path and digital audio equipment.

say, the signal degradation caused by modulation, transmission and reception for FM, and the degradation caused by cutting, mastering, pressing and record playback for records. There may be some basis for saying that the deterioration is slight, but some years ago the sound quality available from records and FM was fairly uniform. Then with the advent of metal tape and evaporation tape, the sound quality available from tape recorders improved greatly, so that today three secondary sound sources of comparable quality are available. As a result, there will probably be many vicissitudes in the fortunes of secondary sound sources. Thus, the state of the manufacturers involved will be fairly dreadful, although the hi-fi enthusiast will conversely be very well provided for.

However, the move towards digital equipment will not stop at the industrial sphere and secondary source production, and in fact, it ought not to stop there. With the PCM tape recorder as a starting point, once this equipment has been introduced, the sphere of application will gradually widen more and more. As shown in Fig. 9-1, secondary source provision will become wholly digital, and we can hope to see appropriate systems introduced into the home to convert these sources into analog form for the listener's enjoyment. This would be only one development, but once realized, it will lead to a complete revolution in the basic equipment supplied as audio systems. This development is really only a matter of time. However, there will no doubt be an intermediate period when both analog and digital software is being used by the consumer. Thus, the consumer will gradually augment his existing stereo equipment with newly developed digital audio equipment of various types. The change from analog to digital will, therefore, be a gradual process (Fig. 9-2).

There are, however, two or three hurdles which must be overcome before secondary sound sources are fully produced using digital audio equipment, and before digital audio equipment starts to be added to existing sound systems. Before this equipment becomes totally accepted, there must first of all be some sort of standardization of secondary sound sources. The records and tapes produced must be available worldwide.

At the moment, much effort is being put into the establishment of world standardization, and these efforts are continually being intensified. However, this standardization for digital equipment is a very difficult process when compared to that of analog. In the case of analog equipment, the articles of standardization relate only to basic parts, and so it is therefore possible to arrive at fixed parameters relating to improvements in sound quality and operation caused by changes in mechanics and devices. Standardization for digital audio equipment covers a wide variety of complex matters, and matters which can be decided upon with ease are limited to sound quality and

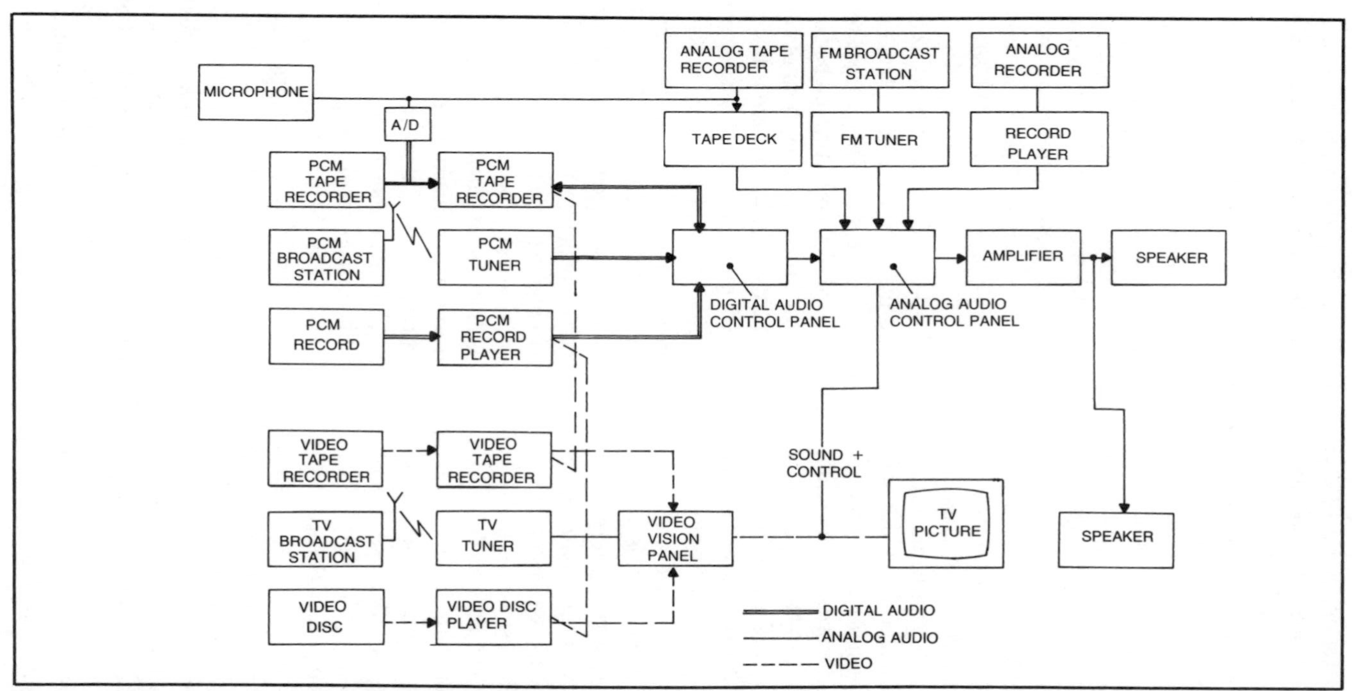

Fig. 9-2. A general plan of record and reproduction systems.

operation. Because digital audio is still in the process of being developed, if subjects for standardization are not debated fully, there is a distinct possibility that, in future, these agreements would become inapplicable. In addition, it is also difficult to gain agreement from all interested manufacturers on the way digital audio technology should be applied in future products. There are a great many obstacles in the way of digital audio standardization.

The next problem is that it is not sufficient merely to state the superiority of digital over analog in terms of sound quality secondary sound sources for the audio world. New equipment must be cheaper than the analog equipment, and also offer superior facilities before it can be considered for use as equipment to produce secondary sound sources.

By agreement, the design of the digital Compact Disc is 12 cm in diameter, which will allow more than one hour of 16 bit, two channel PCM stereo to be recorded on one side. On a record the same size as conventional analog records (30 cm), it would be possible to have one hour of 4 channel PCM recording, one hour of video and 2 channel PCM, or 13 hours of 2 channel PCM stereo. If special materials and techniques are used, then it would be possible not only to have a reproduction disc, but a record and reproduction disc.

It would also be advantageous to have a reproduction unit which is capable of reproducing video discs as well as PCM audio. It is thus very interesting to try and imagine how the various combinations of software, hardware, stereo reproduction, etc. may be combined as a system in creating a marketable product. It may eventually result even in the production of programs which have a profound effect on people's lives. The production of such revolutionary software, however, requires a disproportionate capital investment in record production and in production processes. If we compare the costs with those of an analog record, however, there is no particular reason why digital records should be disproportionately more expensive. With regard to the players also, it is obvious that the digital circuits will be constructed in the form of LSI, thus price and operation should be comparable to analog systems.

At present, the PCM processors marketed for use in combination with a domestic VTR are more expensive than the highest quality open reel tape recorder, besides which, editing is much easier using an analog tape recorder. However, with the advent of LSI, the digital circuits for these processors will of course be created in integrated form, and the price, weight and size of these units will be reduced to a mere fraction of their present size. Once this development is complete, it will not take long for these processors to exceed the popularity of open reel equipment.

If we examine a C90 compact cassette, a format widely used by hi-fi owners, the area used for recording is 0.4 m^2, and this is an area acceptable for use in recording 16 bit stereo digital information. If, for example, we were to make a PCM recording in a manner only slightly more sophisticated than at present, we would be able to make a 45 minute recording on a standard C90 compact cassette. The prerequisites to achieve this feat are thin film heads, tape which can hold twice the maximum recording density possible at present, the development of an effective sound processing system, and economical peripheral structures. Even so, we have to have such aims today so that we can strive to achieve them tomorrow. In addition, there is the possibility of combining stereo PCM with home computer systems, which opens up yet another sphere of speculation.

PCM broadcasts require a transmission bandwidth of 2 MHz, and therefore, a transmission system which has the same bandwidth as a TV wave is necessary. The preliminary experiments for this are already complete, and it has been confirmed that there are no intrinsic technical problems involved. The next major problem is deciding what form these broadcasts should take, what concrete steps can be taken and how radio waves should be apportioned. Once these matters have been clarified, field tests can be started, leading to full-scale experimental broadcasts, so that the extent of the service areas and the effects of interference can be investigated. Over the years, transmission frequencies have exhibited a tendency to creep upwards. The main problem associated with higher bandwidths was that although an exclusive bandwidth guaranteed comparatively good performance, there was a tendency to feel that the resulting dynamic range would not be as wide as required. However, PCM is an ideal answer to these problems of finding a suitable transmission medium for audio signals at high frequencies. It is, in particular, the answer to SHF transmission. From the users point of view, the major advantage is that a PCM tuner would use only a high frequency reception circuit and a D/A converter, so there is no reason why it should differ much in price or shape from an ordinary FM tuner. It would, in fact, have the advantage of offering better performance, since the distortions peculiar to analog FM would disappear.

If we look at PCM in this way, it is clear that it possesses the merits of better sound quality and lower price over its analog competitors, as well as offering the possibility of new systems produced by new combinations of functions. We are just beginning to appreciate how the human voice may be digitally combined with music, and it is interesting to speculate on what kind of equipment could be produced, and how research could be forwarded to combine these new ideas with existing digital audio systems. Digital reverberators, originally designed for the industrial market, as well as volume control, sound adjustment and mixing could be brought into the domestic sphere, and combined together in some form of digital control panel. Such a unit would fill the role of an "operations center" for all the digital equipment used.

There are, however, a few more problems associated with the spread of domestic digital equipment into the home environment. Firstly, because the secondary source material is recorded in a digital form, provided that sufficient care is taken while dubbing to keep the number of dropouts to a minimum, it is possible to make a very high quality copy with little sound degradation compared to the original. When dubbing using analog equipment, there is a 3 dB drop in dynamic range each time copying is carried out, and there is also a corresponding increase in wow and flutter and distortion. Thus, the original analog sound source has, as it were, an intrinsic value. In the case of digital, however, there is no loss of sound quality, so, unless some effective method of dubbing prevention is employed, there is a distinct possibility that companies supplying secondary sources would very shortly become bankrupt. This is, in fact, a very difficult problem to solve satisfactorily.

The next point to be taken into consideration is that the sound quality available from the three basic secondary sound sources, disc, tape and broadcast, would be more or less identical because the sampling frequencies and quantization bits would of necessity be more or less fixed. This may seem to be beneficial rather than otherwise from the consumer's point of view, but might in fact pose a danger to the suppliers of software. In the present analog system, it is already possible to make good quality copies of disc, tape or broadcast material. The disc has the advantage of good quality allied to immediate track access, while despite recent advances, tape is still inferior to disc in terms of quality and track access, although it does have the advantages of easy operation, and the ability to record as well as reproduce.

Broadcast material on FM is not inferior to disc, but it also has the advantages of volume and price. The reason why these three sources co-exist at the present time is precisely because each source offers different benefits and disadvantages. However, in a digital environment, all three sources would have the same advantage of excellent sound quality. Furthermore, if it was feasible to produce discs which could record as well as reproduce, the situation of the tape production houses would become very shaky indeed. In this situation, it would be necessary indeed to take steps to maintain the balance between these three sources of software, by putting more emphasis on ease of operation and facilities. However, even though all sorts of dire situations may be imagined, it is still necessary to investigate all possibilities as thoroughly as possible, and create an atmosphere for present discussion of future possibilities.

THE FUTURE OF ANALOG SYSTEMS

Despite the fact that part of the audio system may be digital, the voices and music filling the air are still analog quantities, and the ears and brain which make sense of them are also analog. The most striking fact about the audio field is that, after all, part of it will have to remain analog. The various problems associated with sound reproduction which existed ten years ago have now been analyzed, and the most serious problems in sound deterioration have now been solved by adding digital technology to tape recorders and the FM broadcast chain. Now is the time to examine the sound reproduction chain afresh and reconsider the various points which may give rise to difficulties in the future.

Let us look first of all at methods of sound collection. The noise level for studios used for music and voice pick-up, and the standard for measurement and insulation are based on a measurement standard which indicates a number of "noise criteria" (NC) based on a financial viewpoint. Recording studios and voice broadcast studios are measured as NC15 to 20 dB; while a TV studio is NC25 dB. As matters stand at the moment, these figures pose no particular problem. This, however, is primarily because these figures adequately cover tape recorder noise. Now that PCM equipment is available, it is vital to reconsider studio noise levels because it is highly probable that even NC15 dB will pose a problem for noise spectra.

Owing to recent tendencies in sound, and methods of pick-up and level adjustment, the microphone has effectively been pushed to its maximum limits at the time of sound pick-up. Much attention has been paid to extending the maximum sound pressure level possible for sound pick-up, so that pick-up of sound pressure levels up to 135 dB can be effectively accomplished with a high frequency distortion ratio of less than 1%. However, it is now necessary to revise these parameters to give a 10 dB increase in noise levels, air current noise and vibration noise. There will undoubtedly be problems associated with demanding a dynamic range in excess of 100 dB from microphone pre-amplifiers. Such demands for higher microphone performance will naturally follow from the use of PCM equipment, which entails lower noise levels at higher frequencies.

Equipment for sound quality adjustment, sound level attenuation and the addition of reverberation which is at present analog, will very shortly, become digital because of demands for increased dynamic range. As the process continues, it will become natural to include A/D converters and LSI circuits in the actual body of the microphone. As a result, programs could be transmitted in PCM form, and then re-transmitted after passing through digital sound level adjustment equipment. There is no way that one can accurately predict this type of development, but one can imagine that the present mechanical oscillations in a microphone could be transformed into electrical oscillations so that a direct digital signal could be more readily obtained.

Now let us move to the reproduction chain. As a result of dramatically increasing the dynamic range, the most pressing new requirement for the reproduction chain for secondary musical sources now that the sound quality has been increased will of necessity be an increase in the maximum output sound level. At present, no-one is investigating the re-design of speaker capacity to give higher performance levels, and although some effort is being put into increasing the maximum output level for amplifiers, this would, in fact, have to be 500 watts or more to give a satisfactory result. A speaker driven with this level of power would, therefore, need to be re-designed to incorporate new materials, a new structure to deal with the maximum input levels as well as non-linear distortion. It would even be desirable to add some kind of cooling system to deal with the dramatic temperature increases resulting from such high power handling.

It is probably best to concentrate on speaker systems as the best method for improving audio performance in future. It will be necessary to create a pistonic movement covering the whole, wide playback frequency by introducing materials such as titanium, boron, or developing high rigidity alloys of these materials for the diaphragm section, and by designing flat, honeycomb structures for the baffle board. This type of re-design will reduce the nonlinear distortion in the oscillating sections, but it will also be necessary to develop new materials for the frame and cabinet, and strengthen the weak places so that the non-oscillating sections are also improved in quality. All these design improvements will mean that there will be a great increase in sound quality when compared with the situation ten years ago. However, there are a great many difficulties involved in the attempt to improve the sound quality available from speakers. Yet, as these points are gradually resolved, they will lead to an improvement in sound quality.

Once PCM technology is added to microphones, we can expect to see analog speakers introduced which include parallel developments, such as the conversion of sound vibrations into digital electronic oscillations. Whatever the sound quality of such speakers, it is fascinating to imagine what form they will actually take.

Furthermore, when we can reproduce excellent quality sound at the necessary high output levels, with a noise level close to zero, it will be necessary to face the next difficulty; a suitable room to hear the sound produced. In general, there will be many occasions when these listening rooms are, from the point of view of a domestic listening environment, lacking in terms of the consideration which should be given to domestic listening and noise distribution. The problems would be that the sound levels would be too low for a Japanese style room, but too high for a western style room, and there would be trouble with long pass echoes and flutter echoes. There would, therefore, be a likelihood that where the listening conditions differ from the ideal, there would be times when a satisfactory result could not be obtained. In addition to this, it will soon be necessary to speedily counter domestic noise levels, because with the dramatic increase in maximum sound output, noise pollution of the environment could become a serious problem unless sufficient care is taken in insulation and soundproofing.

As mentioned above, the application of PCM technology to the audio field as a measure to gain improvement in tape recorder performance may eventually effect a change in equipment such as microphones and speakers which are at present difficult to make in a digital form. It may even affect methods and behavior in areas such as studios and listening rooms which it is impossible to "digitize" except in the very long term. In the final analysis, apart from the changes in production of secondary source material, no-one can guarantee that these effects will not stretch to altering the very nature of the software produced.

The essence of audio is that developments in one area of the audio system invariably lead to changes in other areas. So PCM is playing the role of catalyst. We are the lucky beneficiaries of these new developments.

Fig. 9-3. World's first digital audio CD broadcast.

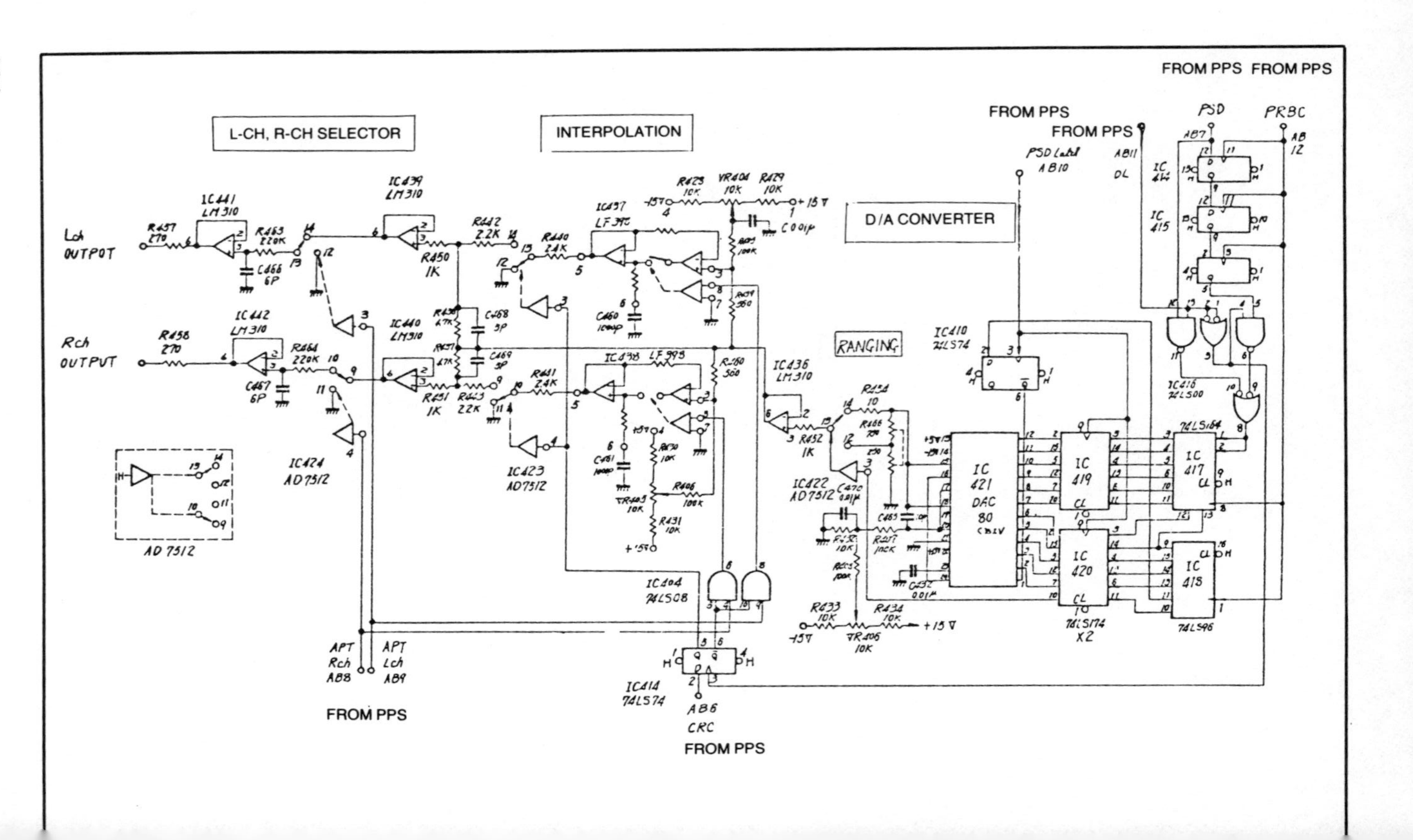
L-CH, R-CH SELECTOR
INTERPOLATION
D/A CONVERTER
RANGING
FROM PPS
FROM PPS
FROM PPS
FROM PPS
FROM PPS
FROM PPS
FROM PPS
Lch OUTPUT
Rch OUTPUT
PSD
PRBC
AD7512

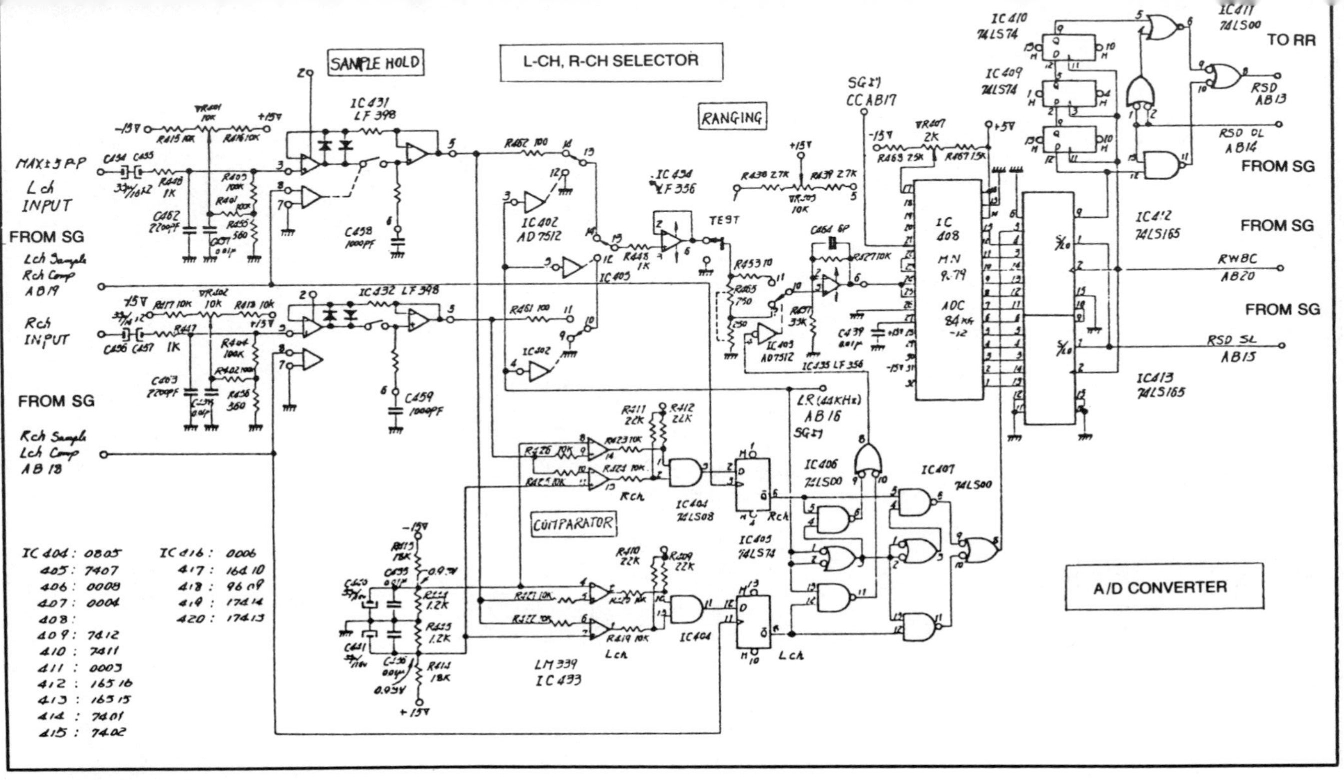

Fig. 9-4. A/D, D/A circuit.

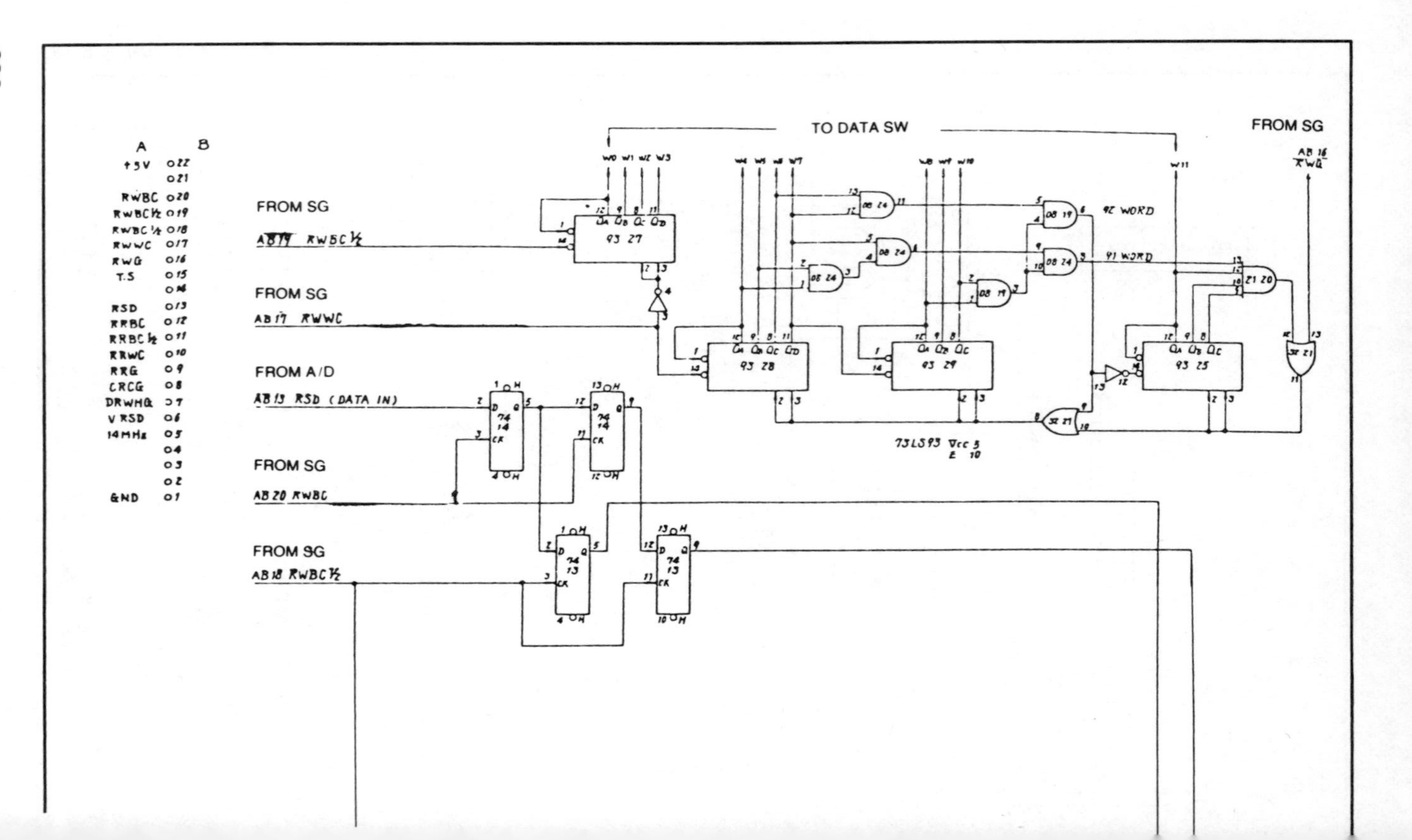
TO DATA SW
FROM SG
AB 16 RWG
FROM SG
AB17 RWBC½
FROM SG
AB17 RWWC
FROM A/D
AB13 RSD (DATA IN)
FROM SG
AB20 RWBC
FROM SG
AB18 RWBC½
92 WORD
91 WORD
74LS93 Vcc 5 E 10
A B
+5V 022
021
RWBC 020
RWBC½ 019
RWBC¼ 018
RWWC 017
RWG 016
T.S 015
014
RSD 013
RRBC 012
RRBC½ 011
RRWC 010
RRG 09
CRCG 08
DRWHG 07
V RSD 06
14MHz 05
04
03
02
GND 01

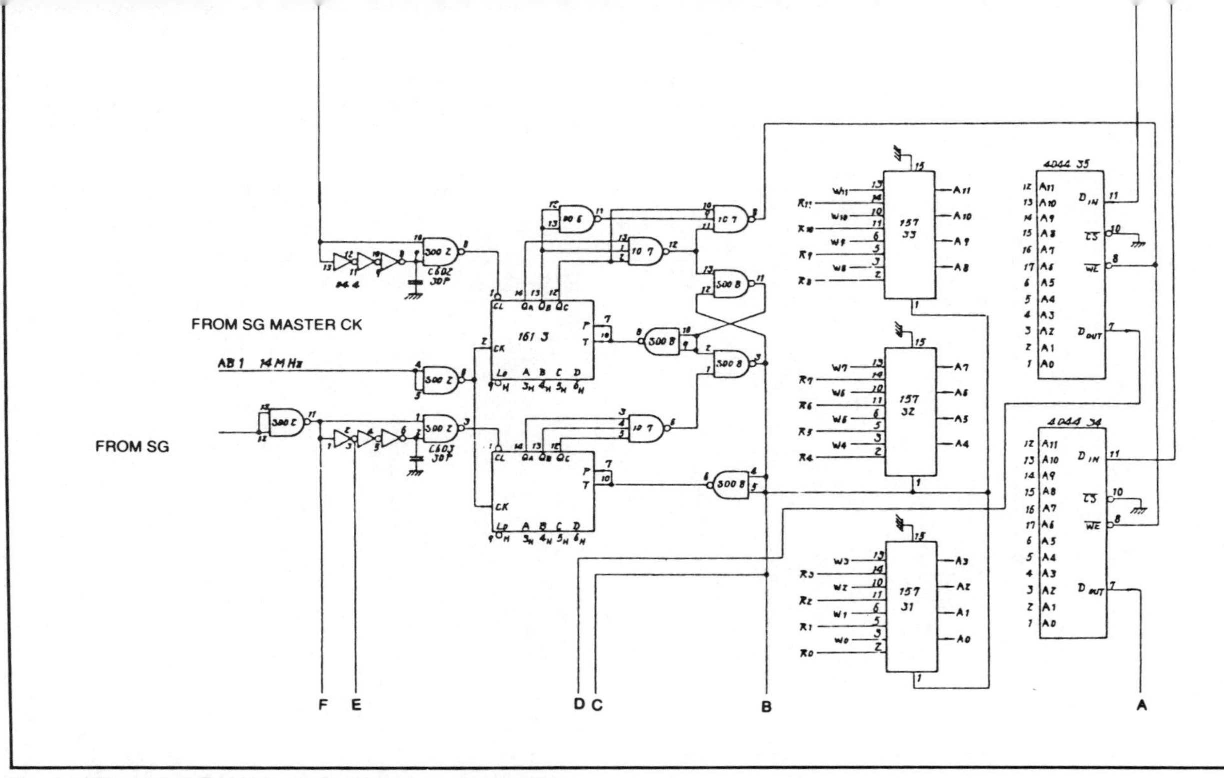

Fig. 9-5. Recording RAM circuit (continued on page 288).

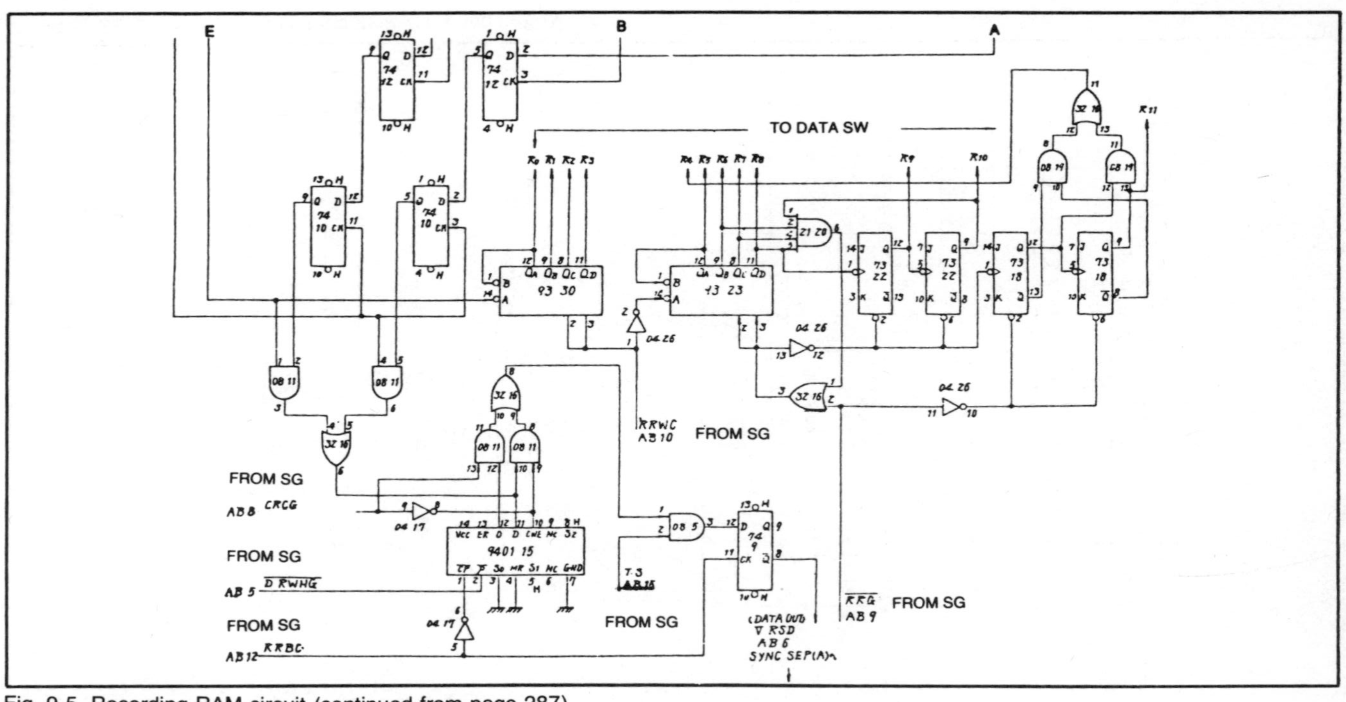

Fig. 9-5. Recording RAM circuit (continued from page 287).

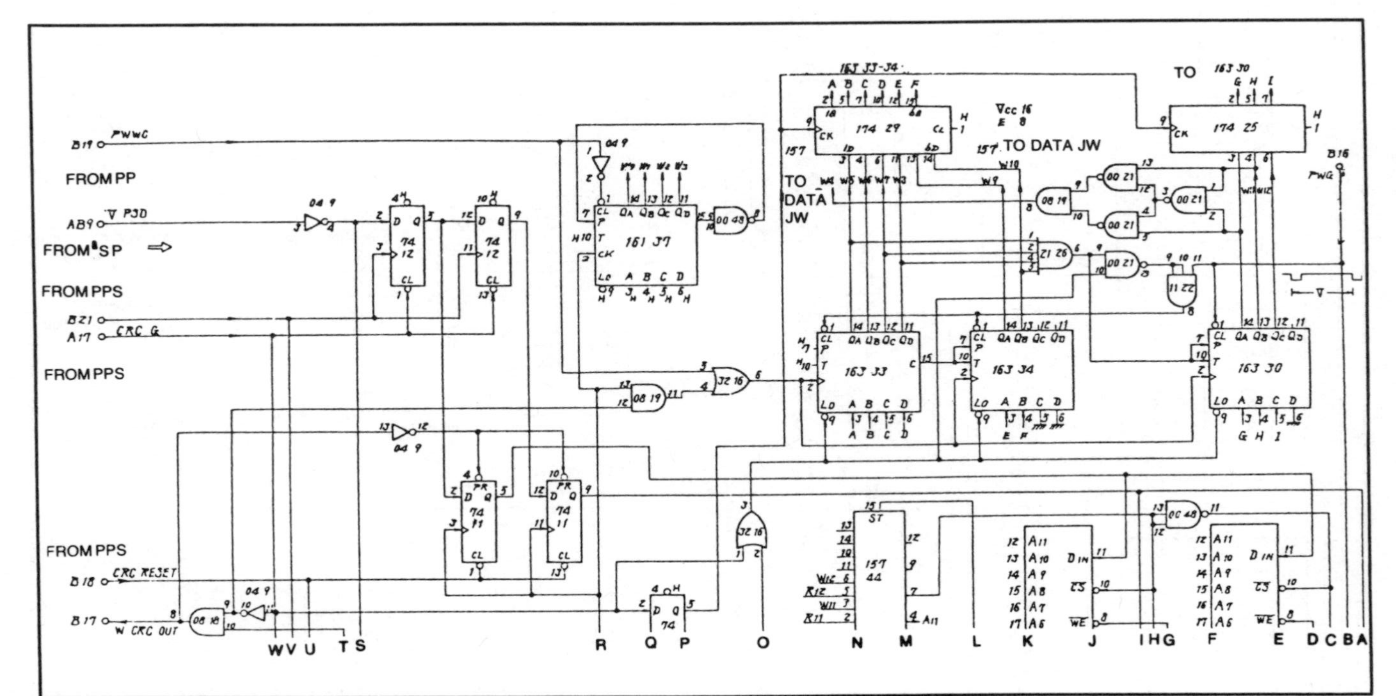

Fig. 9-6. Playback RAM circuit (continued on page 290).

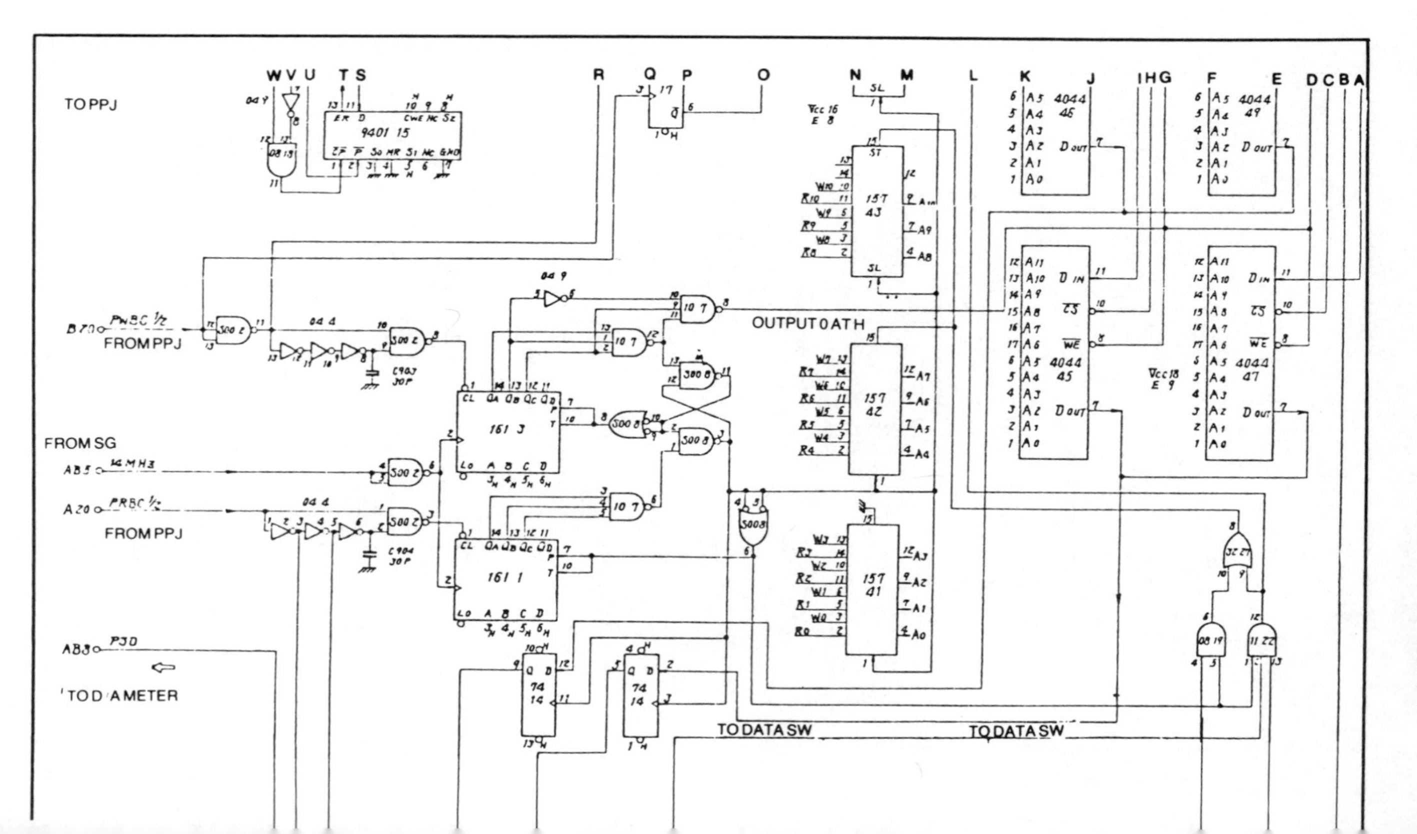
TO PPJ
W V U T S R Q P O N M L K J I H G F E D C B A
FROM PPJ
FROM SG
FROM PPJ
OUTPUT DATH
TO DATA SW
TO DATA SW
TO DAMETER

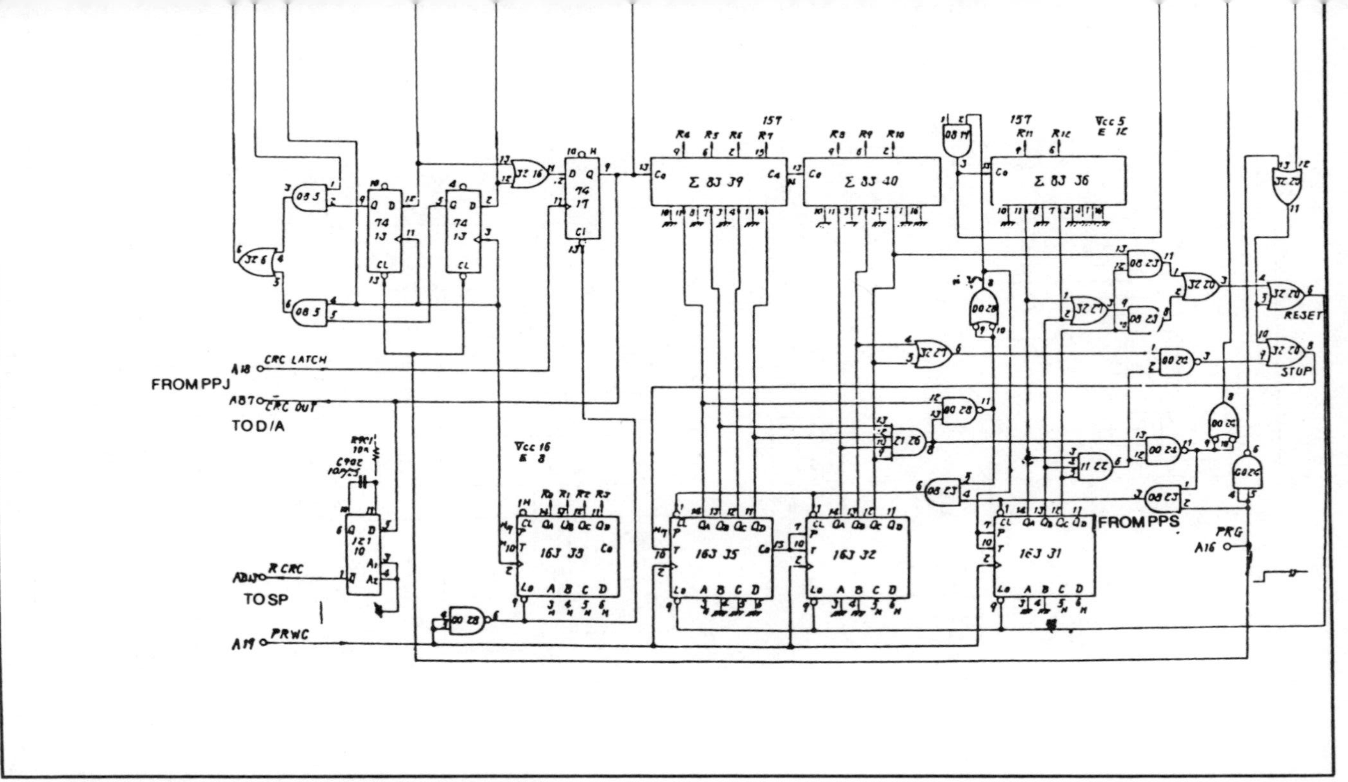

Fig. 9-6. Playback RAM circuit (continued from page 289).

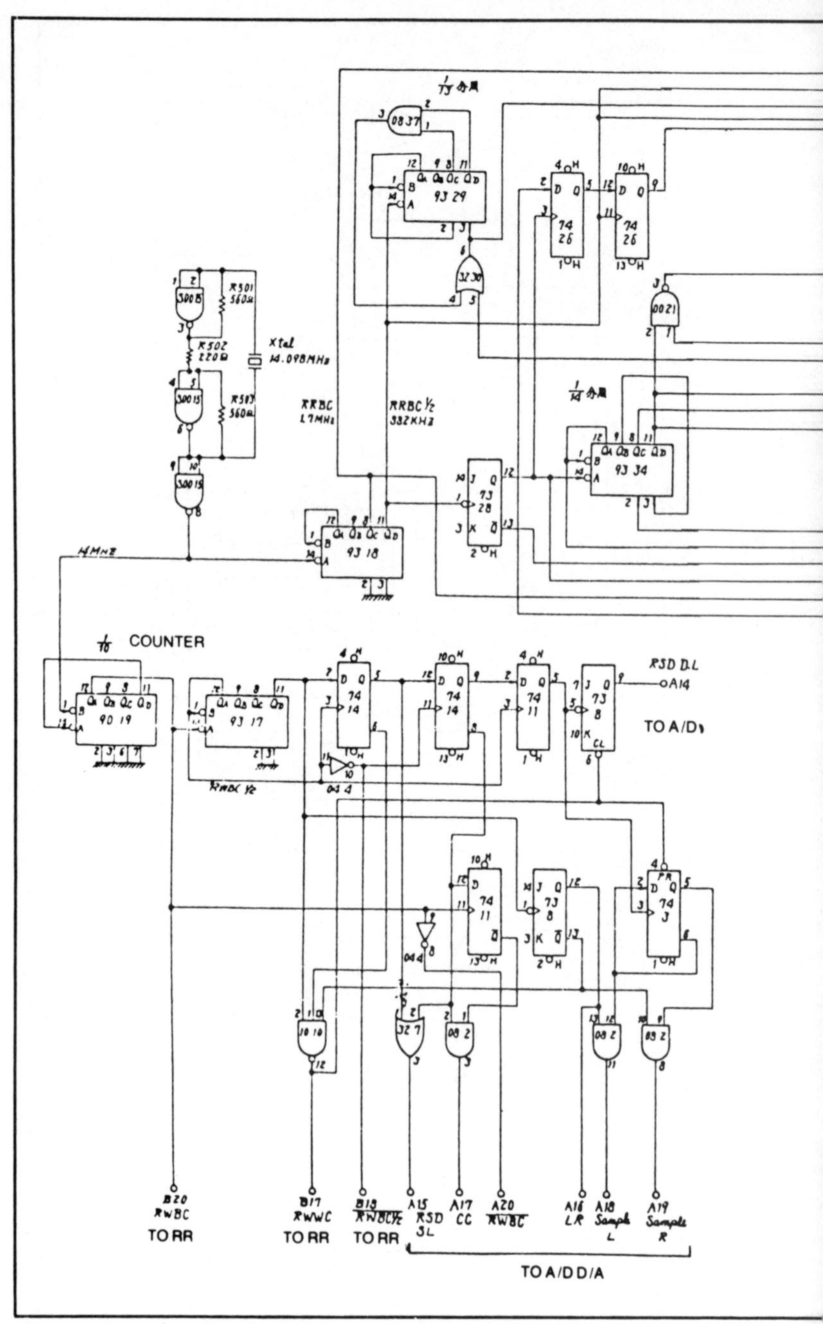

Fig. 9-7. Snyc generator circuit (continued on page 294).

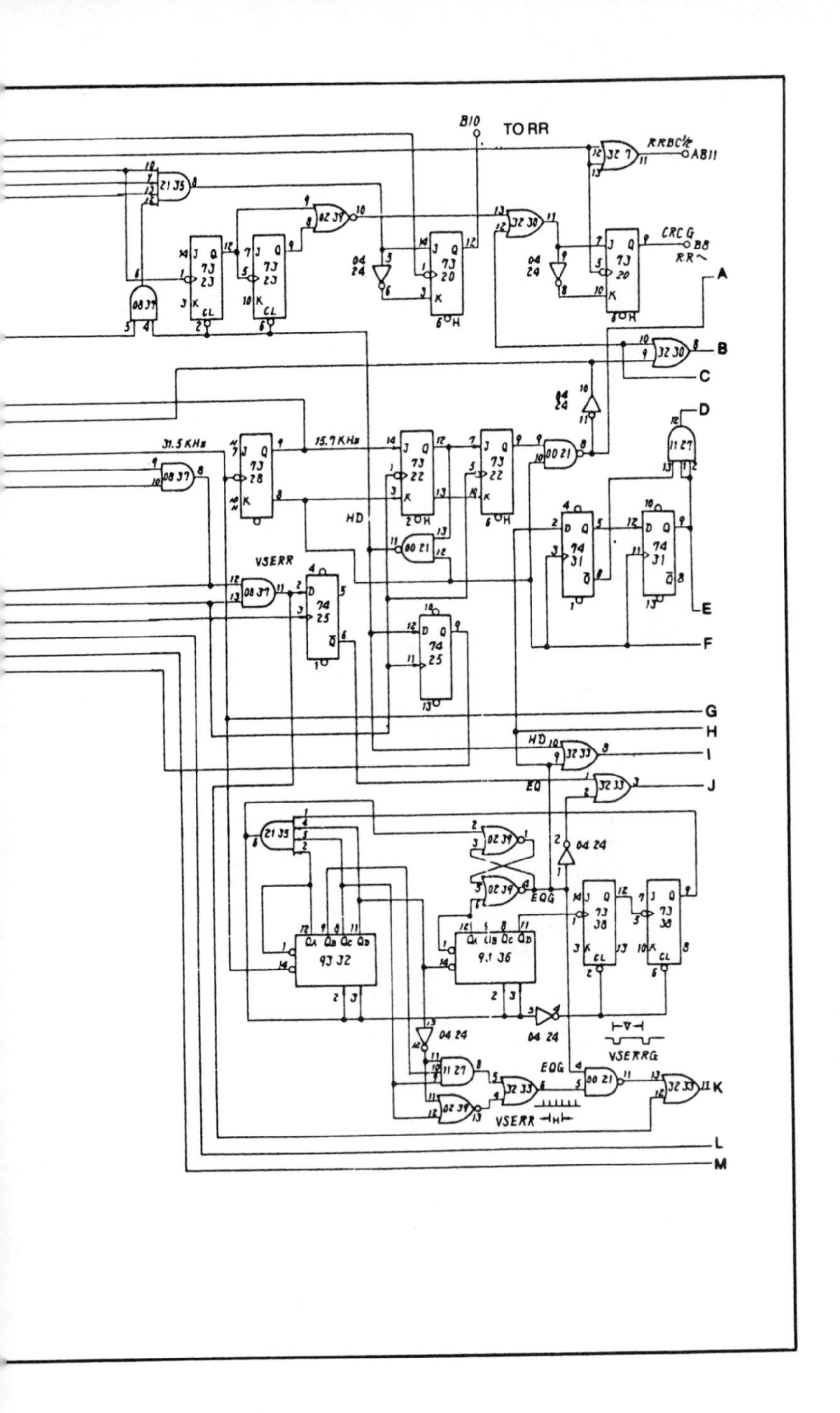
B10
TO RR
RRBC½
AB11
CRC G
B8
RR
A
B
C
D
E
F
G
H
I
J
K
L
M
31.5 KHz
15.7 KHz
HD
VSERR
EQ
EQG
VSERRG

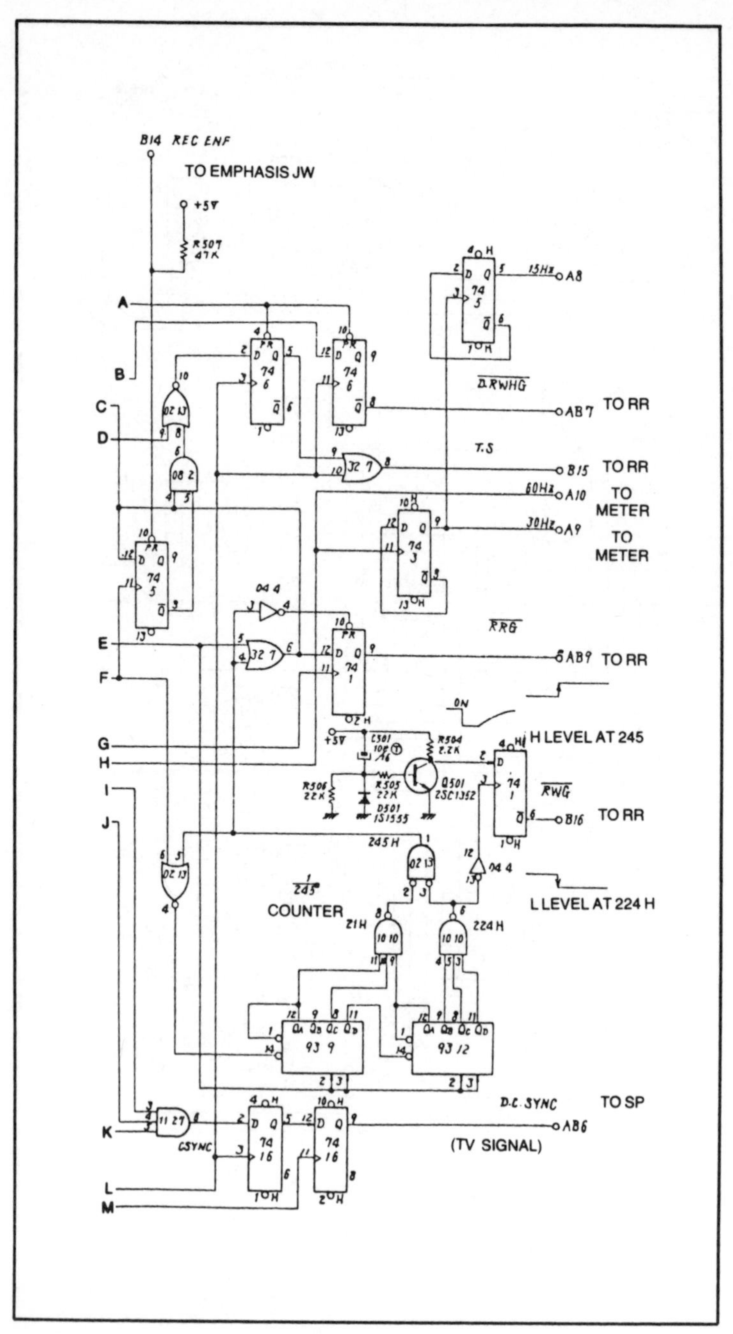

Fig. 9-7. Sync generator circuit (continued from page 293).

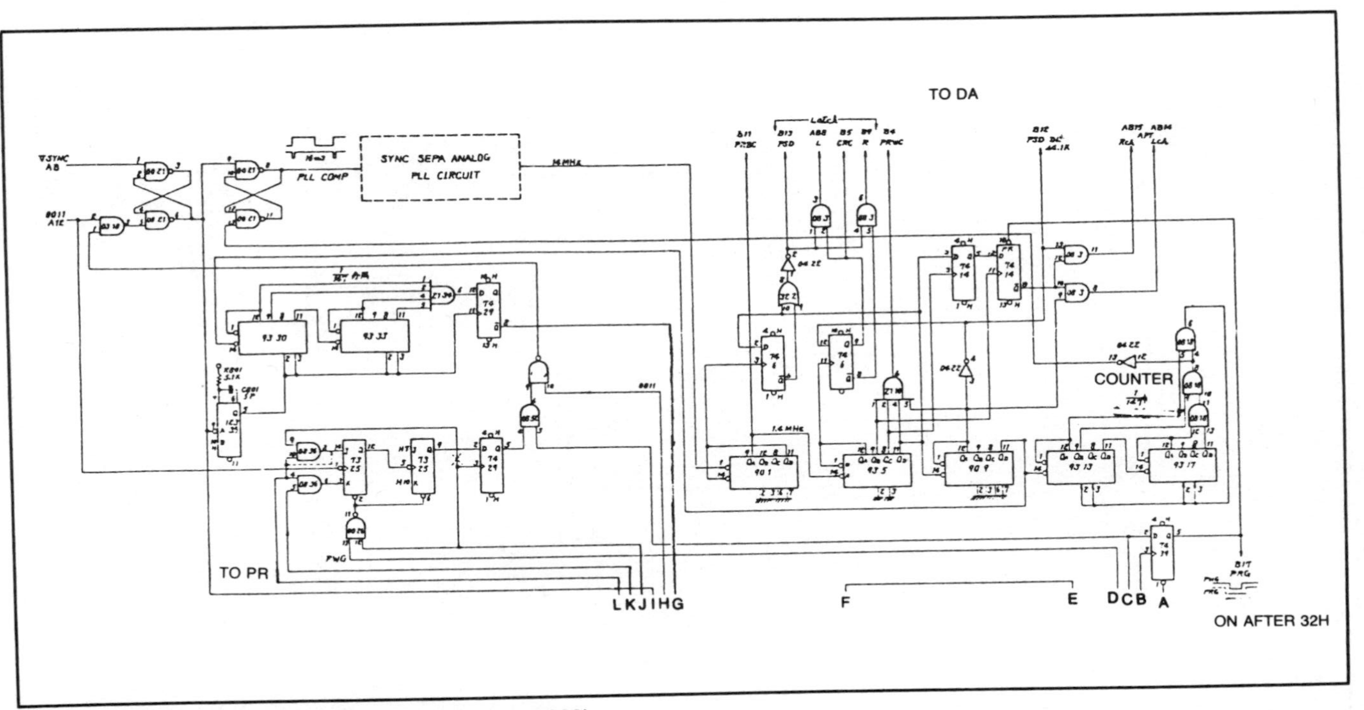

Fig. 9-8. Sync separator (D) circuit (continued on page 296).

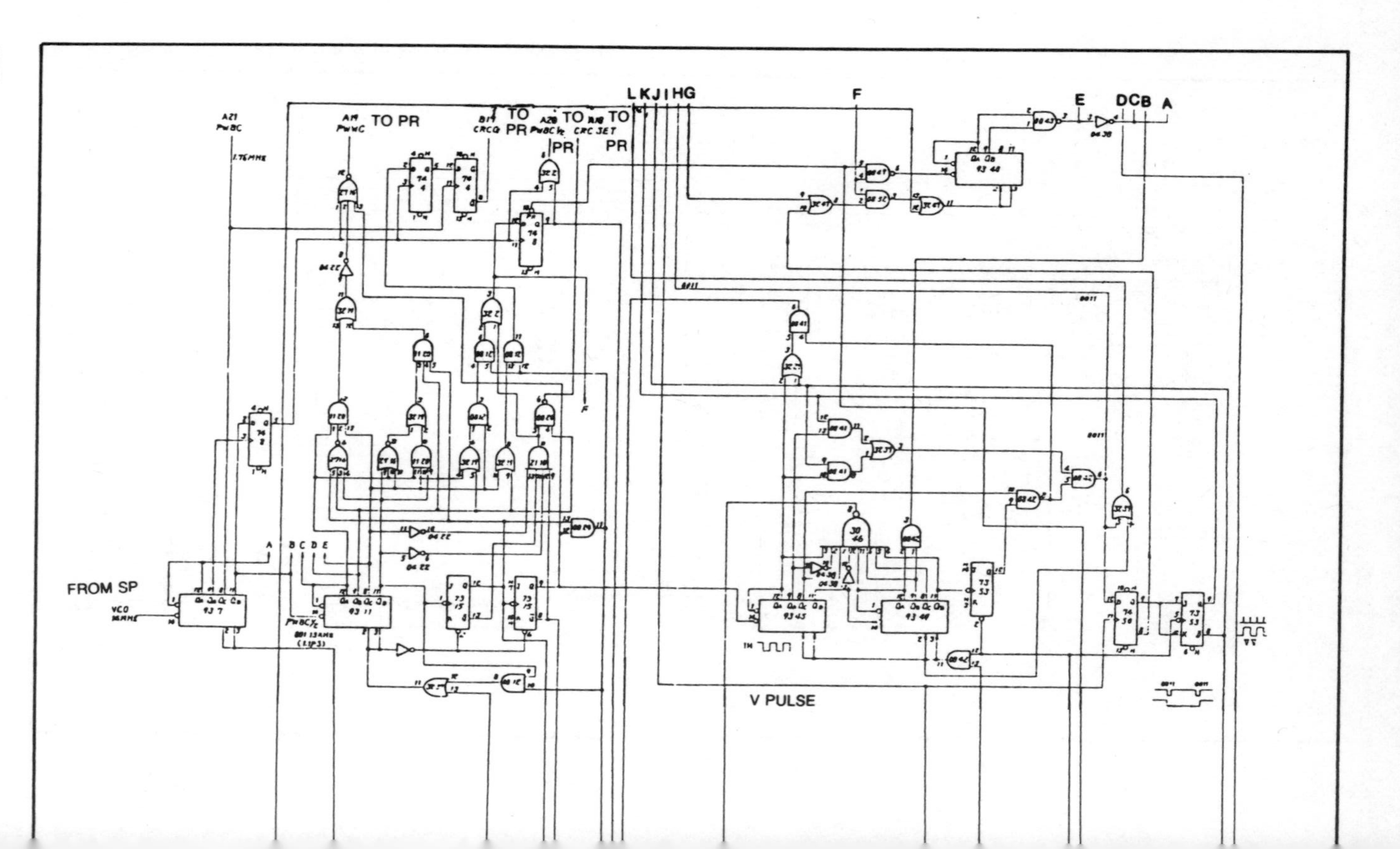

LKJIHG
F
E
DCB A
TO PR
TO PR
TO PR
TO PR
FROM SP
A B C D E
V PULSE

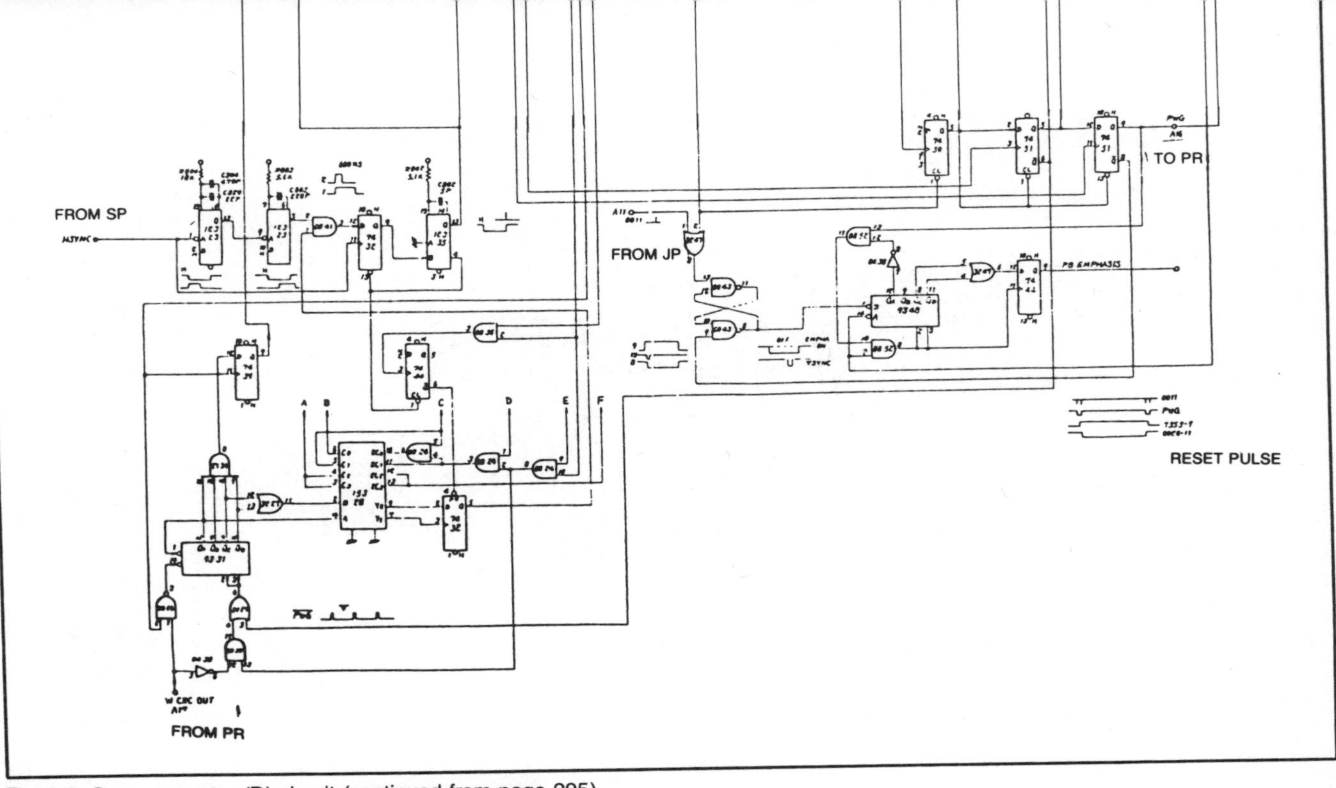

Fig. 9-8. Sync separator (D) circuit (continued from page 295).

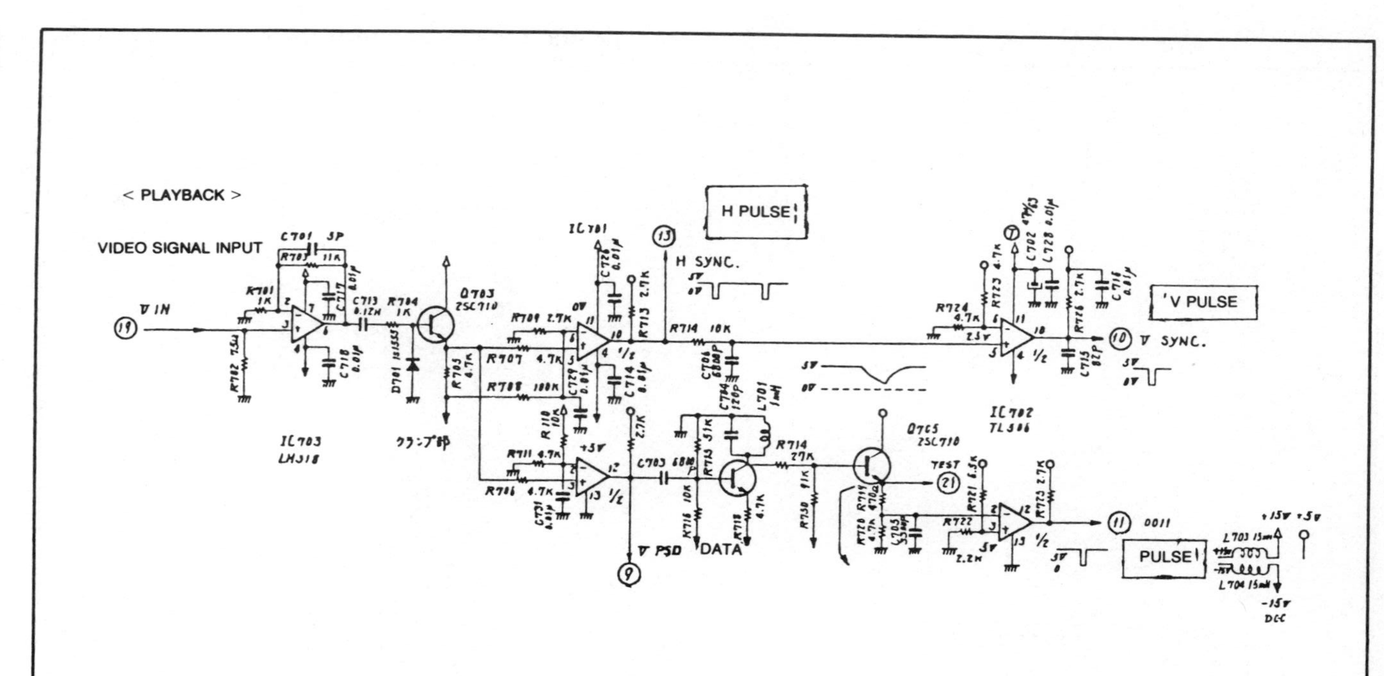
< PLAYBACK >
VIDEO SIGNAL INPUT
V IN
H PULSE
H SYNC.
'V PULSE
V SYNC.
V PSD
DATA
TEST
PULSE
IC701
IC702
TL506
IC703
LM318
Q703
2SC710
Q705
2SC710

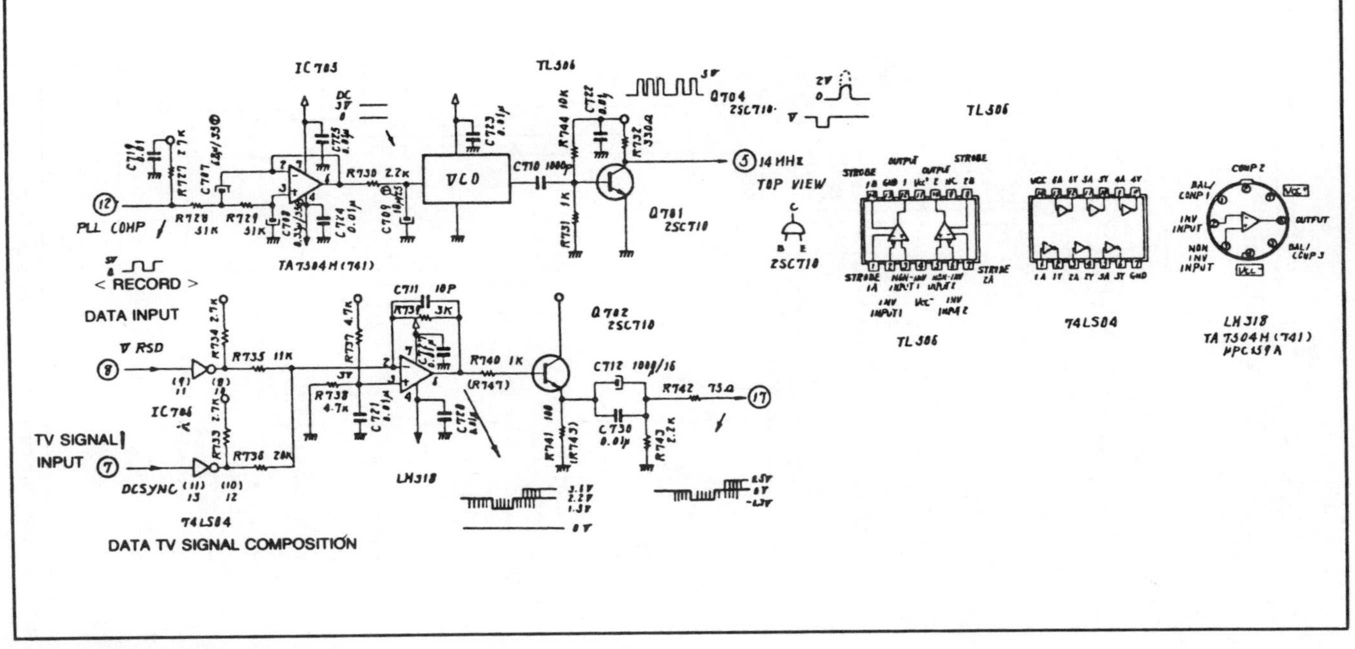

Fig. 9-9. Sync separator (A) circuit.

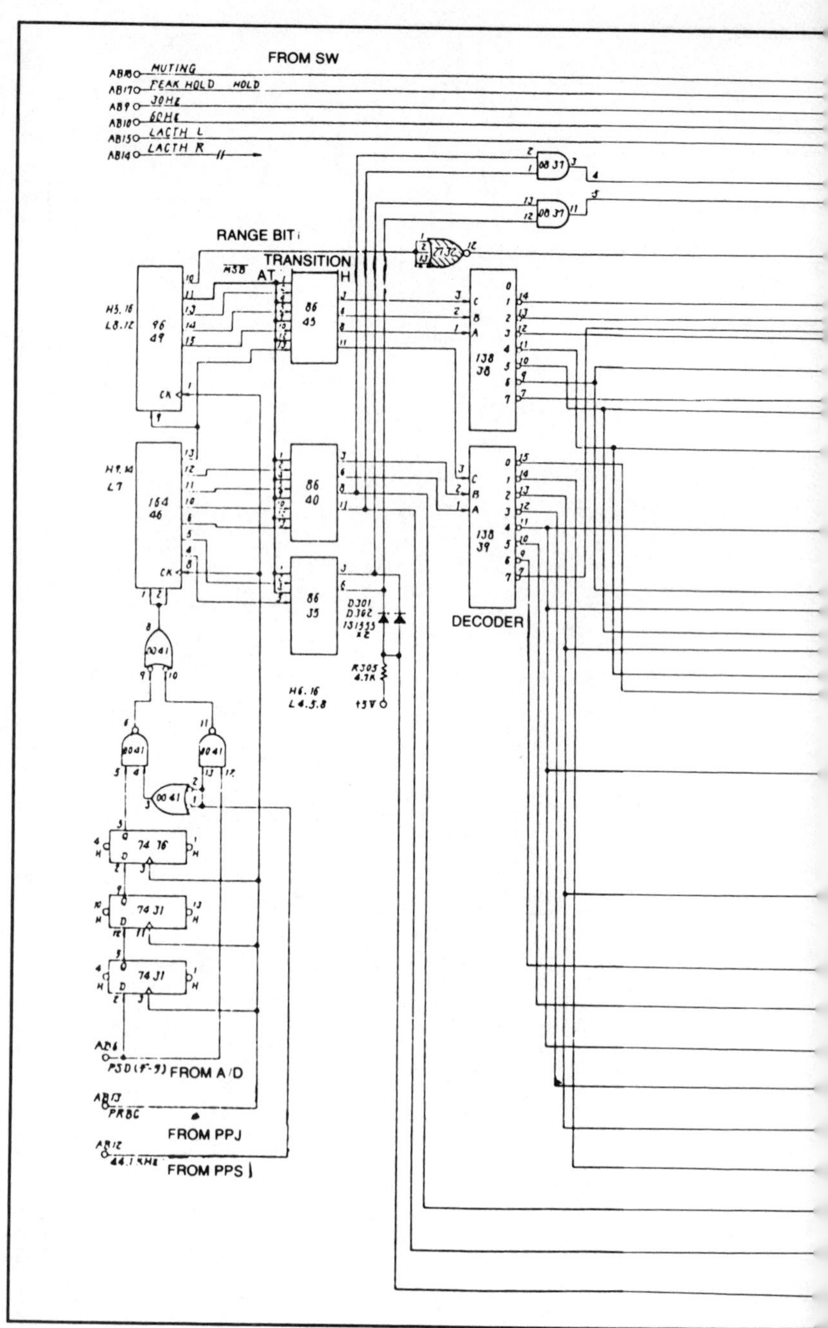

Fig. 9-10. Peak meter circuit (continued on page 302).

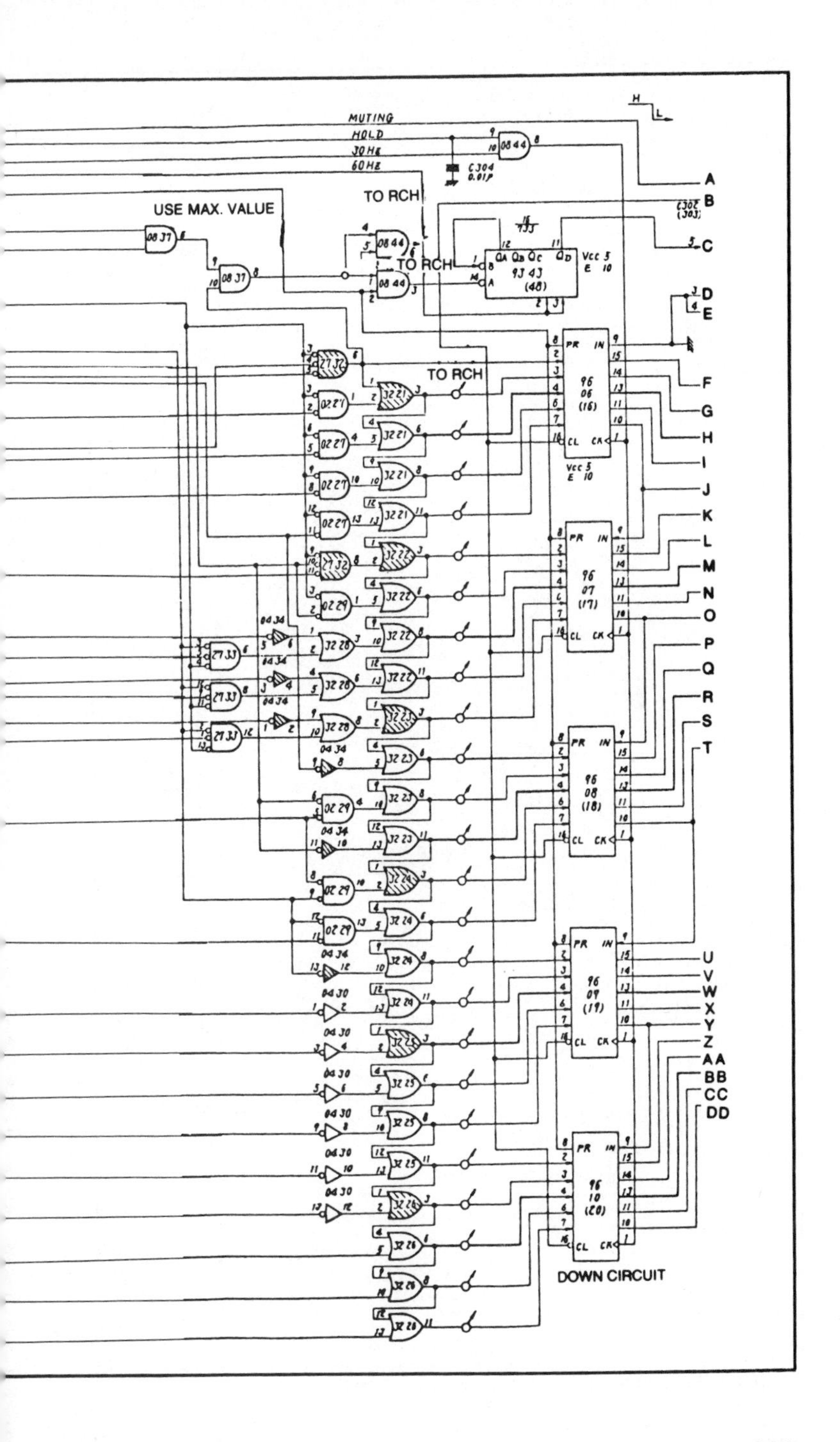
MUTING
HOLD
30Hz
60Hz
USE MAX. VALUE
TO RCH
TO RCH
TO RCH
A
B
C
D
E
F
G
H
I
J
K
L
M
N
O
P
Q
R
S
T
U
V
W
X
Y
Z
AA
BB
CC
DD
DOWN CIRCUIT

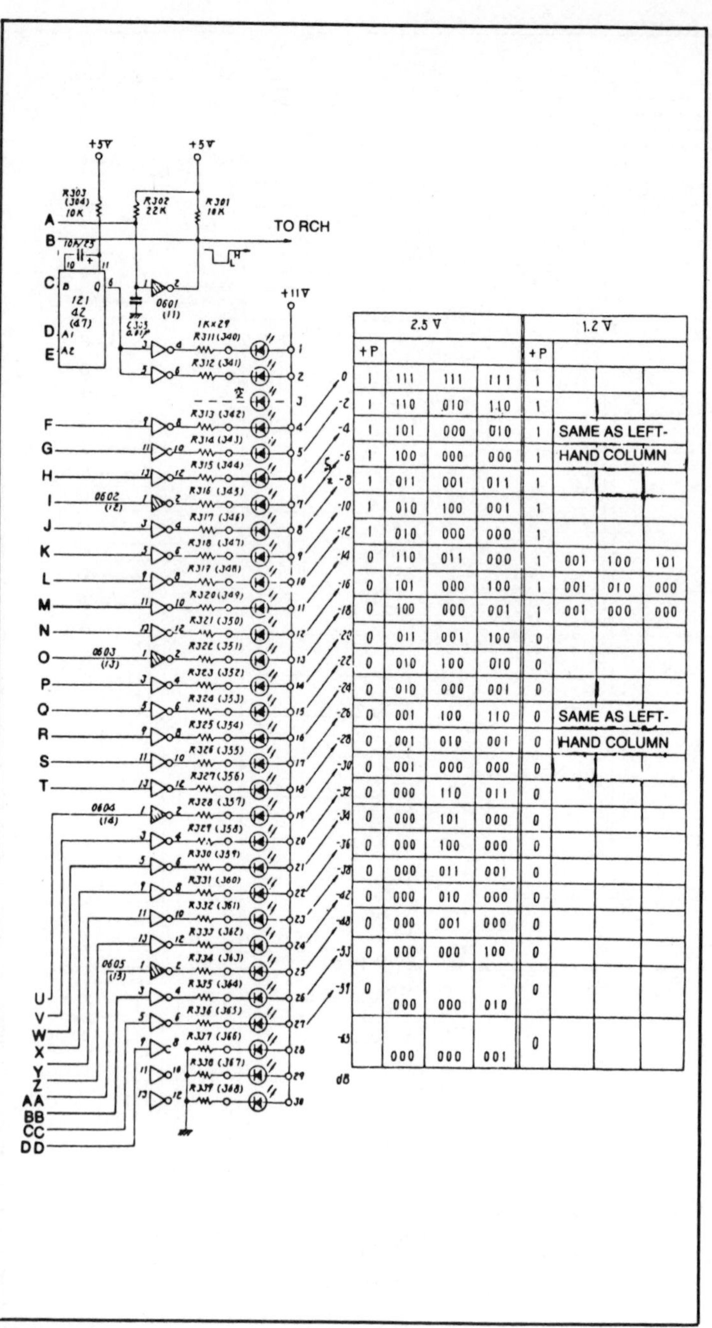

dB	2.5 V				1.2 V			
	+P				+P			
0	1	111	111	111	1			
-2	1	110	010	110	1			
-4	1	101	000	010	1	SAME AS LEFT-		
-6	1	100	000	000	1	HAND COLUMN		
-8	1	011	001	011	1			
-10	1	010	100	001	1			
-12	1	010	000	000	1			
-14	0	110	011	000	1	001	100	101
-16	0	101	000	100	1	001	010	000
-18	0	100	000	001	1	001	000	000
-20	0	011	001	100	0			
-22	0	010	100	010	0			
-24	0	010	000	001	0			
-26	0	001	100	110	0	SAME AS LEFT-		
-28	0	001	010	001	0	HAND COLUMN		
-30	0	001	000	000	0			
-32	0	000	110	011	0			
-34	0	000	101	000	0			
-36	0	000	100	000	0			
-38	0	000	011	001	0			
-42	0	000	010	000	0			
-48	0	000	001	000	0			
-53	0	000	000	100	0			
-59	0	000	000	010	0			
-63		000	000	001	0			

Fig. 9-10. Peak meter circuit (continued from page 307).

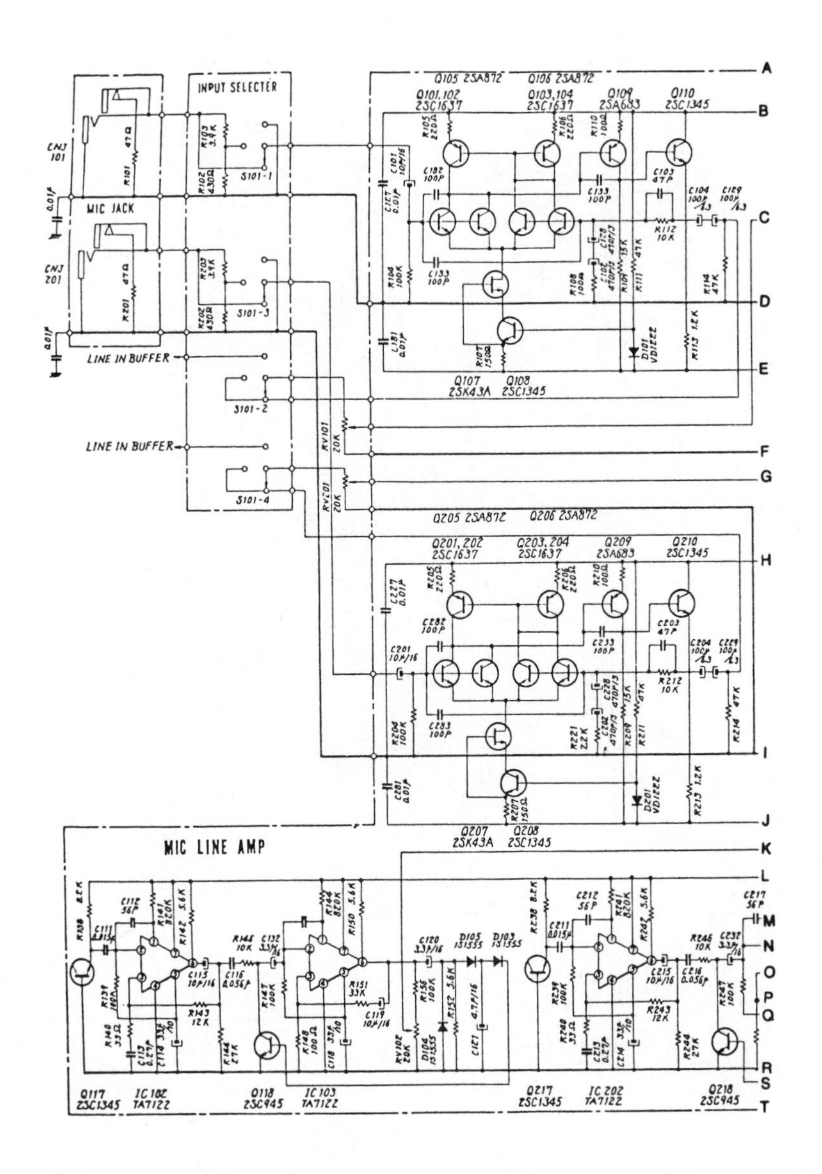

Fig. 9-11. Audio amp circuit.

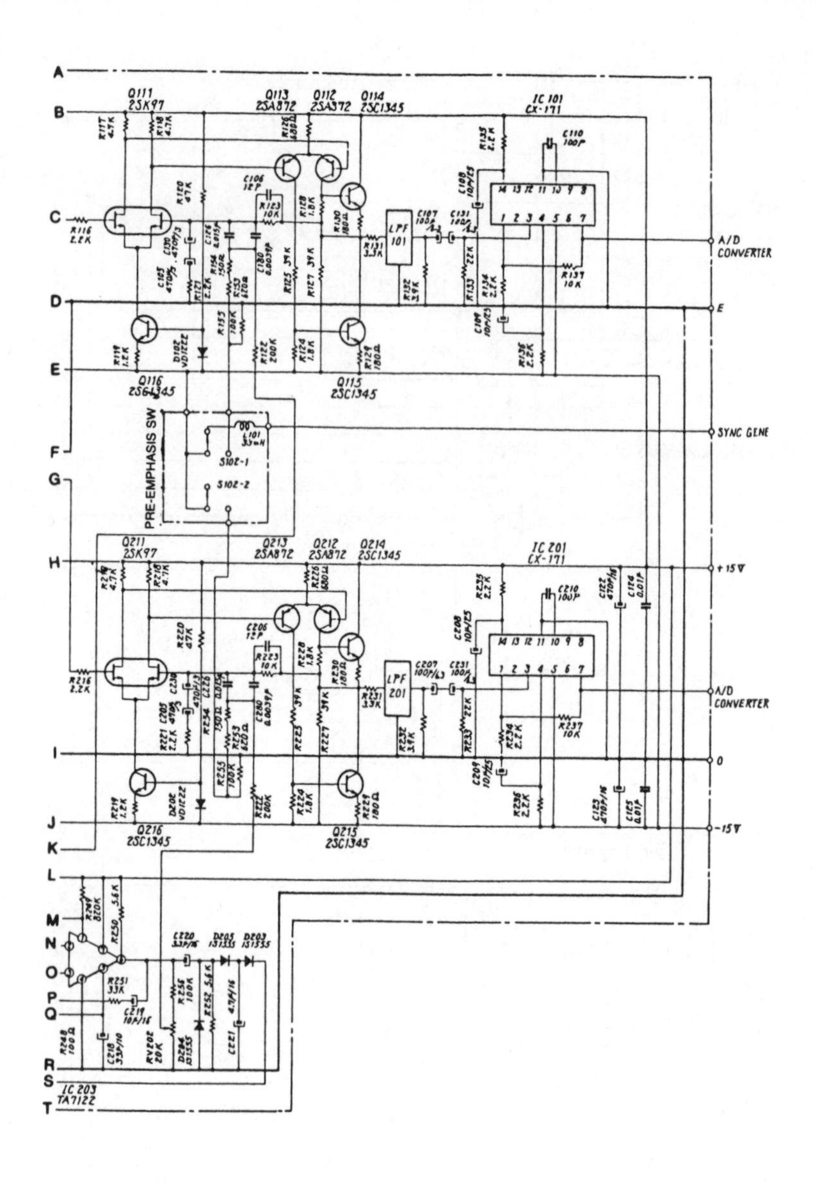

Fig. 9-11. Audio amp circuit. (Continued from page 303).

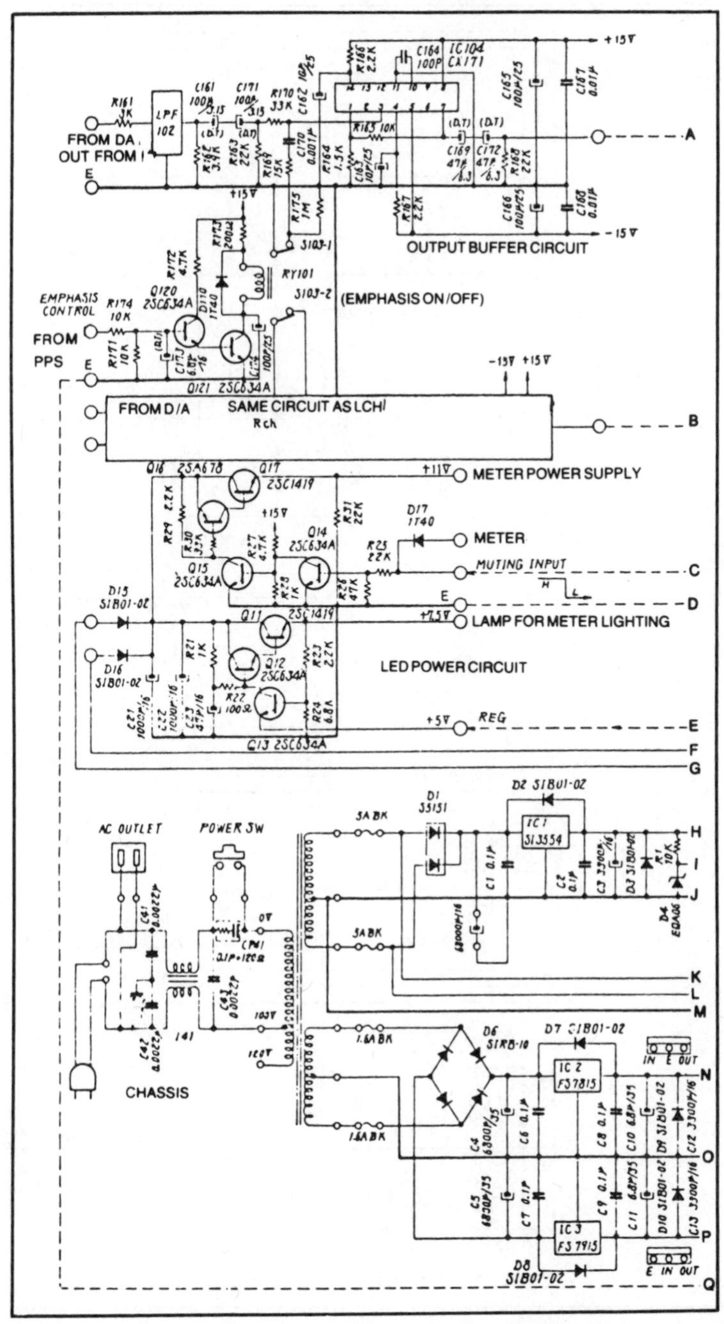

Fig. 9-12. Power supply and buffer circuit (continued on page 306).

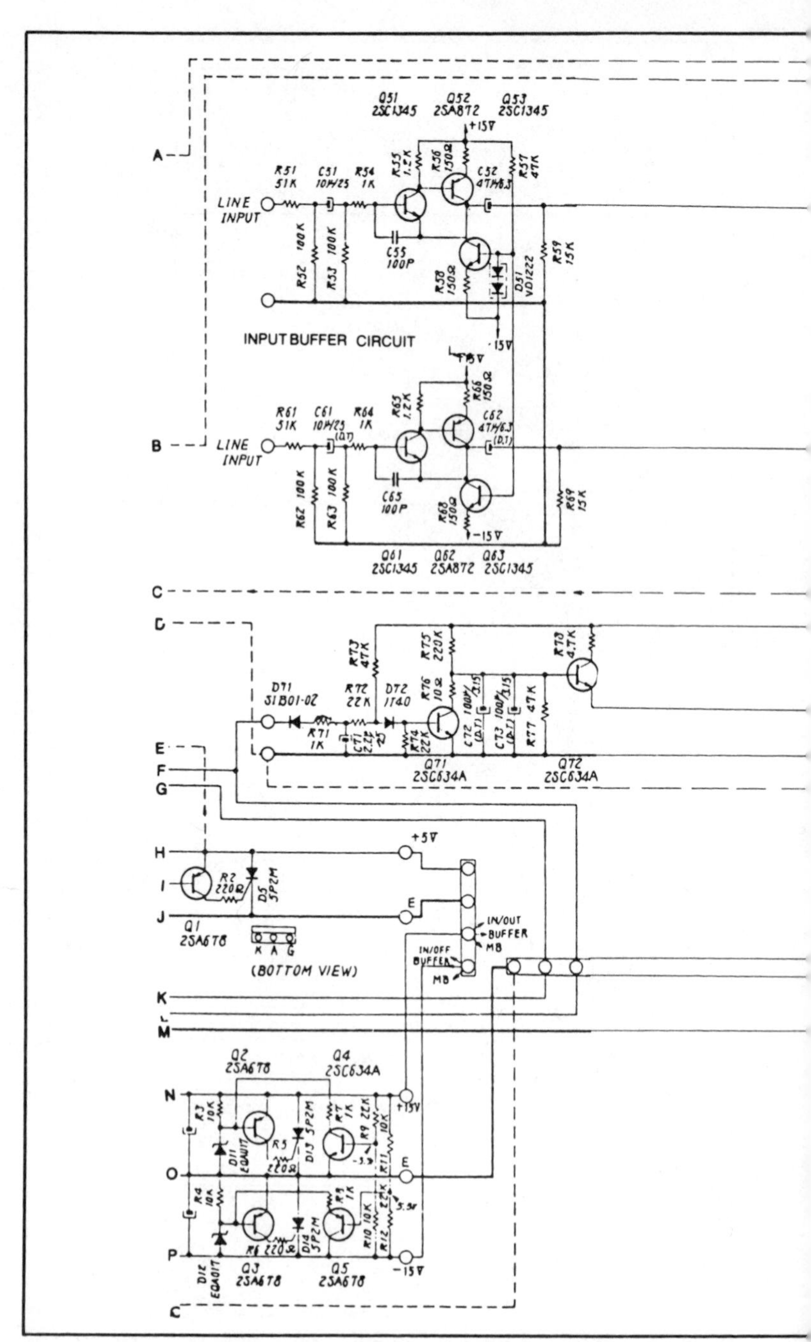

Fig. 9-12. Power supply and buffer circuit (continued from page 305).

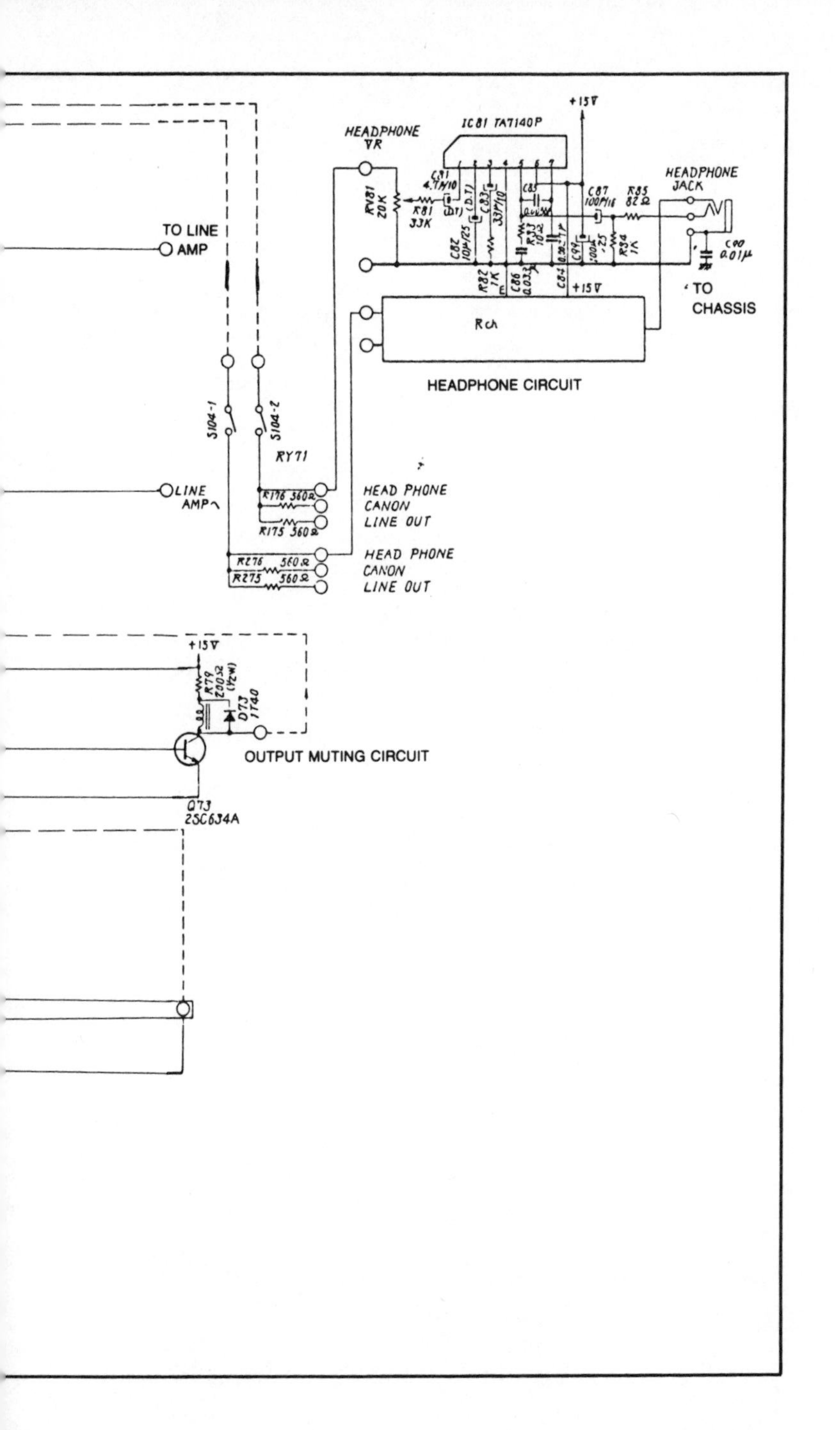
TO LINE AMP
HEADPHONE VR
IC81 TA7140P
+15V
HEADPHONE JACK
RV81 20K
R81 33K
C81 4.7M/10
C87 100M/16
R85 82Ω
C90 0.01μ
TO CHASSIS
Rch
HEADPHONE CIRCUIT
S104-1
S104-2
RY71
LINE AMP
R176 560Ω
R175 560Ω
HEAD PHONE
CANON
LINE OUT
R276 560Ω
R275 560Ω
HEAD PHONE
CANON
LINE OUT
+15V
R79
D73
OUTPUT MUTING CIRCUIT
Q73
2SC634A

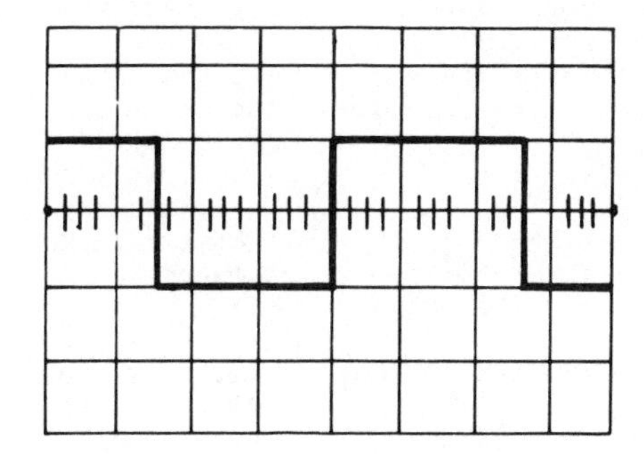

Index

D

E

F